Lecture Notes in Computer Science 807

Edited by G. Goos and J. Hartmanis

Advisory Board: W. Brauer D. Gries J. Stoer

Maxime Crochemore, Dan Gusfield (Eds.)

Combinatorial
Pattern Matching

5th Annual Symposium, CPM 94
Asilomar, CA, USA, June 5-8, 1994
Proceedings

Springer-Verlag
Berlin Heidelberg New York
London Paris Tokyo
Hong Kong Barcelona
Budapest

Maxime Crochemore   Dan Gusfield (Eds.)

Combinatorial Pattern Matching

5th Annual Symposium, CPM 94
Asilomar, CA, USA, June 5-8, 1994
Proceedings

Springer-Verlag

Berlin Heidelberg New York
London Paris Tokyo
Hong Kong Barcelona
Budapest

Series Editors

Gerhard Goos
Universität Karlsruhe
Postfach 69 80
Vincenz-Priessnitz-Straße 1
D-76131 Karlsruhe, Germany

Juris Hartmanis
Cornell University
Department of Computer Science
4130 Upson Hall
Ithaca, NY 14853, USA

Volume Editors

Maxime Crochemore
Institut Gaspard Monge, Université de Marne la Vallée
F-93160 Noisy le Grand, France

Dan Gusfield
Department of Computer Science, University of California
Davis, CA 95616-8692, USA

CR Subject Classification (1991):F.2.2, I.5.4, I.5.0, I.7.3, H.3.3, E.4, G.2.1, J.3

ISBN 3-540-58094-8 Springer-Verlag Berlin Heidelberg New York
ISBN 0-387-58094-8 Springer-Verlag New York Berlin Heidelberg

CIP data applied for

© Springer-Verlag Berlin Heidelberg 1994
Printed in Germany

Typesetting: Camera-ready by author
SPIN: 10131146   45/3140-543210 - Printed on acid-free paper

Foreword

The papers contained in this volume were presented at the fifth annual symposium on Combinatorial Pattern Matching, held June 5 - 8, 1994 at Asilomar, California. They were selected from 41 abstracts submitted in response to the call for papers.

Combinatorial Pattern Matching addresses issues of searching and matching of strings and more complicated patterns such as trees, regular expressions, extended expressions, etc. The goal is to derive non-trivial combinatorial properties for such structures and then to exploit these properties in order to achieve superior performances for the corresponding computational problems.

In recent years, a steady flow of high-quality scientific study of this subject has changed a sparse set of isolated results into a full-fledged area of algorithmics. This area is expected to grow even further due to the increasing demand for speed and efficiency that comes especially from molecular biology and various genome projects, but also from areas such as information retrieval, pattern recognition, compiling, data compression, and program analysis. The stated objective of annual CPM gatherings is to bring together active researchers for an intensive exchange of information about current and future research in combinatorial pattern matching.

The general organisation and orientations of CPM Conferences are coordinated by a Steering Committee composed of A. Apostolico, M. Crochemore, Z. Galil, and U. Manber.

The first four meetings were held at the University of Paris (1990), at the University of London (1991), at the University of Arizona (Tucson, 1992), and at the University of Padova (1993). After the first meeting, a selection of the papers appeared as a special issue of *Theoretical Computer Science*. The proceedings of the third and fourth meetings appeared as volumes 644 and 684 of the present series.

External referees who helped with the selection of papers for CPM 94 are gratefully acknoledged.

Line Fonfrède and Jean-Louis Barrière (Institut Gaspard Monge, France) provided technical help during the selection process. Debbie Chadwick (Computer Science Department, U.C. Davis) helped with secretarial and administrative work. The conference was supported in part by the National Science Foundation and the University of California, Davis. The efforts of all are gratefully acknowledged.

Program Committee

M. Crochemore, *PC-Chair*	D. S. Hirschberg
A. Ehrenfeucht	E. W. Myers
A. S. Fraenkel	I. Simon
Z. Galil	E. Ukkonen
D. Gusfield, *Chair*	M. S. Waterman

Table of Contents

Session 3: Combinatorial aspects

Session 4: More Bio-Informatics

A Space Efficient Algorithm for Finding the Best Non-overlapping Alignment Score

Gary Benson*

Department of Mathematics, University of Southern California

Abstract. Repeating patterns make up a significant fraction of DNA and protein molecules. These repeating regions are important to biological function because they may act as catalytic, regulatory or evolutionary sites and because they have been implicated in human disease. Additionally, these regions often serve as useful laboratory tools for such tasks as localizing genes on a chromosome and DNA fingerprinting. In this paper, we present a *space efficient* algorithm for finding the maximum alignment score for any two substrings of a single string T under the condition that the substrings do not overlap. In a biological context, this corresponds to the largest repeating region in the molecule. The algorithm runs in $O(n^2 \log^2 n)$ time and uses only $O(n^2)$ space.

1 Introduction

DNA and proteins are long linear molecules made up of several kinds of individual units. In DNA, there are four kinds of units (*bases* or *nucleotides*); in proteins there are 20 kinds of units (*amino acids*). Because of their linear structure, these molecules can be thought of as strings over a finite alphabet.

Repeating patterns make up a significant fraction of DNA and protein molecules. The exact function of many of these repeating regions is unknown. In some cases (*e.g.* the protein collagen), the repetition produces a structural attribute. But, in many others, the repetition may function as a catalytic, regulatory or evolutionary site. For example, the centromeric region of DNA controls the movement of the chromosome during cell division. This region termed a *satellite* consists of many contiguous copies of a species specific pattern and may serve as a protein binding site.

In still other cases, repeating regions have been implicated in human disease. A region consisting of a three nucleotide repeat on the human X chromosome is sometimes replicated incorrectly, causing the number of repeats to balloon from 50 to hundreds or thousands. Individuals with this defect suffer from fragile-X mental retardation. Several other diseases are also now known to have their basis in huge expansions of different trinucleotide repeats.

* Supported by NSF grants DMS-87-20208, DMS-90-05833 and NIH grant GM-36230.

Besides their importance in understanding protein and DNA function, repeating regions are useful laboratory tools. For example, the number of copies of a pattern at a particular site on a chromosome is often variable among individuals (*polymorphic*). Such polymorphic regions are helpful in localizing genes to specific regions of the chromosome and also in determining the probability of a match between two samples of genetic material (DNA fingerprinting).

Given their importance and given the exponential growth in the size of the DNA and protein databases, efficient methods for detecting repeating regions are required.

Due to the action of evolutionary mutation, repeating regions rarely consist of exact repeats. Rather they are approximate repeats contaminated with substitutions, deletions and insertions. It is thus natural to consider approximate string matching techniques when designing algorithms for detecting repeats.

Let $T = t_1 \cdots t_n$ and $W = w_1 \cdots w_m$ be two strings over an alphabet Σ. Let $T[i, j] = t_i \cdots t_j$ and $W[g, h] = w_g \cdots w_h$ be two substrings. An *alignment* of $T[i, j]$ and $W[g, h]$ is a sequence Q of *edit operations* [5] that transforms substring $T[i, j]$ into substring $W[g, h]$. The allowed operations are: insert a symbol into $T[i, j]$, delete a symbol from $T[i, j]$ and replace a symbol in $T[i, j]$ with a (possibly identical) symbol in $W[g, h]$. If a weighting function δ is defined for each possible edit operation [9], then, we can compute a *score* for an alignment by adding the weights assigned to each operation in Q.

In the *global alignment problem*, we seek the optimal cost alignment for T and W. In the *local alignment problem*, initial deletions and terminal insertions have zero cost. This has the effect of permitting a global alignment for any two substrings $T[i, j]$ and $W[g, h]$. Either problem can be solved in $O(nm)$ time by dynamic programming. Typically, in the biological domain, δ is negative for all operations except replacement of similar symbols and the object is to maximize the alignment score.

In this paper, we consider the problem of finding the maximum alignment score for any two substrings of a single string T under the condition that the substrings do not overlap. That is, the maximum alignment score between two substrings $T[g, h]$ and $T[i, j]$ such that $g \leq h < i \leq j$. In a biological context, this corresponds to the largest repeating region in the molecule.

In [6], Miller observed that for a general weighting function δ, the problem can be solved in $O(n^3)$ time and $O(n^2)$ space by a modification of the Smith-Waterman algorithm [8]. That time was improved by Kannan and Myers [2] to $O(n^2 \log^2 n)$ in a rather complicated recursive algorithm. Unfortunately, their algorithm requires $O(n^2 \log n)$ space. They considered reducing the space to $O(n^2)$ to be an important open problem.

In a similar vein, Landau and Schmidt [4] gave an algorithm for identifing approximate *tandem* or contiguous repeats. Their algorithm uses a very restricted weighting function for the edit operations. Either insertions and deletions have infinite negative weight (Hamming distance) or each edit operation has a weight of one (edit distance). The time for their algorithm is $O(kn \log(\frac{n}{k}))$ for a Hamming distance of at most k and $O(kn \log k \log n)$ for an edit distance of at most k. Note that this matches the time of the Kannan-Myers algorithm when the edit distance is at most n.

The **main contribution of this paper** is a new, **space efficient algorithm** for finding the maximum alignment score for two non-overlapping substrings of a sequence T. Our algorithm is **simpler** than the algorithm of [2], uses the same time and **uses only** $O(n^2)$ **space**.

The remainder of the paper is organized as follows. In section 2, we formally define our problem. In section 3 we briefly discuss *edit graphs* and several algorithms by other authors that we use as subroutines. In section 4 we present an overview of our algorithm. In section 5 we introduce the idea of ranks and show how they can be built and used efficiently and in section 6 we present a new algorithm satisfying the time and space bounds we claim.

2 Problem description

Let $T = t_1 \cdots t_n$ be a sequence and δ be a weighting function. Let $S([g, h], [i, j])$ be the best alignment score for substrings, $T[g, h]$ and $T[i, j]$. We seek to find

$$H = \max_{1 \leq g \leq h < i \leq j \leq n} \{S([g, h], [i, j])\}$$

that is, the maximum alignment score for two *non-overlapping* substrings of T. Miller [6] calls such non-overlapping regions *twins* and we adopt this nomenclature.

Wolog, we will assume that n is a power of 2. This can be accomplished by padding the sequence if necessary. Although in this paper, we only discuss finding the best score, we can additionally find the substrings and once the substrings are determined, the alignment can be computed in time and space $O(n^2)$.

3 Preliminaries

The best local alignment score for a string T versus itself can be computed by the Smith-Waterman [8] dynamic programming algorithm. If we exclude the trivial alignment of T with itself, the resulting alignment consists of two (possibly overlapping) substrings of T. Let $S(i, j) = S([*, i], [*, j])$ be the best scoring alignment between any substring ending at T_i and any substring ending at T_j.

 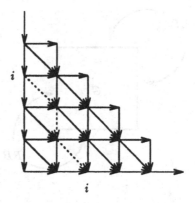

a   b

Fig. 1.

a) edit graph for the scoring matrix
b) the lower diagonal and a twin (dotted) bounded by i.

The recurrence is simple:

$$S(i,j) = \max \begin{cases} S(i, j-1) + \delta(\text{ insert } T_j) \\ S(i-1, j) + \delta(\text{ delete } T_i) \\ S(i-1, j-1) + \delta(\text{replace } T_i \text{ with } T_j) \\ 0 \end{cases}$$

The final option, which restricts the scores to non-negative values, permits *local alignment*, that is, the starting indices of the substrings are not fixed.

Note that computing a single entry in the scoring matrix requires knowing the value of only three other entries. Because of this, the scoring matrix can be viewed as a *weighted edit graph* [3, 7] where the entries are the nodes and the weights δ are assigned to the edges (figure 1). An alignment consists of a path through the edit graph and its score is the sum of the edge weights along the path.

Throughout the remainder of this paper, we will think of our problem in terms of finding high scoring paths in the edit graph.

As has been observed [2], a path in the edit graph is a twin iff it lies entirely within a rectangle bounded by row i and column i for some i, $1 < i < n$ and because of this, we need consider only the lower diagonal of the edit graph.

In our algorithm, we will use several other algorithms as subroutines. They are: 1) the DIST table construction algorithm from [1], 2) the Propagate algorithm from [2] and 3) the Twins algorithm from [6]. In the remainder of this section, we give a brief overview of each algorithm.

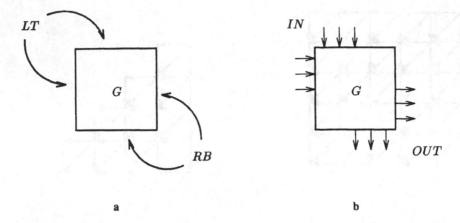

a   b

Fig. 2.

a) The DIST table for G contains best scores from every node in LT to every node in RB

b) Algorithm Propagate finds the best score $OUT(y)$ for every node $y \in RB$ given initial values $IN(x)$ on every node $x \in LT$.

3.1   DIST table construction

Let G be an $n \times m$ edit graph. Let LT be the set of nodes on the left and top edges of G. Similarly, let RB be the set of nodes on the right and bottom edges of G. We seek to construct a table $DIST_G[x, y]$ that gives the best score through G from every node $x \in LT$ to every node $y \in RB$ (figure 2). In [1], it was shown that the time to construct such a table is $O((n+m)^2 \log(n+m))$. The construction is done recursively by dividing G into four equal sized subgraphs, finding the tables for each subgraph and then combining the tables. Combining takes $O((n+m)^2)$ time and relies on the easily proven fact [1] that the best scoring path for one pair of nodes can not cross the best scoring path for another pair of nodes.

3.2   Propagate Algorithm

Let G be an $n \times m$ edit graph as above. Suppose we assign a value $IN(x)$ to each node $x \in LT$. We want to determine the following maximum values $OUT(y)$:

$$\forall y \in RB, \quad OUT(y) = \max_{x \in LT}\{IN(x) + DIST_G[x, y]\}$$

that is, the best value that comes out of y given the initial values on the x (figure 2).

Using the same property of non-crossing best scoring paths, [2] show how to compute $OUT(y)$ in time $O(n \log m + m)$. The method is to first find the best score for y_{mid}, the middle $y \in RB$ and then to recursively find the best scores for all y to the right of y_{mid} and to the left of y_{mid}.

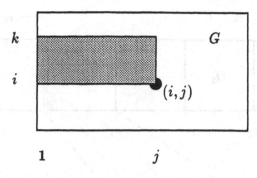

Fig. 3.

$C(i, j, k)$ is the best score from any node in the shaded region to node (i, j).

3.3 Twins Algorithm

Let G be an $n \times m$ edit graph as above. We want to compute the best score to every node (i, j) beginning anywhere in rows k to i, $k \leq i$ (figure 3). That is

$$C(i, j, k) = \max_{k \leq x \leq i;\ 1 \leq y \leq j} \{S([y, j], [x, i])\}$$

Miller [6] observed that $C(i, j, k)$ can be calculated by using the algorithm of [8]. Simply keep a list of scores of size at most n at each node of G and compute $S([*, j], [x, i])$ separately for each value of x between k and i. The best score $C(i, j, k)$ is then the max of the scores on the list for node (i, j). The scores can be restricted to non-overlapping alignments by requiring that $k \geq j$. The time is $O(n^2 m)$.

4 Algorithm Overview

Our algorithm will work as follows (figure 6):

1. Divide the edit graph into rectangles and, within each, use dynamic programming [8] to find the best score to every node. The rectangles are arranged so that alignments are non-overlapping. This handles some, but not all possible non-overlapping alignments.
2. Adjacent rectangles form an upper and lower *panel* where a non-overlapping alignment can begin and end. Find the best remaining non-overlapping scores from nodes in a panel $[k, j]$ of rows to nodes in a panel $[k, j]$ of columns. Note that the union of these panels can be divided into an upper rectangle A, a lower rectangle B and a triangle T.
   (a) For alignments with at least one end in either rectangle A or B, we jump best values through a series of DIST tables.
   (b) For alignments *strictly within* the triangle T, we recurse.

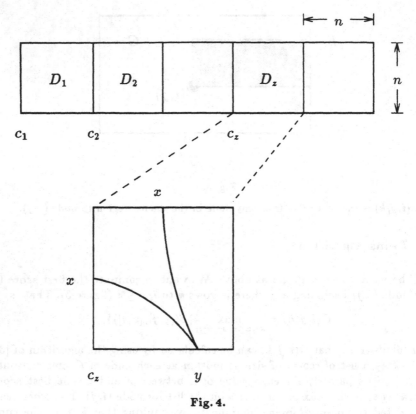

Fig. 4.

Computing the values $OUT(y)$ with a rank of DIST tables.

5 Ranks of Tables

In this section, we show how to efficiently jump a set of scores through a series of DIST tables. Let G be an $n \times m$ edit graph with $n \leq m$. Let there be $\frac{m}{n}$ consecutive $n \times n$ DIST tables $D_1, \ldots D_{\frac{m}{n}}$ covering G. (See figure 4.) We call such a collection of tables a *rank*. The *size* of a rank is the length of a side of a DIST table in the rank. Here the rank has size n.

We begin by showing that a set of IN values can be efficiently jumped across G using the rank of tables. A similar idea appears in algorithm Mesh_propagate in [2].

> **Lemma 1.** *Let G be an $n \times m$ edit graph with $n \leq m$. Let a size n rank of DIST tables cover G as above. Let each node $x \in LT_G$ have an associated score $IN(x)$. Then the best score to each element $y \in RB_G$ that originates anywhere in LT_G can be computed in time $O(m \log n)$.*

Proof. (See figure 4.) Let c_z be the column on the left side of DIST table D_z and let RB_z be the set of nodes on the right or bottom edges of D_z. We show by induction that the best score from LT_G to an element $y \in RB_z$ is:

$$\forall y \in RB_z, \quad OUT(y) = \max_{x \in LT_z} \{VALUE(x) + D_z[x, y]\}$$

Where

$$VALUE(x) = \begin{cases} IN(x) & \text{if } x \in LT_G \\ OUT(x) & \text{if } x \in RB_{z-1} \end{cases}$$

That is, the best score to a node $y \in RB_z$ consists of the maximum sum of 1) the score in node $x \in LT_z$ and 2) the edge-to-edge score of x to y. Clearly, this is true for D_1. Inductively, suppose it is true for D_k. In D_{k+1}, the best score comes either from the nodes x on the top of D_{k+1} or from the nodes to the left of c_{k+1}. The definition correctly selects a maximum from the nodes on the top. For a node x to the left, note that a best scoring path can be partitioned into a best scoring path from x to c_{k+1} and a best scoring path from c_{k+1} to y. Again, the definition correctly selects a maximum.

To compute the best scores on row n, we compute for each table D_z in order for $z = 1, 2, \ldots$, the best scores to RB_z using the algorithm Propagate [2]. The time for one table is $O(n \log n)$. There are $\frac{m}{n}$ tables so the total time is $O(m \log n)$. ∎

Lemma 2. *The space required to store a rank of DIST tables is $O($the area that the tables cover$)$.*

Proof. In a rank of size k, each table is square, so the edge to edge information occupies $O(k^2)$ space, which is $O($the area covered by the table$)$. ∎

5.1 Building the ranks

Next we consider how to efficiently build and discard a sequence of DIST table ranks. Consider a block of n rows in an edit graph. The DIST table ranks are built for each row *in order* from n up to 1 so that at each row i there are at most $\log n$ ranks between rows i and n. When we build the ranks for row i, we assume that the ranks for row $i+1$ are already available. For illustrative purposes, label each row i with $n - i$ (figure 5). In the binary representation of each label, the ones indicate exactly the number and size of ranks between rows i and n. For example, for row n with label 0, there are 0 ranks. For row $n - 7$ with label $7 = 0111_2$, there are three ranks, the bottom rank of size 4, the next rank of size 2 and the top rank of size 1. Notice that for row $n - 8$, the next row up, the

ROW   LABEL

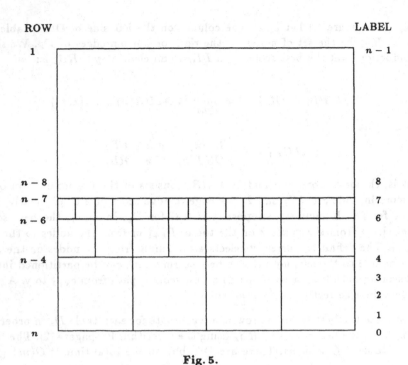

Fig. 5.

Three ranks lie between rows $n - 7$ and n.

label is $8 = 1000_2$ and there is a single rank of size 8 between this row and row n.

The order of the construction of the ranks can be determined by examining the label bits from right to left, stopping after the first 1 is reached. Each time we build a rank from smaller ranks, the smaller ranks are discarded. Consider, again, the case of row $n - 8$. Reading the label from right to left, we encounter three zeros. The first indicates a rank between rows $n - 8$ and $n - 7$. The next a rank between rows $n - 8$ and $n - 6$. The last indicates a rank between rows $n - 8$ and $n - 4$. The 1 indicates a rank between rows $n - 8$ and $n - 0$, and this final rank is the only one not discarded.

Lemma 3. *Let G be an $n \times m$ edit graph with $n \leq m$. A sequence of DIST table ranks as described above can be constructed in time $O(nm \log n)$ and space $O(nm)$.*

Proof. By Lemma 2, the space for a table is O(the area it covers). Since smaller DIST tables are discarded, the tables are non-overlapping. Therefore, the space is at most $O(nm)$. Each table in a rank is square and is constructed from four smaller tables as in the algorithm of [1]. The time to build all the tables is equivalent to the time to build $\frac{m}{n}$ tables of size $n \times n$ or $O(mn \log n)$. ∎

5.2 Jumping Values

Finally, for an edit graph G of size $n \times m$ we show how to efficiently calculate, for each k, the best score to every node $x \in RB_G$ beginning anywhere in rows k to n.

The values can be computed by the following algorithm which is implemented with the ranks construction just described:

Algorithm Jumps

1. For each node $x \in RB_G$ create a list $VALUES(x)$. Initialize the score at the top of each list to be zero.
2. For each row k from n up to 1 do
   (a) Build the rank tables for row k.
   (b) Starting with $IN(y) = 0$ for every node y in row k, compute $OUT(x)$ for every node $x \in RB_G$ by jumping values across the ranks using the method of lemma 5.1. (Nodes in RB_G in rows $1, \ldots, k-1$ get no values from this computation.)
   (c) At each node $x \in RB_G$, compare $OUT(x)$ with the score at the top of $VALUES(x)$. If $OUT(x)$ is larger, add it to the top of $VALUES(x)$. Otherwise, discard $OUT(x)$ and add a duplicate of the previous score to the top. (This ensures that the scores are in non-increasing order from the top.)

Let a node $x \in RB_G$ be in row k. The final scores in $VALUES(x)$ are the best scores to node x. They are ordered so that the best score from rows 1 to k is on top, the best score from rows 2 to k is next, etc. The important point here is that the filling of $VALUES(x)$ in step 2(c) serves as a checkpoint to guarantee that the scores are non-increasing from the top.

Theorem 4. *Let G be an $n \times m$ edit graph with $n \leq m$. Using algorithm Jumps, the values $C(n, y, k)$ in row n ($\forall y, k, 1 \leq y \leq m, 1 \leq k \leq n$) and $C(x, m, k)$ in column m ($\forall x, k, 1 \leq x \leq n, 1 \leq k \leq x$) can be calculated in time $O(mn \log^2 n)$ and space $O(nm)$.*

Proof. Time: Building the ranks takes total time $O(mn \log n)$ (lemma 3). Transferring values across one rank takes $O(m \log n)$ time (lemma 1). Each row i must transfer it's values across at most $\log n$ ranks for a time of $O(m \log^2 n)$ per row. The time for all n rows is therefore $O(mn \log^2 n)$.

Space: The space to store the ranks is $O(nm)$ (lemma 3). Each of the $n + m$ lists $VALUES(x)$ holds at most n scores. The total space is therefore $O(mn)$. ∎

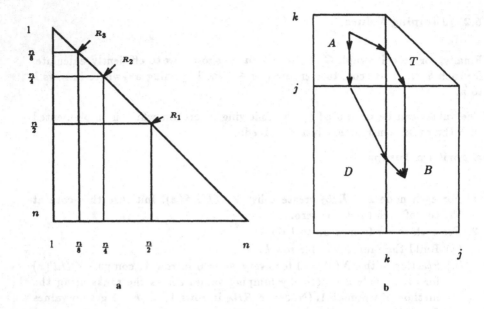

Fig. 6.

a) Partition indices and rectangles.
b) step 1, scores from nodes in A to nodes in B.

6   An $O(n^2)$ Space Algorithm

Here, we outline the $O(n^2)$ space and $O(n^2 \log^2 n)$ time algorithm. It has some of the flavor of an $n^{2.5} \log^{0.5} n$ time algorithm described in [2]. We adopt some of the nomenclature from that paper.

Algorithm Best Scores

- Partition the interval $[n, 1]$ into $\log n$ *panels* P_i of geometrically decreasing size (figure 6). The panels are $[n, \frac{n}{2}], [\frac{n}{2}, \frac{n}{4}], \ldots, [2, 1]$ with partition indices $n/2^i$. Each partition index defines a rectangle R_i ($T(1, \frac{n}{2^i})$ versus $T(\frac{n}{2^i}, n)$) of the edit graph.

- Run the Smith-Waterman algorithm [8] in each rectangle R_i of the edit graph. Each node saves its best score.

- For each panel, P_i, $i = \log n, \ldots, 1$ do the following:
  Let $[k, j] = [\frac{n}{2^i}, \frac{n}{2^{i+1}}]$ and associate three parts of the edit graph with the panel: 1) the rectangular part, A, of rows k to j, 2) the rectangular part, B, of columns k to j, and 3) the triangular part, T. Also, associate a DIST table rank D.

At this point, we have to compute four classes of scores.

(1) Scores that originate in A and end in T.
(2) Scores that originate in T and end in B.
(3) Scores that originate in A and end in B.
(4) Scores that originate and end in T.

1. **Class (2) and (3):** Run algorithm Jumps on the $\frac{n}{2^i}$ *rows* in A and T combined to get the best scores from nodes in A and T to row j. Although Jumps is described for rectangles, we can use appropriate weightings on the edges outside the lower triangle to preclude using those edges.
   Using DIST table rank D, jump the values on row j at the bottom of A to column k. We now have lists $VALUES(x)$ for every node in the LT border of B.
   Run a variation of algorithm Jumps on the $\frac{n}{2^i}$ *columns* in B. This time, we use Jumps to carry a set of scores on the LT border of B to a column in B. The variation builds the ranks for columns from k to j (*i.e.* from k to $k+1$, from k to $k+2$, etc.). At each iteration i, the top score in each list $VALUES(x)$ is removed and used as the value $IN(x)$.

2. **Class (1):** Similar to step 1 above. Run algorithm Jumps on the $\frac{n}{2^i}$ *rows* in A to get the best scores from nodes in A to column k. Then run the variation of Jumps on the $\frac{n}{2^i}$ *columns* in T. (During iteration i, an appropriately large negative value can be used as $IN(x)$ for x in rows $1, \ldots, i-1$.)

3. **Class (4):** Recursively run Best Scores on the triangle T.

4. Each node picks the best of
   1) the scores computed in steps 1, 2 or 3 above and
   2) the score from the Smith-Waterman algorithm.

- Pick the best score over all the nodes.

Theorem 5. *Algorithm Best Scores runs in time $O(n^2 \log^2 n)$ and space $O(n^2)$ assuming the DIST table ranks D for all the panels can be constructed in these bounds.*

Proof. The algorithm operates on a sequence of triangles of geometrically decreasing area. Excluding the DIST table ranks D, the time and space recursions are:

$$T(n) \le t(n) + \sum_{i=1}^{\log n} T\left(\frac{n}{2^i}\right)$$

$$S(n) \le s(n) + \sum_{i=1}^{\log n} S\left(\frac{n}{2^i}\right)$$

where $t(n)$ and $s(n)$ are respectively (within just the largest triangle), the time for the Smith-Waterman algorithm in the rectangles plus the time for algorithm Jumps (steps 1 and 2) and the space for the ranks and the lists.

For the largest rectangle, the time for Smith-Waterman is $O(n^2)$ and the time for Jumps is $O(n^2 \log^2 n)$. Since the rectangles and panels are of geometrically decreasing size, $t(n) = O(n^2 + n^2 \log^2 n)$. Similarly, $s(n)$ is $O(n^2)$ since the DIST tables we construct within the panels (for steps 1 and 2) never overlap. The solution then for $T(n)$ is $O(n^2 \log^2 n)$ and for $S(n)$ is $O(n^2)$. $\blacksquare$

7 Constructing the DIST Table Ranks D

In this section, we show how to construct the DIST table ranks D in the time and space bounds of Theorem 5.

> **Theorem 6.** *The DIST table ranks D can be constructed in $O(n^2 \log n)$ time and $O(n^2)$ space.*

Proof. Consider first the original lower diagonal graph. We start with the rank for the smallest panel $P_{\log n}$ Note each table in this rank has size 2. We can build this rank, and then complete algorithm Best Scores for the smallest panel. Then, *and this is the key point*, we can use the rank for panel $P_{\log n}$ to build the rank for the next largest panel $P_{\log n - 1}$ and then discard the rank for panel $P_{\log n}$. We will never use it again. Each table in the new rank has size 4, so we can use the algorithm of [1] as in lemma 3. Continuing on in this way, the time for all the tables is $O(n^2 \log n)$, the time to build the largest rank. We discard smaller tables after we use them to build the next largest tables, so the tables are non-overlapping. Since the recursively processed triangles are geometrically decreasing in area, the total space is $O(n^2)$. $\blacksquare$

8 Acknowledgement

I would like to thank two of the referees for spotting an error in the original description of algorithm Best Scores and for making several useful comments.

References

1. A. Apostolico, M.J. Atallah, L.L. Larmore, and S. Mcfaddin, "Efficient Parallel Algorithms for String Editing and Related Problems," *SIAM J. Comput.*, 19, pp 968-988, 1990.

2. S. Kannan and E. Myers, "An Algorithm for Locating Non-overlapping Regions of Maximum Alignment Score," *Fourth Annual Symposium on Combinatorial Pattern Matching*, pp 74-86, 1993.

3. Z.M. Kedem and H. Fuchs, "On finding several shortest paths in certain graphs," *Proc. 18th Allerton Conference on Communication, Control and Computing*, pp 677-683, October 1980.

4. G. Landau and J. Schmidt, "An algorithm for Approximate Tandem Repeats," *Fourth Annual Symposium on Combinatorial Pattern Matching*, pp 120-133, 1993.

5. V.I. Levenshtein, "Binary codes capable of correcting deletions, insertions and reversals," *Soviet Phys. Dokl.*, 10, pp707-710, 1966.

6. W. Miller, "An algorithm for locating a repeating region," *manuscript*, 1992.

7. E. Myers, "An $O(ND)$ difference algorithm and its variants," *Algorithmica*, 1, pp 251-266, 1986.

8. T.F. Smith and M.S. Waterman, "Identification of common molecular sequences," *J. Mol. Biol.*, 147, pp 195-197, 1981.

9. R.A. Wagner and M.J. Fisher, "The string-to-string correction problem," *J. ACM*, 21, pp 168-173, 1974.

The Parameterized Complexity
of Sequence Alignment and Consensus

Hans Bodlaender[1], Rodney G. Downey[2], Michael R. Fellows[3], Harold T. Wareham[3]

[1] Computer Science Department, Utrecht University, P.O. Box 80.089, 3508 TB Utrecht, the Netherlands, hansb@cs.ruu.nl
[2] Mathematics Department, Victoria University, P.O. Box 600, Wellington, New Zealand, downey@math.vuw.ac.nz
[3] Computer Science Department, University of Victoria, Victoria, British Columbia V8W 3P6, Canada, mfellows@csr.uvic.ca

Abstract. The LONGEST COMMON SUBSEQUENCE problem is examined from the point of view of parameterized computational complexity. There are several different ways in which parameters enter the problem, such as the number of sequences to be analyzed, the length of the common subsequence, and the size of the alphabet. Lower bounds on the complexity of this basic problem imply lower bounds on a number of other sequence alignment and consensus problems. At issue in the theory of parameterized complexity is whether a problem which takes input (x, k) can be solved in time $f(k) \cdot n^\alpha$ where α is independent of k (termed *fixed-parameter tractability*. It can be argued that this is the appropriate asymptotic model of feasible computability for problems for which a small range of parameter values cover important applications — a situation which certainly holds for many problems in biological sequence analysis. Our main results show that: (1) The LONGEST COMMON SUBSEQUENCE (LCS) parameterized by the number of sequences to be analyzed is hard for $W[t]$ for all t. (2) The LCS problem problem, parameterized by the length of the common subsequence, belongs to $W[P]$ and is hard for $W[2]$. (3) The LCS problem parameterized both by the number of sequences and the length of the common subsequence, is complete for $W[1]$. All of the above results are obtained for unrestricted alphabet sizes. For alphabets of a fixed size, problems (2) and (3) are fixed-parameter tractable. We conjecture that (1) remains hard.

1 Introduction

The computational problem of finding the longest common subsequence of a set of k strings (the LCS problem) has been studied extensively over the last twenty years (see [Hir83,IF92] and references). This problem has many applications. When $k = 2$, the longest common subsequence is a measure of the similarity of two strings and is thus useful in in molecular biology, pattern recognition, and text compression [San72,LF78,Mai78]. The version of LCS in which the the number of strings is unrestricted is also useful in text compression [Mai78], and is a special case of the multiple sequence alignment and consensus subsequence discovery problems in molecular biology [Pev92,DM93a,DM93b].

To date, most research has focused on deriving efficient algorithms for the LCS problem when $k = 2$ (see [Hir83,IF92] and references). Most of these algorithms are based on the dynamic programming approach [PM92], and require quadratic time. Though the k-unrestricted LCS problem is NP-complete [Mai78], certain of the algorithms for the $k = 2$ case have been extended to yield algorithms that require $O(n^{(k-1)})$ time and space, where n is the length of the longest of the k strings (see [IF92] and references; see also [Bae91]).

In this paper, we analyze the *Longest common subsequence* problem from the point of view of parameterized complexity theory introduced in [DF92]. The parameterizations of the *Longest Common Subsequence* problem that we consider are defined as follows.

LONGEST COMMON SUBSEQUENCE (LCS-1, LCS-2 and LCS-3)
Input: A set of k strings X_1, ..., X_k over an alphabet Σ, and a positive integer m.
Parameter 1: k (We refer to this problem as LCS-1.)
Parameter 2: m (We refer to this problem as LCS-2.)
Parameter 3: (k, m) (We refer to this problem as LCS-3.)
Question: Is there a string $X \in \Sigma^*$ of length at least m that is a subsequence of X_i for $i = 1, ..., k$?

Our results are summarized in the following table.

Table 1. The Fixed-Parameter Complexity of the LCS Problem

Problem	Fixed	Alphabet Size $|\Sigma|$	
		Unbounded	Fixed
LCS-1	k	W[t]-hard, $t \geq 1$	?
LCS-2	m	W[2]-hard	FPT
LCS-3	k, m	W[1]-complete	FPT

In §2 we give some background on parameterized complexity theory. In §3 we detail the proof that LCS-3 is complete for W[1]. This implies that LCS-1 and LCS-2 are W[1]-hard, results which can be improved by further arguments to show that LCS-1 is hard for $W[t]$ for all t, and that LCS-2 is hard for $W[2]$. Concretely, none of these three parameterized versions of LCS is thus fixed-parameter tractable unless the well-known (and apparently resistant) k-CLIQUE problem is fixed-parameter tractable.

2 Parameterized Computational Complexity

The theory of parameterized compuational complexity is motivated by the observation that many NP-complete problems take as input two objects, for example, perhaps a graph G and and integer k. In some cases, e.g., VERTEX COVER, the problem can be solved in linear time for every fixed parameter value, and is well-solved for problems with $k \leq 20$. For other problems, for example CLIQUE and MINIMUM DOMINATING SET we have the contrasting situation where the best known algorithms are based on brute force, essentially, and require time $\Omega(n^k)$. If $P = NP$ then all three of these problems are fixed-parameter tractable. The theory of parameterized computational complexity explores the apparent qualitative difference between these problems (for fixed parameter values). It is particularly relevant to problems where a small range of parameter values cover important applications — this is certainly the case for many problems in computational biology. For these the theory offers a more sensitive view of tractability vs. apparent intractability than the theory of NP-completeness.

2.1 Parameterized Problems and Fixed-Parameter Tractability

A *parameterized problem* is a set $L \subseteq \Sigma^* \times \Sigma^*$ where Σ is a fixed alphabet. For convenience, we consider that a parameterized problem L is a subset of $L \subseteq \Sigma^* \times N$. For a parameterized problem L and $k \in N$ we write L_k to denote the associated fixed-parameter problem $L_k = \{x | (x, k) \in L\}$.

Definition 1. We say that a parameterized problem L is (uniformly) *fixed-parameter tractable* if there is a constant α and an algorithm Φ such that Φ decides if $(x, k) \in L$ in time $f(k)|x|^\alpha$ where $f : N \to N$ is an arbitrary function.

2.2 Problem Reductions

A direct proof that a problem such as MINIMUM DOMINATING SET is not fixed-parameter tractable would imply $P \neq NP$. Thus a completeness program is reasonable.

Definition 2. Let A, B be parameterized problems. We say that A is (uniformly many:1) *reducible* to B if there is an algorithm Φ which transforms (x, k) into $(x', g(k))$ in time $f(k)|x|^\alpha$, where $f, g : N \to N$ are arbitrary functions and α is a constant independent of k, so that $(x, k) \in A$ if and only if $(x', g(k)) \in B$.

It is easy to see that if A reduces to B and B is fixed parameter tractable then so too is A. It is important to note that there are two ways in which parameterized reductions differ from familiar P-time reductions: (1) the reduction may be polynomial in n, but (for example) exponential in the parameter k, and (2) the slice A_k must be mapped to a single slice $B_{g(k)}$ (unlike NP-completeness reductions which may map k to $k' = n - k$, for example).

2.3 Complexity Classes

The classes are intuitively based on the complexity of the circuits required to check a solution, or alternatively, the "natural logical depth" of the problem.

Definition 3. A Boolean circuit is of *mixed type* if it consists of circuits having gates of the following kinds.
(1) *Small gates: not* gates, *and* gates and *or* gates with bounded fan-in. We will usually assume that the bound on fan-in is 2 for *and* gates and *or* gates, and 1 for *not* gates.
(2) *Large gates: and* gates and *or* gates with unrestricted fan-in.

Definition 4. The *depth* of a circuit C is defined to be the maximum number of gates (small or large) on an input-output path in C. The *weft* of a circuit C is the maximum number of large gates on an input-output path in C.

Definition 5. We say that a family of decision circuits F has *bounded depth* if there is a constant h such that every circuit in the family F has depth at most h. We say that F has *bounded weft* if there is constant t such that every circuit in the family F has weft at most t. The *weight* of a boolean vector x is the number of 1's in the vector.

Definition 6. Let F be a family of decision circuits. We allow that F may have many different circuits with a given number of inputs. To F we associate the parameterized circuit problem $L_F = \{(C, k) : C$ accepts an input vector of weight $k\}$.

Definition 7. A parameterized problem L belongs to $W[t]$ if L reduces to the parameterized circuit problem $L_{F(t,h)}$ for the family $F(t, h)$ of mixed type decision circuits of weft at most t, and depth at most h, for some constant h.

Definition 8. A parameterized problem L belongs to $W[P]$ if L reduces to the circuit problem L_F, where F is the set of all circuits (no restrictions).

We designate the class of fixed-parameter tractable problems FPT.
The framework above describes a hierarchy of parameterized complexity classes

$$FPT \subseteq W[1] \subseteq W[2] \subseteq \cdots \subseteq W[P]$$

for which there are many natural hard or complete problems [DF92].
For example, all of the following problems are now known to be complete for $W[1]$: SQUARE TILING, INDEPENDENT SET, CLIQUE, and BOUNDED POST CORRESPONDENCE PROBLEM, k-STEP DERIVATION FOR CONTEXT-SENSITIVE GRAMMARS, VAPNIK-CHERVONENKIS DIMENSION, and the k-STEP HALTING PROBLEM FOR NONDETERMINISTIC TURING MACHINES [CCDF93,DEF93,DFKHW93]. Thus, any one of these problems is fixed-parameter tractable if and only if all of the others are; and none of the problems for which we here prove W hardness results are fixed-parameter tractable unless all of these are also. DOMINATING SET is complete for $W[2]$ [DF92]. Fixed parameter tractability for DOMINATING SET, or any other $W[2]$-hard problem implies fixed parameter tractability for all problems in $W[1]$ mentioned above, and all other problems in $W[2] \supseteq W[1]$.

3 The Reductions

Theorem 9. *LCS-3 is complete for* $W[1]$.

Proof. Membership in $W[1]$ can be seen by a reduction to WEIGHTED CNF SATISFIABILITY for expressions having bounded clause size. By padding with new symbols or by repeating some of the X_i, we can assume for convenience (with polynomially bounded blow-up) that $k = m$. The idea is to use a truth assignment of weight k^2 to indicate the k positions in each of the k strings of an instance of LCS-3 that yield a common subsequence of length k.

The details are as follows. Let $X_1, \ldots, X_k$ be an instance of LCS-3. By a trivial padding with symbols having only a single occurence we may assume that the strings X_i are all of length n. Let $a[i,j]$ denote the jth symbol of X_i. Let $B = \{b[i,j,r] : 1 \leq i \leq k, 1 \leq j \leq n, 1 \leq r \leq k\}$ be a set of boolean variables. The interpretation we intend for the variable $b[i,j,r]$ is that the rth symbol $x[r]$ of a length k common subsequence $X = x[1] \cdots x[k]$ occurs as the symbol $a[i,j]$ in the string X_i, that is, $x[j] = a[i,j]$. Let B_i be the set of elements $b[i,j,r]$ with first index i.

Let $E = E_1 E_2 E_3$ be the boolean expression over the set of variables B where

$$E_1 = \prod_{i=1}^{k} \prod_{r=1}^{k} \prod_{1 \leq j < j' \leq n} (\neg b[i,j,r] + \neg b[i,j',r])$$

$$E_2 = \prod_{1=1}^{k} \prod_{j=1}^{n} \prod_{1 \leq r < r' \leq k} (\neg b[i,j,r] + \neg b[i,j,r'])$$

$$E_3 = \prod_{r=1}^{k} \prod_{1 \leq i < i' \leq k} \prod_{1 \leq j \leq j' \leq n} \prod_{a[i,j] \neq a[i',j']} (\neg b[i,j,r] + \neg b[i',j',r])$$

We claim that E has a weight k^2 truth assignment if and only if the X_i have a common subsequence of length k. It is easy to verify that a truth assignment corresponding to a length k common subsequence according to our intended interpretation of the boolean variables satisfies E. For the converse direction, suppose τ is a weight k^2 truth assignment that satisfies E. The clauses of E_1 insure (by the Pigeonhole Principle) that no more than k variables of B_i are set *true* for $i = 1, \ldots, k$. Consequently there must be exactly k variables set to *true* in each B_i, and since E_2 is satisfied, these must indicate k distinct positions in X_i according to our interpretation. The clauses of E_3 insure that the corresponding subsequence symbols in the k strings are the same.

To show $W[1]$-hardness we reduce from CLIQUE. Let $G = (V, E)$ be a graph for which we wish to determine whether G has a k-clique. We show how to construct a family $\mathcal{F}_G$ of $k' = f(k)$ sequences over an alphabet Σ that have a common subsequence of length $k'' = g(k)$ if and only G contains a k-clique. Assume for convenience that the vertex set of G is $V = \{1, \ldots, n\}$.

The Alphabet We first describe the alphabet $\Sigma = \Sigma_1 \cup \Sigma_2 \cup \Sigma_3 \cup \Sigma_4$. We refer to these as *vertex symbols* (Σ_1), *edge symbols* (Σ_2), *vertex position symbols* (Σ_3), and *edge position symbols* (Σ_4).

$$\Sigma_1 = \{\alpha[p, q, r] : 1 \leq p \leq k, \ 0 \leq q \leq 1, \ 1 \leq r \leq n\}$$

$$\Sigma_2 = \{\beta[i, j, q, u, v] : 1 \leq i < j \leq k, \ 0 \leq q \leq 1, \ 1 \leq u < v \leq n, \ uv \in E\}$$

$$\Sigma_3 = \{\gamma[p, q, b] : 1 \leq p \leq k, \ 0 \leq q \leq 1, 0 \leq b \leq 1\}$$

$$\Sigma_4 = \{\delta[i, j, q, b] : 1 \leq i < j \leq k, \ 0 \leq q \leq 1, \ 0 \leq b \leq 1\}$$

We will use the following shorthand notation to refer to various subsets of Σ. The notation indicates which indices are held fixed to some value, with "*" indicating that the index should vary over its range of definition in building the set. For example, $\Sigma_1[p, *, r] = \{\alpha[p, q, r] : 0 \leq q \leq 1\}$ is the set of two elements with the first and third indices fixed at p and r, respectively.

An Example of a Clique Representation The sequences in $\mathcal{F}$ are constructed in such a way that the k-cliques in G (considered with vertices in ascending order) are in 1:1 correspondence with the common subsequences of length k''. It will be useful in motivating the construction to consider an example of this intended correspondence. Consider a graph having a 3-clique on the vertices $\{a, b, c\}$.

This 3-clique would be represented by the following common subsequence $\sigma(a, b, c)$, which we describe according to a hierarchy of factorizations. (Exponential notation indicates repetition of a symbol.)

$$\sigma(a, b, c) = \langle \text{first vertex} \rangle \langle \text{second vertex} \rangle \langle \text{third vertex} \rangle$$

where

$$\langle \text{first vertex} \rangle = \langle \text{vertex 1} \rangle \langle \text{edge (1,2)} \rangle \langle \text{edge (1,3)} \rangle \langle \text{vertex 1 echo} \rangle$$

$$\langle \text{second vertex} \rangle = \langle \text{vertex 2} \rangle \langle \text{edge (1,2) echo} \rangle \langle \text{edge (2,3)} \rangle \langle \text{vertex 2 echo} \rangle$$

$$\langle \text{third vertex} \rangle = \langle \text{vertex 3} \rangle \langle \text{edge (1,3) echo} \rangle \langle \text{edge (2,3) echo} \rangle \langle \text{vertex 3 echo} \rangle$$

and where the constituent subsequences over Σ are

$$\langle \text{vertex 1} \rangle = \gamma[1, 0, 0]^w \alpha[1, 0, a] \gamma[1, 0, 1]^w$$

$$\langle \text{edge (1,2)} \rangle = \delta[1, 2, 0, 0]^w \beta[1, 2, 0, a, b] \delta[1, 2, 0, 1]^w$$

$$\langle \text{edge (1,3)} \rangle = \delta[1, 3, 0, 0]^w \beta[1, 3, 0, a, c] \delta[1, 3, 0, 1]^w$$

$$\langle \text{vertex 1 echo} \rangle = \gamma[1, 1, 0]^w \alpha[1, 1, a] \gamma[1, 1, 1]^w$$

$$\langle \text{vertex 2} \rangle = \gamma[2, 0, 0]^w \alpha[2, 0, b] \gamma[2, 0, 1]^w$$

$$\langle \text{edge (1,2) echo} \rangle = \delta[1, 2, 1, 0]^w \beta[1, 2, 1, a, b] \delta[1, 2, 1, 1]^w$$

$$\langle \text{edge (2,3)} \rangle = \delta[2, 3, 0, 0]^w \beta[2, 3, 0, b, c] \delta[2, 3, 0, 1]^w$$

$$\langle \text{vertex 2 echo} \rangle = \gamma[2, 1, 0]^w \alpha[2, 1, b] \gamma[2, 1, 1]^w$$

$$\langle \text{vertex 3} \rangle = \gamma[3, 0, 0]^w \alpha[3, 0, c] \gamma[3, 0, 1]^w$$

$$\langle \text{edge (1,3) echo} \rangle = \delta[1, 3, 1, 0]^w \beta[1, 3, 1, a, c] \delta[1, 3, 1, 1]^w$$

$$\langle \text{edge (2,3) echo} \rangle = \delta[2, 3, 1, 0]^w \beta[2, 3, 1, b, c] \delta[2, 3, 1, 1]^w$$

$$\langle \text{vertex 3 echo} \rangle = \gamma[3, 1, 0]^w \alpha[3, 1, c] \gamma[3, 1, 1]^w$$

In the above, the position symbols are repeated $w = w(k)$ times for reasons useful for the correctness argument concerning the reduction.

The Target Parameters There are $f_1(k) = 2k + k(k-1) = k^2 + k$ position symbols (in Σ_3 and Σ_4). We take $w = f_1(k)^2 + 1$, $k' = f_1(k) + 2$, and $k'' = (w+1)f_1(k)$.

Symbol Subsets and Operations It is convenient to introduce a linear ordering on Σ that corresponds to the "natural" order in which the various symbols occur, as illustrated by the example above. We can achieve this by defining a "weight" on the symbols of Σ and then ordering the symbols by weight.

Let $N = 2kn$ (a value conveniently larger than k and n). Define the *weight* $||a||$ of a symbol $a \in \Sigma$ by

$$||a|| = \begin{cases} pN^6 + qN^5 + r & \text{if } a = \alpha[p,q,r] \in \Sigma_1 \\ q'iN^6 + qjN^6 + q'N^4 + q'jN^3 + qiN^3 + uN + v & \text{if } a = \beta[i,j,q,u,v] \in \Sigma_2 \\ pN^6 + qN^5 + bN^2 & \text{if } a = \gamma[p,q,b] \in \Sigma_3 \\ q'iN^6 + qjN^6 + q'N^4 + q'jN^3 + qiN^3 + bN^2 & \text{if } a = \delta[i,j,q,b] \in \Sigma_4 \end{cases}$$

where $q' = (q-1)^2$.

Define a linear order on Σ by $a < b$ if and only if $||a|| < ||b||$. The reader can verify that, assuming $a < b < c$, the symbols of the example sequence $\sigma(a,b,c)$ described above occur in ascending order.

For $a, b \in \Sigma$, $a < b$, we define the *segment* $\Sigma(a,b)$ to be $\Sigma(a,b) = \{e \in \Sigma : a \le e \le b\}$, and we define similarly the segments $\Sigma_i(a,b)$.

If Γ is a finite set of symbols, then it is easy to see that there is a "universal" string $(m\Gamma) \in \Gamma^*$ of length $m|\Gamma|$ that contains as a subsequence every string of length at most m over Γ, for example, by running through the symbols in Γ m times. We will use the notation $(m\Gamma)$ to refer to any choice of such a string. Where m is unimportant except that it be "large enough" (with the understanding that this means also "not too large") we may write $(*\Gamma)$ for convenience.

If $\Gamma \subseteq \Sigma$, let $(\uparrow \Gamma)$ be the string of length $|\Gamma|$ which consists of one occurence of each symbol in Γ in ascending order, and let $(\downarrow \Gamma)$ be the string of length $|\Gamma|$ which consists of one occurence of each symbol in Γ in descending order.

String Gadgets We next describe some "high level" component subsequences for the construction. In the following let $\updownarrow$ denote either $\uparrow$ or $\downarrow$. Product notation is interpreted as refering to concatenation.

Vertex and Edge Selection Gadgets

$$\langle \updownarrow \text{ vertex } p \rangle = \gamma[p,0,0]^w (\updownarrow \Sigma_1[p,0,*]) \gamma[p,0,1]^w$$

$$\langle \updownarrow \text{ vertex } p \text{ echo} \rangle = \gamma[p,1,0]^w (\updownarrow \Sigma_1[p,1,*]) \gamma[p,1,1]^w$$

$$\langle \updownarrow \text{ edge } (i,j) \rangle = \delta[i,j,0,0]^w (\updownarrow \Sigma_2[i,j,0,*,*]) \delta[i,j,0,1]^w$$

$$\langle \updownarrow \text{ edge } (i,j) \text{ echo} \rangle = \delta[i,j,1,0]^w (\updownarrow \Sigma_2[i,j,1,*,*]) \delta[i,j,1,1]^w$$

$$\langle \updownarrow \text{ edge } (i,j) \text{ from } u \rangle = \delta[i,j,0,0]^w (\updownarrow \Sigma_2[i,j,0,u,*]) \delta[i,j,0,1]^w$$

$$\langle \updownarrow \text{ edge } (i,j) \text{ to } v \rangle = \delta[i,j,1,0]^w (\updownarrow \Sigma_2[i,j,1,*,v]) \delta[i,j,1,1]^w$$

Control and Selection Assemblies

$$\langle \updownarrow \text{ control } p \rangle = \langle \updownarrow \text{ vertex } p \rangle \left(\prod_{s=1}^{p-1} \langle \updownarrow \text{ edge } (s,p) \text{ echo} \rangle \right)$$

$$\cdot \left(\prod_{s=p+1}^{k} \langle \updownarrow \text{ edge } (p,s) \rangle \right) \langle \updownarrow \text{ vertex } p \text{ echo} \rangle$$

$$\langle \uparrow \text{ choice } p \rangle = \prod_{x=1}^{n} \left(\gamma[p,0,0]^w \alpha[p,0,x] \gamma[p,0,1]^w \prod_{t=1}^{p-1} \langle \uparrow \text{ edge } (t,p) \text{ to } x \rangle \right.$$

$$\left. \cdot \prod_{t=p+1}^{k} \langle \uparrow \text{ edge } (p,t) \text{ from } x \rangle \gamma[p,1,0]^w \alpha[p,1,x] \gamma[p,1,1]^w \right)$$

$$\langle \downarrow \text{ choice } p \rangle = \prod_{x=n}^{\text{down to } 1} \left(\gamma[p,0,0]^w \alpha[p,0,x] \gamma[p,0,1]^w \prod_{t=1}^{p-1} \langle \downarrow \text{ edge } (t,p) \text{ to } x \rangle \right.$$

$$\left. \cdot \prod_{t=p+1}^{k} \langle \downarrow \text{ edge } (p,t) \text{ from } x \rangle \gamma[p,1,0]^w \alpha[p,1,x] \gamma[p,1,1]^w \right)$$

Edge Symbol Pairing Gadget

$$\langle \text{edge } (i,j) \text{ from } u \text{ to } v \rangle = \beta[i,j,0,u,v] (*\Sigma(\delta[i,j,0,1], \delta[i,j,1,0])) \beta[i,j,1,u,v]$$

The Reduction We may now describe the reduction. The instance of LCS-3 consists of strings which we may consider as belonging to three subsets: *Control*, *Selection* and *Check*. The two strings in the *Control* set are

$$X_1 = \prod_{t=1}^{k} \langle \uparrow \text{ control } t \rangle$$

$$X_2 = \prod_{t=1}^{k} \langle \downarrow \text{ control } t \rangle$$

The $2k$ strings in the *Selection* set are, for $p = 1, ..., k$

$$Y_p = \left(\prod_{t=1}^{p-1} \langle \uparrow \text{ control } t \rangle \right) \langle \uparrow \text{ choice } p \rangle \left(\prod_{t=p+1}^{k} \langle \uparrow \text{ control } t \rangle \right)$$

$$Y_p' = \left(\prod_{t=1}^{p-1} \langle \downarrow \text{ control } t \rangle \right) \langle \downarrow \text{ choice } p \rangle \left(\prod_{t=p+1}^{k} \langle \uparrow \text{ control } t \rangle \right)$$

The $2\binom{k}{2} = k(k-1)$ strings in the *Check* set are, for $1 \le i < j \le k$

$$Z_{i,j} = \left(\prod_{t=1}^{i-1} \langle\uparrow \text{ control } t\rangle\right) \langle\uparrow \text{ vertex } i\rangle \left(\prod_{s=1}^{i-1} \langle\uparrow \text{ edge } (s,i) \text{ echo}\rangle\right) \left(\prod_{s=i+1}^{j-1} \langle\uparrow \text{ edge } (i,s)\rangle\right)$$

$$\cdot \; \delta[i,j,0,0]^w \prod_{\substack{1 \le u < v \le n \\ uv \in E}}^{\text{lex}\uparrow} \langle\text{edge } (i,j) \text{ from } u \text{ to } v\rangle$$

$$\cdot \; \delta[i,j,1,1]^w \left(\prod_{s=i+1}^{j-1} \langle\uparrow \text{ edge } (s,j) \text{ echo}\rangle\right) \left(\prod_{s=j+1}^{k} \langle\uparrow \text{ edge } (j,s)\rangle\right)$$

$$\cdot \; \langle\uparrow \text{ vertex } j \text{ echo}\rangle\langle \prod_{t=j+1}^{k} \langle\uparrow \text{ control } t\rangle$$

$$Z'_{i,j} = \left(\prod_{t=1}^{i-1} \langle\downarrow \text{ control } t\rangle\right) \langle\downarrow \text{ vertex } i\rangle \left(\prod_{s=1}^{i-1} \langle\downarrow \text{ edge } (s,i) \text{ echo}\rangle\right) \left(\prod_{s=i+1}^{j-1} \langle\downarrow \text{ edge } (i,s)\rangle\right)$$

$$\cdot \; \delta[i,j,0,0]^w \prod_{\substack{1 \le u < v \le n \\ uv \in E}}^{\text{lex}\downarrow} \langle\text{edge } (i,j) \text{ from } u \text{ to } v\rangle$$

$$\cdot \; \delta[i,j,1,1]^w \left(\prod_{s=i+1}^{j-1} \langle\downarrow \text{ edge } (s,j) \text{ echo}\rangle\right) \left(\prod_{s=j+1}^{k} \langle\downarrow \text{ edge } (j,s)\rangle\right)$$

$$\cdot \; \langle\downarrow \text{ vertex } j \text{ echo}\rangle\langle \prod_{t=j+1}^{k} \langle\downarrow \text{ control } t\rangle$$

We comment that the key difference between $Z_{i,j}$ and $Z'_{i,j}$ is that in $Z_{i,j}$ the edge symbol pairing gadgets occur in increasing lexicographic order, and in $Z'_{i,j}$ the gadgets are in decreasing lexicographic order.

Proof of Correctness Where S_1 and S_2 are strings of symbols, let $l(S_1, S_2)$ denote the maximum length of a common subsequence of S_1 and S_2.

In the Control Strings X_1 and X_2 we distinguish certain substrings that we term *positions*. Note that both of these strings are formed as the concatenation of four different kinds of substrings: $\langle\text{vertex}\rangle$, $\langle\text{vertex echo}\rangle$, $\langle\text{edge}\rangle$ and $\langle\text{edge echo}\rangle$, and that each of these "vertex and edge selection" substrings begins and ends with a matched pair of substrings of repeated symbols from Σ_3 (in the case of vertex selection), or from Σ_4 (in the case of edge selection). These matched pairs of position symbol substrings determine a *position* — note that these position symbol substrings (and therefore the positions defined) occur in the same order in X_1 and X_2. Thus there are $k(2 + k - 1) = k^2 + k$ positions.

Between a matched pair of position symbol substrings in X_1 there is a set of symbols in increasing order that we will term a *set of (vertex or edge) stairs*, and in X_2 in the corresponding position there occurs the same set of symbols in decreasing order. The proof of the following claim is trivial.

Claim 1. Suppose Σ is a linearly ordered finite alphabet, and that $S \uparrow$ is the string consisting of the symbols of Σ in increasing order, and that $S \downarrow$ is the symbols of Σ in decreasing order. Then $l(S \uparrow, S \downarrow) = 1$. $\qquad\square$

Claim 2. A common subsequence C of the control sequences X_1 and X_2 of maximum length l satisfies the conditions: (1) $l = k''$, and (2) C consists of the position symbol substrings (common to X_1 and X_2) together with one symbol in each position defined by these substrings.

Proof. It is clear that $l \geq k''$ because there are many different common subsequences of length k'' consisting of all the position symbol substrings (which are the same in X_1 and X_2) together with a single choice of vertex or edge symbol in each position. Now suppose there is a common subsequence C of length greater than k'' and fix attention on subsequences C_1 of X_1 and C_2 of X_2 that are isomorphic to C (for the reason that C might occur in more than one way as a subsequence). Then C_1 must contain two vertex or edge symbols ϵ_1 and ϵ_2 that occur on the same set of stairs in X_1. By Claim 1, these two symbols, considered now in C_2, cannot occur on the same set of stairs in X_2. This implies that any position symbols between ϵ_1 and ϵ_2 in X_2 do not belong to C_2. Consequently, there are at least $2w$ position symbols of X_2 that do not occur in $C = C_2$. But in order for the length of C to be at least k'', this means that C must contain more than $f_1(k)^2$ vertex and edge symbols. By the Pigeonhole Principle, there must therefore be a set of stairs in X_1 that contains $m > f_1(k)$ vertex or edge symbols of C_1. By Claim 1, no more than one of the corresponding symbols in C_2 can occur on any set of stairs in X_2, and therefore X_2 must have at least m sets of stairs, a contradiction. This establishes (1), and furthermore shows that no two symbols of a common subsequence of length k'' can occur on the same set of stairs. Thus (2) may also be concluded by observing that there must be at least one vertex or edge symbol from each set of stairs, else the length of C would be less than k''. $\qquad\square$

By Claim 2, if C is a common subsequence of of X_1 and X_2 of length k'', we may refer unambiguously to the vertices and edges represented in the various positions of C. In particular, note that these positions occur in k *vertex units*, each of which consists of an *initial vertex position*, followed by $k - 1$ *edge* and *edge echo positions* and concluding with a *terminal vertex echo position*. If uv is an edge of the graph with $u < v$, then we refer to u as the *initial vertex* and to v as the *terminal vertex* of the edge.

Claim 3. If C is a subsequence of length k'' common to the Control and Selection sets, then in each vertex unit: (1) the vertex u represented in the initial vertex position is also represented in the terminal vertex echo position, (2) each edge represented in an edge echo position has terminal vertex u, and (3) each edge represented in an edge position has initial vertex u.

Proof. Suppose C is a subsequence of length k'' common to X_1 and X_2. We argue that if C is also common to Y_p and Y_p' then the statements of the Lemma are satisfied for the pth vertex unit. Let C_p and C_p' denote specific subsequences of Y_p and Y_p', respectively, with $C = C_p = C_p'$.

The strings Y_p and Y_p' differ from the Control strings X_1 and X_2, respectively, only in the replacement of a $\langle \updownarrow \text{ control } p \rangle$ gadget with a $\langle \updownarrow \text{ choice } p \rangle$ gadget. In particular, the position symbols in the other constituent substrings occur in the same way in all four strings, and so by Claim 2, C_p (C_p') must include all of the length w position symbol substrings in Y_p (Y_p') occuring outside of $\langle \uparrow \text{ choice } p \rangle$ ($\langle \downarrow \text{ choice } p \rangle$). Furthermore, C_p must contain precisely two vertex symbols α and α', appropriately positioned, from $\langle \uparrow \text{ choice } p \rangle$, and C_p' must contain the same two (and no other) vertex symbols from $\langle \downarrow \text{ choice } p \rangle$.

The subsequence of Y_p consisting of all the vertex symbols in $\langle \uparrow \text{ choice } p \rangle$ is the vertex index increasing sequence

$$S = \prod_{x=1}^{n} (\alpha[p,0,x]\alpha[p,1,x])$$

and the subsequence of Y_p' consisting of all the vertex symbols in $\langle \downarrow \text{ choice } p \rangle$ is the vertex index decreasing sequence

$$S' = \prod_{x=n}^{1} (\alpha[p,0,x]\alpha[p,1,x])$$

The only possibility for α and α' to be common to S and S' is for α and α' to represent the same vertex u, that is, $\alpha = \alpha[p,0,u]$ and $\alpha' = \alpha[p,1,u]$. This establishes (1).

Consider the position symbols occuring in Y_p between α and α' in C_p, and occuring in Y_p' between α and α' in C_p'. Since these must occur in C (by Lemma 2) and this can happen in only one way, all of these position symbols must belong to C_p and C_p', respectively. This insures (2) and (3). $\qquad\qquad \square$

The length w substrings of the position symbols $\delta[i,j,0,0]$ and $\delta[i,j,0,1]$ in C define the $(i,j)^{th}$ edge position in the ith vertex unit and the length w substrings of the position symbols $\delta[i,j,1,0]$ and $\delta[i,j,1,1]$ in C define the $(i,j)^{th}$ edge echo position in the jth vertex unit. We term these a *corresponding pair* of edge and edge echo positions.

Claim 4. If C is a subsequence of length k'' common to the Control, Selection and Check sets, then for each corresponding pair of an edge position and an edge echo position, the same edge must be represented in the two positions.

Proof. Suppose C is a subsequence of length k'' common to the Control and Selection sets. We argue that if C is also common to $Z_{i,j}$ and $Z_{i,j}'$ then Lemma holds for the $(i,j)^{th}$ corresponding pair of positions. Let $C_{i,j}$ and $C_{i,j}'$ denote specific subsequences of $Z_{i,j}$ and $Z_{i,j}'$ isomorphic to C.

It is convenient to consider $Z_{i,j}$ (and similarly $Z'_{i,j}$) under the factorization $Z_{i,j} = Z_{i,j}(1)Z_{i,j}(2)Z_{i,j}(3)$ where

$$Z_{i,j}(2) = \prod_{\substack{1 \le u < v \le n \\ uv \in E}}^{\text{lex}\uparrow} \langle \text{edge } (i,j) \text{ from } u \text{ to } v \rangle$$

and where $Z_{i,j}(1)$ and $Z_{i,j}(3)$ are the appropriately defined prefix and suffix (respectively) of $Z_{i,j}$.

Since none of the position symbols in $Z_{i,j}(1)$ or $Z_{i,j}(3)$ occur in $Z_{i,j}(2)$, all of the position symbols in $Z_{i,j}(1)$ and $Z_{i,j}(3)$ must belong to $C_{i,j}$. Similarly, all of the position symbols in $Z'_{i,j}(1)$ and $Z'_{i,j}(3)$ must belong to $C'_{i,j}$. This implies, by Lemma 2, that $C_{i,j} \cup Z_{i,j}(2) = C'_{i,j} \cup Z'_{i,j}(2)$ begins with a symbol $\beta[i,j,0,u,v]$ and ends with a symbol $\beta[i,j,x,y]$. We argue that necessarily $u = x$ and $v = y$.

From the fact that $\beta[i,j,1,x,y]$ follows $\beta[i,j,0,u,v]$ in $Z_{i,j}(2)$, and from the construction of the latter in increasing lexicographic order, we may deduce that (u,v) precedes (x,y) lexicographically. Similarly, since $Z'_{i,j}(2)$ is constructed in decreasing lexicographic order, we obtain that (x,y) precedes (u,v), and therefore $(x,y) = (u,v)$. □

We now argue the correctness of the reduction as follows. If G has a k-clique, then it is easily seen that there is a common subsequence of length k'' in which the k vertex units represent the vertices of the clique, and the edge and edge echo positions within each vertex unit represent the edges incident on the represented vertex of the unit in increasing lexicographic order. (Each edge is thus represented twice, in the vertex units corresponding to its endpoints, first in an edge position in the initial vertex unit, and second in an edge echo position in the terminal vertex unit.)

Conversely, suppose there is a common subsequence C of length k''. By Claims 2 and 3, C represents a sequence of k vertices of G. That these must be a clique in G follows from Claim 4 and the definition of the "edge from" and "edge to" gadgets, which restrict the edges represented in a vertex unit to those present in the graph and for which the vertex is, respectively, initial or terminal. That completes the proof. □

Theorem 9 implies immediately that LCS-1 and LCS-2 are hard for $W[1]$, but it is possible to say more about the parameterized complexity of these problems.

Theorem 10. *LCS-1 is hard for $W[t]$ for all t.*

The reduction which establishes this is quite complicated and will appear elsewhere. Interestingly, it provides the starting point for a number of other important hardness results in parameterized complexity theory, such as the results that TRIANGULATING COLORED GRAPHS, INTERVALIZING COLORED GRAPHS and BANDWIDTH are hard for $W[t]$ for all t [BFH94].

Theorem 11. *LCS-2 is hard for $W[2]$.*

Proof. We reduce from DOMINATING SET. Let $G = (V, E)$ be a graph with $V = \{1, ..., n\}$. We will construct a set S of strings that have a common subsequence of length k if and only if G has a k-element dominating set.

The alphabet for the construction is

$$\Sigma = \{\alpha[i, j] : 1 \leq i \leq k, 1 \leq j \leq n\}$$

We use the following notation for important subsets of the alphabet.

$$\Sigma_i = \{\alpha[i, j] : 1 \leq j \leq n\}$$

$$\Sigma[t, u] = \{\alpha[i, j] : (i \neq t) \text{ or } (i = t \text{ and } j \in N[u])\}$$

The set S consists of the following strings.

Control Strings

$$X_1 = \prod_{i=1}^{k} (\uparrow \Sigma_i)$$

$$X_2 = \prod_{i=1}^{k} (\downarrow \Sigma_i)$$

Check Strings

For $u = 1, ..., n$:

$$X_u = \prod_{i=1}^{k} (\uparrow \Sigma[i, u])$$

To see that the construction works correctly, first note that by Claim 1 of the proof of Theorem 9, it follows easily that any sequence C of length k common to both control strings must consist of exactly one symbol from each Σ_i in ascending order. Thus to such a sequence C we may associate the set V_C of vertices represented by C: if $C = \alpha[1, u_1] \cdots \alpha[k, u_k]$, then $V_C = \{u_i : 1 \leq i \leq k\} = \{x : \exists i\ \alpha[i, x] \in C\}$.

We argue that if C is also a subsequence of the check strings $\{X_u\}$, then V_C is a dominating set in G. To this end, let $u \in V(G)$ and fix a substring C_u of X_u with $C_u = C$.

Claim. For some index j, $1 \leq j \leq k$, the symbol $\alpha[j, u_j]$ occurs in the $(\uparrow \Sigma[j, u])$ portion of X_u, and thus $u_j \in N[u]$ by the definition of $\Sigma[j, u]$.

We argue by induction on k. The case of $k = 1$ is clear. For the induction step, there are two cases: (1) the first $k - 1$ symbols of C_u occur in the prefix $(\uparrow \Sigma[1, u]) \cdots (\uparrow \Sigma[k - 1, u])$ of X_u, and the induction hypothesis immediately yields the Claim, or (2) the symbol $\alpha[k-1, u_{k-1}]$ occurs in the $(\uparrow \Sigma[k, u])$ portion of $C_u \cap X_u$. In case (2), this implies that the symbol $\alpha[k, u_k]$ of $C = C_u$ also occurs in the $(\uparrow \Sigma[k, u])$ part of X_u.

By the Claim, if C is a subsequence of the *Control* and *Check* strings, then every vertex of G has a neighbor in V_C, that is, V_C is a dominating set in G.

Conversely, if $D = \{u_1, ..., u_k\}$ is a k-element dominating set in G with $u_1 < \cdots < u_k$, then the sequence $C = \alpha[1, u_1] \cdots \alpha[k, u_k]$ is easily seen to be common to the strings of S. $\qquad\square$

4   Conclusions

Our results suggest that the general LCS problem is not fixed-parameter tractable when either k or m are fixed. It is important to note, however, that our results here apply only to the version of the problem where the size of the alphabet is unbounded. Since many applications involve fixed-size alphabets, the question of whether LCS-1 remains hard for W for a fixed alphabet size is very interesting. We have recently been able to show that LCS remains hard when parameterized by both the number of strings and the alphabet size.

Our results also have implications for the fixed-parameter tractability of the multiple sequence alignment and consensus subsequence discovery problems in molecular biology. This is so because the LCS problem is a special case of each of these problems. The problem of aligning k sequences is often re-stated as that of finding a minimal-cost path between two vertices in a particular type of edge-weighted k-dimensional graph [Pev92]. The LCS problem can be stated in this form using the edge-weighting in Section 3 of [Pev92], and is hence a restriction of the multiple sequence alignment problem (albeit, that version of the problem which allows arbitrary alignment evaluation functions). The LCS problem is shown to be a restriction of the consensus subsequence problem in Section 3 of [DM93b]. By the results of this paper, the general multiple sequence alignment (consensus subsequence discovery) problem is W[t]-hard for all t (W[2]-hard), and hence unlikely to be fixed-parameter tractable, when the number of sequences and the cost of the alignment (length of the consensus subsequence) are fixed.

Fixed-parameter complexity analysis may be relevant to many computational problems in biology. Many of these problems are known either to be NP-complete in general, e.g. evolutionary tree estimation by parsimony, character compatibility and distance-matrix fitting criteria (see [War93] and references), or to require time $O(n^k)$ when k is fixed, such as multiple sequence alignment using the SP or evolutionary tree alignment evaluation functions [Pev92]. To solve such problems in practice, investigators must often settle for suboptimal solutions obtained by algorithms that are fast but are either approximate or solution-constrained [KS83,San85,Pev92,Gus93,War93]. For instances of such problems, critical parameters such as the number of sequences or taxa are often small but nontrivial, e.g., $5 \leq k \leq 20$. These are precisely the situations in which fixed-parameter algorithms might be useful. Apart from showing for some problems that such algorithms are unlikely to exist by analyses such as presented in this paper, such results can be viewed as clarifying the contribution that each parameter makes to a problem's complexity, and may thus suggest computation-saving constraints that may yet yield restricted versions of these problems of feasible complexity.

References

[Bae91] R. A. Baeza-Yates. Searching subsequences. *Theoretical Computer Science* 78 (1991), 363–376.

[BFH94] H. Bodlaender, M. Fellows and M. Hallett. Beyond NP-completeness for problems of bounded width: hardness for the W hierarchy. To appear, *Proceedings fo the ACM Symposium on the Theory of Computing*, 1994.

[CCDF93] L. Cai, J. Chen, R. Downey and M. Fellows. The parameterized complexity of short computations and factorization. University of Victoria, Technical Report, Department of Computer Science, July, 1993.

[DEF93] R. Downey, P. Evans and M. Fellows. Parameterized learning complexity. *Proc. Sixth ACM Workshop on Computational Learning Theory (COLT)*, pp. 51–57, ACM Press, 1993.

[DF92] R. Downey and M. Fellows. Fixed-parameter intractability (extended abstract). In *Proceedings of the Seventh Annual Conference on Structure in Complexity Theory*, pp. 36–49, IEEE Computer Society Press, Los Alamitos, CA, 1992.

[DFKHW93] R. Downey, M. Fellows, B. Kapron, M. Hallett and H.T. Wareham. The parameterized complexity of some problems in logic and linguistics. Workshop on Recursion Theory and Complexity in Logic, Vancouver, B.C., Canada, October, 1993, and University of Victoria, Technical Report, Department of Computer Science, July, 1993.

[DM93a] W. H. E. Day and F. R. McMorris. Discovering consensus molecular sequences. In O. Opitz, B. Lausen, and R. Klar (eds.) *Information and Classification – Concepts, Methods, and Applications*, pp. 393–402, Springer-Verlag, Berlin, 1993.

[DM93b] W. H. E. Day and F. R. McMorris. The computation of consensus patterns in DNA sequences. *Mathematical and Computer Modelling* 17 (1993), 49–52.

[Gus93] D. Gusfield. Efficient methods for multiple sequence alignment with guaranteed error bounds. *Bulletin of Mathematical Biology* 55 (1993), 141–154.

[Hir83] D. S. Hirschberg. Recent results on the complexity of common subsequence problems. In D. Sankoff and J. B. Kruskal (eds.) *Time Warps, String Edits, and Macromolecules: The Theory and Practice of Sequence Comparison*, pp. 325–330, Addison-Wesley, Reading, MA, 1983.

[IF92] R. W. Irving and C. B. Fraser. Two algorithms for the longest common subsequence of three (or more) strings. In A. Apostolico, M. Crochemore, Z. Galil, and U. Manber (eds.) *Proceedings of the Third Annual Symposium on Combinatorial Pattern Matching*, pp. 214–229, Lecture Notes in Computer Science no. 644, Springer-Verlag, Berlin, 1992.

[KS83] J. B. Kruskal and D. Sankoff. An anthology of algorithms and concepts for sequence comparison. In D. Sankoff and J. B. Kruskal (eds.) *Time Warps, String Edits, and Macromolecules: The Theory and Practice of Sequence Comparison*, pp. 265–310, Addison-Wesley, Reading, MA, 1983.

[LF78] S. Y. Lu and K. S. Fu. A sentence-to-sentence clustering procedure for pattern analysis. *IEEE Transactions on Systems, Man, and Cybernetics* 8 (1978), 381–389.

[Mai78] D. Maier. The complexity of some problems on subsequences and supersequences. *Journal of the ACM* 25 (1978), 322–336.

[PM92] W. R. Pearson and W. Miller. Dynamic programming algorithms for biological sequence comparison. *Methods in Enzymology* 183 (1992), 575–601.

[Pev92] P. A. Pevzner. Multiple alignment, communication cost, and graph matching. *SIAM Journal on Applied Mathematics* 52 (1992), 1763–1779.

[San72] D. Sankoff. Matching comparisons under deletion/insertion constraints. *PNAS* 69 (1972), 4–6.

[San85] D. Sankoff. Simultaneous solution of the RNA folding, alignment, and protosequence problems. *SIAM Journal on Applied Mathematics* 45 (1985), 810–825.

[War93] H. T. Wareham. *On the Computational Complexity of Inferring Evolutionary Trees*, M.Sc. Thesis, Technical Report no. 9301, Department of Computer Science, Memorial University of Newfoundland, 1993.

Computing all Suboptimal Alignments in Linear Space†

Kun-Mao Chao

Department of Computer Science and Engineering,
The Pennsylvania State University,
University Park, PA 16802, USA.

Abstract. Recently, a new compact representation for suboptimal alignments was proposed by Naor and Brutlag (1993). The kernel of that representation is a minimal directed acyclic graph (DAG) containing all suboptimal alignments. In this paper, we propose a method that computes such a DAG in space *linear* to the graph size. Let F be the area of the region of the dynamic-programming matrix bounded by the suboptimal alignments and W the maximum width of that region. For two sequences of lengths M and N, it is shown that the worst-case running time is $O(MN + F \log \log W)$. To exploit the computed DAG, we employ a variant of Aho-Corasick pattern matching machine (Aho and Corasick, 1975) to locate all occurrences of specified patterns, and then find a path in the DAG that maximizes the sum of the scores of the non-overlapping patterns occurring in it. An example illustrates the utility.

1. Introduction

Biologically significant alignments are not necessarily mathematically optimized. It has been shown that sometimes the neighborhood of an optimal alignment reveals additional interesting biological features (Waterman and Byers, 1985; Saqi and Sternberg, 1991). Besides, the most strongly conserved regions can be effectively located by inspecting the range of variation of suboptimal alignments (Vingron and Argos, 1990; Zuker, 1991; Chao et al., 1993). While rigorous statistical analysis for the mean and variance of an optimal alignment score is not yet available, suboptimal alignments have been successfully used to informally estimate the significance of an optimal alignment.

However, it is essentially impractical to enumerate all suboptimal alignments since the number could be enormous. Therefore, a more compact representation of all suboptimal alignments is indispensable. A 0-1 matrix can be used to indicate if a pair of positions is in some suboptimal alignment or not (Vingron and Argos, 1990; Zuker, 1991). As pointed out by Naor and Brutlag (1993), this approach misses some connectivity information among those pairs of positions. They then used a set of "canonical" suboptimal alignments to represent all suboptimal alignments. The kernel of that representation is a minimal directed acyclic graph (DAG) containing all suboptimal alignments. Although their work was based on a simple scoring scheme, it is also applicable for affine gap penalties. ("Affine" means that a gap of length k is penalized $\alpha + k \times \beta$, i.e., it costs α to open up a gap plus β for each symbol in the gap.)

Traditional dynamic-programming algorithms for sequence comparison require quadratic space, and hence are infeasible for long protein or DNA sequences. Fortunately, quadratic-time, linear-space methods have been successfully designed to conquer this problem (Hirschberg, 1975; Myers and Miller, 1988).

†This work was supported by grant R01 LM05110 from the National Library of Medicine.

In this paper, we propose a method that computes the DAG representing all suboptimal alignments. The space requirement is linear to the size of the DAG. The time, however, is out-put-sensitive. Let F be the area of the region of the dynamic-programming matrix bounded by the suboptimal alignments and W the maximum width of that region. For two sequences of lengths M and N, it is shown that the worst-case running time is $O(MN + F \log \log W)$.

To exploit the computed DAG, we employ a variant of Aho-Corasick pattern matching machine (Aho and Corasick, 1975) to locate all occurrences of specified patterns, and then find a path in the DAG that maximizes the sum of the scores of the non-overlapping patterns occurring in it. This is useful in delivering a more "meaningful" alignment. For instance, if there is more than one optimal alignment, we would prefer the one revealing more motifs of interest.

The rest of the paper is organized as follows. In Section 2, we present a relatively simple linear-space algorithm for computing the DAG in time $O(MN + F \log W)$. In Section 3, the algorithm is refined to compute the DAG in time $O(MN + F \log \log W)$. In Section 4, we discuss an algorithm that finds a path in the computed DAG with the maximum pattern score. In Section 5, an example illustrates the utility. Section 6 discusses some future research directions.

2. A Simple Linear-Space Algorithm for Computing the DAG

Given two sequences $A = a_1 a_2 \cdots a_M$ and $B = b_1 b_2 \cdots b_N$, an *alignment* of A and B is obtained by introducing dashes into the two sequences such that the lengths of the two resulting sequences are identical and no column contains two dashes. Let Σ denote the input symbol alphabet. A score $\sigma(a, b)$ is defined for each $(a, b) \in \Sigma \times \Sigma$. A gap of length k is penalized $\alpha + k \times \beta$. The score of an alignment is the sum of σ scores of all columns with no dashes minus the penalties of the gaps.

It is helpful to think of an alignment as a path in the alignment graph, $G_{A,B}$, defined as follows. $G_{A,B}$ is a directed graph with $3(M+1)(N+1)$ nodes, denoted $(i, j)_D$, $(i, j)_I$ and $(i, j)_S$, where $i \in [0, M]$ and $j \in [0, N]$. Table 1 depicts all the edges in $G_{A,B}$:

edge	weight	aligned pair	range
$(i-1, j)_D \rightarrow (i, j)_D$	$-\beta$	$\begin{bmatrix} a_i \\ - \end{bmatrix}$	$i \in [1, M]$ and $j \in [0, N]$
$(i-1, j)_S \rightarrow (i, j)_D$	$-(\alpha + \beta)$	$\begin{bmatrix} a_i \\ - \end{bmatrix}$	$i \in [1, M]$ and $j \in [0, N]$
$(i, j-1)_I \rightarrow (i, j)_I$	$-\beta$	$\begin{bmatrix} - \\ b_j \end{bmatrix}$	$i \in [0, M]$ and $j \in [1, N]$
$(i, j-1)_S \rightarrow (i, j)_I$	$-(\alpha + \beta)$	$\begin{bmatrix} - \\ b_j \end{bmatrix}$	$i \in [0, M]$ and $j \in [1, N]$
$(i-1, j-1)_S \rightarrow (i, j)_S$	$\sigma(a_i, b_j)$	$\begin{bmatrix} a_i \\ b_j \end{bmatrix}$	$i \in [1, M]$ and $j \in [1, N]$
$(i, j)_D \rightarrow (i, j)_S$	0	none	$i \in [0, M]$ and $j \in [0, N]$
$(i, j)_I \rightarrow (i, j)_S$	0	none	$i \in [0, M]$ and $j \in [0, N]$

Table 1. The weights and aligned pairs associated with edges of $G_{A, B}$.

Let s denote $(0, 0)_S$ and t denote $(M, N)_S$. A path is *normal* if and only if it does not contain subpaths of the form $(i-1, j)_D \rightarrow (i-1, j)_S \rightarrow (i, j)_D$ or $(i, j-1)_I \rightarrow (i, j-1)_S \rightarrow (i, j)_I$. It can be shown that alignments of A and B are in one-to-one correspondence with normal s-t paths (Myers and Miller, 1989). Furthermore, define the score of an s-t path P, denoted as

$Score(P)$, to be the sum over the weights of its edges. $Score(P)$ is the score of the alignment corresponding to P.

Suppose we are given a threshold score Δ that does not exceed the optimum score. A Δ-suboptimal path (or Δ-path) is an s-t path with score at least as large as Δ. A Δ-suboptimal grid point (or Δ-point) is a grid point where at least one of its nodes appears in some Δ-path. Obviously, both $(0,0)$ and (M,N) are Δ-points. A Δ-suboptimal edge (or Δ-edge) is an edge that appears in some Δ-path.

Our goal is to compute a directed acyclic graph, denoted by $DAG_\Delta = (V_\Delta, E_\Delta)$, where V_Δ is the set of nodes in all Δ-points and E_Δ is the set of all Δ-edges. In the following, we will show how to construct V_Δ in $O(|V_\Delta|)$ space.

Let $Score^-(i,j)_X$ be the maximum score of any path from s to $(i,j)_X$, where $X \in \{D, I, S\}$. With proper initializations, these scores can be computed by the following recurrence relations (Myers and Miller, 1988):

$$Score^-(i,j)_D = \max\{Score^-(i-1,j)_D - \beta, Score^-(i-1,j)_S - \alpha - \beta\}$$
$$Score^-(i,j)_I = \max\{Score^-(i,j-1)_I - \beta, Score^-(i,j-1)_S - \alpha - \beta\}$$
$$Score^-(i,j)_S = \max\{Score^-(i-1,j-1)_S + \sigma(a_i,b_j), Score^-(i,j)_D, Score^-(i,j)_I\}$$

Similarly, let $Score^+(i,j)_X$ be the maximum score of any path from $(i,j)_X$ to t, where $X \in \{D, I, S\}$. With proper initializations, these scores can be computed by the following recurrence relation:

$$Score^+(i,j)_S = \max\{Score^+(i+1,j+1)_S + \sigma(a_{i+1},b_{j+1}), Score^+(i+1,j)_D - \alpha - \beta,$$
$$Score^+(i,j+1)_I - \alpha - \beta\}$$
$$Score^+(i,j)_D = \max\{Score^+(i+1,j)_D - \beta, Score^+(i,j)_S\}$$
$$Score^+(i,j)_I = \max\{Score^+(i,j+1)_I - \beta, Score^+(i,j)_S\}$$

Define $Score(i,j) = \max\{Score^-(i,j)_X + Score^+(i,j)_X \mid X \in \{D, I, S\}\}$.

Lemma 1. A grid point (i,j) is a Δ-point if and only if $Score(i,j) \geq \Delta$.

Proof. Omitted. $\square$

Let $[T,B] \times [L,R]$ denote the rectangle whose upper left corner is (T,L) and lower right corner is (B,R). We say that $[T,B] \times [L,R]$ contains (i,j) (or (i,j) is in $[T,B] \times [L,R]$) if $T \leq i \leq B$ and $L \leq j \leq R$.

Lemma 2. Let (i,j) be a Δ-point in $[1,M-1] \times [1,N-1]$. At least one of $(i,j-1)$, $(i-1,j)$ and $(i-1,j-1)$ is a Δ-point. At least one of $(i,j+1)$, $(i+1,j)$ and $(i+1,j+1)$ is a Δ-point.

Proof. Omitted. $\square$

Given a rectangle, denoted by Π, let π be the set of Δ-points on Π's boundaries. If π is not empty, let π_{i_1} and π_{i_2} be the minimum and maximum index, respectively, of the rows containing some of π's elements, and let π_{j_1} and π_{j_2} be the minimum and maximum index, respectively, of the columns containing some of π's elements.

Lemma 3. If π is empty, there is no Δ-point in Π. Otherwise, $[\pi_{i_1}, \pi_{i_2}] \times [\pi_{j_1}, \pi_{j_2}]$ contains all Δ-points in Π.

Proof. Suppose there are some Δ-points in Π, and π is empty. Take any such Δ-point. By Lemma 2, we can always trace back from that Δ-point to a boundary Δ-point. A contradiction with the assumption that π is empty.

If π is not empty, we claim $[\pi_{i_1}, \pi_{i_2}] \times [\pi_{j_1}, \pi_{j_2}]$ contains all Δ-points in Π. Indeed, suppose there exists a Δ-point in Π with a row index smaller than π_{i_1}. We can trace back from that Δ-point to a boundary Δ-point with a row index smaller than π_{i_1}, contradicting the assumption that π_{i_1} is the minimum index of the rows that contain some of π's points. Similar arguments apply to π_{i_2}, π_{j_1} and π_{j_2}. It follows that $[\pi_{i_1}, \pi_{i_2}] \times [\pi_{j_1}, \pi_{j_2}]$ contains all Δ-points in Π. □

In fact, it can be shown that $[\pi_{i_1}, \pi_{i_2}] \times [\pi_{j_1}, \pi_{j_2}]$ is the smallest rectangle that contains all Δ-points in Π.

The algorithm for computing all Δ-points is outlined as follows. For each conducted subproblem, the invariant is that $Score^-$ are given for every grid point on the left and upper boundaries, and $Score^+$ are given for every grid point on the right and lower boundaries. With these scores, the $Score^-$ and $Score^+$ for grid points within the subproblem can be computed. Problems with one or two rows or columns, can be solved directly. In general, a larger subproblem is then divided into four non-overlapping subproblems by the middle row and middle column.

To do so, a linear-space forward pass is performed to compute $Score^-$. To maintain the invariant, $Score^-$ are stored in every grid point on the two middle rows and two middle columns. To decide a more accurate range of each subproblem, $Score$ for each grid point on the right and lower boundaries is also determined and stored.

Similarly, a linear-space backward pass is performed to compute $Score^+$. To maintain the invariant, $Score^+$ are stored in every grid point on the two middle rows and two middle columns. $Score$ for each grid point on the left and upper boundaries is also determined and stored.

At this point, the $Score$ for each grid point on the boundaries of the four subrectangles, divided by the middle row and middle column, can be determined in constant time. Take one subrectangle for example, we determine the minimum and maximum indices of the rows and the minimum and maximum indices of the columns that contain at least one Δ-point on the subrectangle's boundaries. Lemma 3 says that the rectangle bounded by these rows and columns contains all Δ-points in the subrectangle. It is therefore enough to consider only the "shrunken" subrectangle. Figure 1 illustrates the approach.

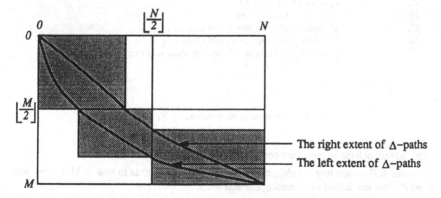

Figure 1. Splitting the problem into subproblems (shaded areas).

Figure 2 gives the pseudo code for constructing V_Δ in linear space. Let $Sub[i]$ be a linked list to store all Δ-points in row i for $0 \le i \le M$. Initially, they are set to be empty. Each time when a Δ-point is found, the function *append* is called to add the point to its Sub list. We assume that $Score^-$ and $Score^+$ are stored in each Δ-point.

```
1.   procedure SUB_OPT(M, N)
2.      { Compute Score⁻ for row 0 and column 0
3.        Compute Score⁺ for row M and column N
4.        for i ← 0 to M do Sub[i] ← φ
5.        sub_opt(0, 0, M, N, initial boundary score vectors)
     }

6.   recursive procedure sub_opt(I₁, J₁, I₂, J₂, boundary score vectors)
        /* Compute all Δ-points in [I₁, I₂] × [J₁, J₂] */
7.      { if I₁ + 1 ≥ I₂ or J₁ + 1 ≥ J₂ then
8.          { Compute and store Score(i, j) for each (i, j) in [I₁, I₂] × [J₁, J₂].
9.            for i ← I₁ to I₂ do
10.               for j ← J₁ to J₂ do { if Score(i, j) ≥ Δ then append(Sub[i], POINT(i, j)) }
11.           return
          }
12.       midI ← ⌊(I₁ + I₂)/2⌋
13.       midJ ← ⌊(J₁ + J₂)/2⌋
14.       A linear-space forward computation is performed to compute Score⁻:
              store Score⁻(i, j) if i = midI or midI + 1, or j = midJ or midJ + 1;
              store Score(i, j) if i = I₂ or j = J₂.
15.       A linear-space backward computation is performed to compute Score⁺:
              store Score⁺(i, j) if i = midI or midI + 1, or j = midJ or midJ + 1;
              store Score(i, j) if i = I₁ or j = J₁.
        /* Divide the problem by row midI and column midJ */
16.       Π₁ ← the set of the grid points on the boundaries of [I₁, midI] × [J₁, midJ]
17.       Π₂ ← the set of the grid points on the boundaries of [I₁, midI] × [midJ + 1, J₂]
18.       Π₃ ← the set of the grid points on the boundaries of [midI + 1, I₂] × [J₁, midJ]
19.       Π₄ ← the set of the grid points on the boundaries of [midI + 1, I₂] × [midJ + 1, J₂]
20.       for k ← 1 to 4 do
21.         { π ← {(i, j) | Score(i, j) ≥ Δ, (i, j) ∈ Πₖ}
22.           if π ≠ φ then
23.             { i₁ ← min{i | (i, j) ∈ π}
24.               j₁ ← min{j | (i, j) ∈ π}
25.               i₂ ← max{i | (i, j) ∈ π}
26.               j₂ ← max{j | (i, j) ∈ π}
27.               Compute Score⁻ for row i₁ and column j₁
28.               Compute Score⁺ for row i₂ and column j₂
29.               sub_opt(i₁, j₁, i₂, j₂, new boundary score vectors);
              }
          }
     }
```

Figure 2. The algorithm for constructing V_Δ in linear space.

The following lemma proves the correctness of the algorithm in Figure 2.

Lemma 4. For each row i, $Sub[i]$ contains only and all Δ-points in row i. Moreover, those points are distinct and linked in increasing column order.

Proof. Omitted. □

36

Space requirement

Theorem 5. The space for the boundary score vectors of all pending subproblems is $O(M + N)$.

Proof. Let $S(m, n)$ denote the worst-case space requirement for the boundary score vectors of all pending subproblems when applying *sub_opt* to a subproblem with m rows and n columns. Since each of its four possible subproblems is solved independently,

$$S(m, n) \le \begin{cases} c(m+n) & \text{for } m \le 2 \text{ or } n \le 2 \\ S(\lceil m/2 \rceil, \lceil n/2 \rceil) + c(m+n) & \text{for } m > 2 \text{ and } n > 2 \end{cases}$$

where c is a constant. It follows $S(M, N) = O(M + N)$. $\square$

Since $|V_\Delta|$ is $\Omega(\max\{M, N\})$, the space for the boundary score vectors and computed Δ-points is $O(|V_\Delta|)$. To see that this dominates the algorithm's space requirements, we need to consider the maximum size of the procedure activation stack, which depends on the maximum recursion depth. The number of rows (and columns) of the problem at a recursive call to *sub_opt* is at most half that of the containing problem (rounded up), so the maximum stack depth is $O(\min\{\log M, \log N\})$.

Time analysis

For each row i, define $L[i]$ and $R[i]$ to be the minimum and maximum index, respectively, of the columns where a Δ-path intersects row i. The band width of row i, $R[i] - L[i] + 1$, is denoted by $W_{row}[i]$. $W_{col}[j]$ is defined in a similar way. W is defined to be $\min\{\max\{W_{row}[i]\}, \max\{W_{col}[j]\}\}$. Let F denote the area of the region of the dynamic-programming matrix bounded by Δ-paths, i.e. $F = \sum_{i=0}^{M} W_{row}[i]$.

Lemma 6. If $R[i] \le midJ$ in the current subproblem, $(i, midJ + 1), \cdots, (i, J_2)$ will not be included in any subsequent subproblem. Similarly, if $L[i] > midJ$ in the current subproblem, $(i, J_1), \cdots, (i, midJ)$ will not be included in any subsequent subproblem.

Proof. Since the right extent of Δ-paths is monotonically increasing, it is easy to see that if $R[i] \le midJ$, $[I_1, i] \times [midJ + 1, J_2]$ does not contain any Δ-points. Either $[I_1, I_2] \times [midJ + 1, J_2]$ does not contain any Δ-points, or the minimum index of the rows that contain some Δ-points in $[I_1, I_2] \times [midJ + 1, J_2]$ is larger than i. In either case, $(i, midJ + 1), \cdots, (i, J_2)$ will not be included in any subsequent subproblem. The case when $L[i] > midJ$ can be proved in a similar way. $\square$

Theorem 7. Let T be the total number of grid points in all the calls to *sub_opt*. $T = O(MN + F \log W)$.

Proof. Let subproblems with no more than two rows or two columns be trivial subproblems. Since each grid point can be included in at most one trivial subproblem, $O(MN)$ grid points are included in such subproblems.

Fix a row i, consider all nontrivial subproblems that include some row i's grid-points. Before reaching the first subproblem with the property $J_1 \le L[i] \le midJ \le R[i] \le J_2$, all its containing subproblems include in total $O(N)$ row i's grid points. This is because all its containing subproblems is either with the property $J_1 \le L[i] \le R[i] < midJ \le J_2$ or $J_1 \le midJ < L[i] \le R[i] \le J_2$ which will truncate half of row i's grid points in the subsequent call (Lemma 6).

The subproblem is further split into at most one subproblem with the property $J_1 \le L[i] \le J_2 \le R[i]$, and at most one with the property $L[i] < J_1 \le R[i] \le J_2$. Now we show that each of them will include $O(N + W_{row}[i] \log W_{row}[i])$ row i's grid points in its subsequent calls. Indeed, consider the subproblem with $J_1 \le L[i] \le J_2 \le R[i]$. If $midJ \ge L[i]$, it is easy to see that all its subsequent subproblems include in total $O(W_{row}[i] \log W_{row}[i])$ row i's grid points. If $midJ < L[i]$, $(i, J_1), \cdots, (i, midJ)$ will be truncated (Lemma 6). Before reaching the subproblem with $midJ \ge L[i]$, those containing subproblems include in total $O(N)$ row i's grid points. Similar arguments apply to the case when $L[i] < J_1 \le R[i] \le J_2$.

It follows that all subproblems include $O(N + W_{row}[i] \log W_{row}[i])$ row i's grid points. Therefore, we have

$$T = O(MN + \sum_{i=0}^{M} W_{row}[i] \log W_{row}[i]) = O(MN + F \log \max \{W_{row}[i]\})$$

In a similar way, we can derive $T = O(MN + F \log \max\{W_{col}[j]\})$. It follows $T = O(MN + F \log W)$.

$\square$

Since $F \le MN$ and $W \le \min \{M, N\}$, $T = O(MN \log \min \{M, N\})$. This remains even when DAG_Δ is sparse because the width of DAG_Δ could be independent of its density. On the other hand, Theorem 7 implies that if $F = O(MN / \log W)$, $T = O(MN)$.

To complete the construction of DAG_Δ, we need to build E_Δ. Let e be an edge from node u to node v. Define $Score(e)$ to be $Score^-(u) + weight(e) + Score^+(v)$. It can be shown that e is a Δ-edge if and only if $Score(e) \ge \Delta$. Obviously, if e is a Δ-edge, both u and v are at some Δ-point. Constructing all Δ-edges from the Sub lists takes $O(|V_\Delta|)$ time.

It should be noted that not every s-t path in DAG_Δ has score at least Δ. However, methods of Waterman and Byers (1985) or Naor and Brutlag (1993) can be applied to DAG_Δ to generate Δ-paths efficiently.

As defined by Naor and Brutlag (1993), an s-t path P is called canonical if there exists an edge e in P such that $Score(e) = Score(P)$. They further showed that canonical Δ-paths can represent all Δ-paths and their number is far less than the number of all Δ-paths. It can be shown that their theorems for canonical paths also hold for DAG_Δ.

3. An Improved Linear-Space Algorithm

For each conducted subproblem, the invariant is that $Score^-$ are given for every grid point on the left and upper boundaries, and $Score^+$ are given for every grid point on the right and lower boundaries. Instead of partitioning a subproblem into four subproblems, we partition it into a different number of subproblems, depending on the recursion depth of a subproblem. Let $L(i)$ and $W(i)$ be the number of rows and columns of a subproblem in recursion depth i, respectively. In general, an $L(i) \times W(i)$ subproblem at recursion depth i is divided into $T^2(i)$ non-overlapping $\dfrac{L(i)}{T(i)} \times \dfrac{W(i)}{T(i)}$ subproblems, where $T(i)$ is determined by the following recurrence relation.

$$T(i) = \begin{cases} b & \text{for } i = 0 \\ T^2(i-1)/2 & \text{for } i > 0 \end{cases}$$

where $b > 2$ is a constant. It can be shown that $T(i) = \dfrac{b^{2^i}}{2^{2^i - 1}}$. $L(i)$ and $W(i)$ are computed as follows.

$$L(i) = \begin{cases} M + 1 & \text{for } i = 0 \\ L(i-1)/T(i-1) & \text{for } i > 0 \end{cases}$$

$$W(i) = \begin{cases} N+1 & \text{for } i=0 \\ W(i-1)/T(i-1) & \text{for } i>0 \end{cases}$$

One can show that $L(i) = \dfrac{2^{2^i-1-i}}{b^{2^i-1}}(M+1)$ and $W(i) = \dfrac{2^{2^i-1-i}}{b^{2^i-1}}(N+1)$.

Theorem 8. The space for the boundary score vectors of all pending subproblems is $O(M+N)$.

Proof. Let $c(n+m)$ be the space required to store boundary score vectors for an $n \times m$ subproblem, where c is a constant. Let $S(k)$ be the total space for the boundary score vectors of all pending subproblems when the recursion depth is k.

$$S(k) = \sum_{i=0}^{k} c(L(i) + W(i))T(i)$$
$$= c \sum_{i=0}^{k} \frac{b}{2^i}(M+N+2)$$
$$\leq 2cb(M+N+2)$$
$$= O(M+N)$$

□

Table 2 illustrates the case when $b = 4$

Recursion depth i	0	1	2	3	4	5	...
$W(i)$	$N+1$	$\frac{1}{2^2}(N+1)$	$\frac{1}{2^5}(N+1)$	$\frac{1}{2^{10}}(N+1)$	$\frac{1}{2^{19}}(N+1)$	$\frac{1}{2^{36}}(N+1)$	...
$T(i)$	2^2	2^3	2^5	2^9	2^{17}	2^{33}	...
$T(i) \times W(i)$	$2^2(N+1)$	$2(N+1)$	$N+1$	$\frac{1}{2}(N+1)$	$\frac{1}{2^2}(N+1)$	$\frac{1}{2^3}(N+1)$	...

Table 2. Rate of growth when $b = 4$.

Lemma 9. The maximum recursion depth is $O(\log\log\min\{M,N\})$.

Proof. The recursive procedure stops at recursion depth i when $L(i) \leq 1$ or $W(i) \leq 1$. Since $L(i) = (\frac{2}{b})^{2^i-1}\frac{1}{2^i}(M+1)$ and $b > 2$, it can be shown that $L(i) \leq 1$ for some $i = c\log\log M$, where c is a constant. Thus, $L(i)$ decreases to 1 in $O(\log\log M)$ steps. Similarly, we can show that $W(i)$ decreases to 1 in $O(\log\log N)$ steps. Therefore, the maximum recursion depth is $O(\log\log\min\{M,N\})$. □

Theorem 10. The total running time is $O(MN\log\log\min\{M,N\})$.

Proof. It takes in total $O(MN)$ time for all the subproblems at the same recursion depth. Lemma 9 shows that the recursion depth is bounded by $O(\log\log\min\{M,N\})$. It follows that the total running time is $O(MN\log\log\min\{M,N\})$. □

Again, Lemma 3 can be applied to reduce the size of each conducted subproblem. With an argument similar to the proof of Theorem 7, we have the following theorem.

Theorem 11. The total running time of the new divide-and-conquer algorithm augmented with shrinking operation described in Lemma 3 is $O(MN + F \log \log W)$.

4. Finding an s-t path in DAG_Δ with the maximum pattern score

This section discusses one way of utilizing DAG_Δ. Given is a set of patterns, where each pattern ω is given a positive score ω_{score}. The pattern score of a path P is defined as the maximum sum of the scores of non-overlapping patterns occurring in P. The goal is to find an s-t path P_Δ in DAG_Δ such that the pattern score of P_Δ is maximum among all s-t paths in DAG_Δ. Furthermore, if there are more than one s-t paths maximizing the pattern score, $Score(P_\Delta)$ is maximum among all such paths.

A pattern ω is said to *occur* at Δ-point (i, j) if $a_{i-|\omega|+1} a_{i-|\omega|+2} \cdots a_i = b_{j-|\omega|+1} b_{j-|\omega|+2} \cdots b_j = \omega$, and $(i-|\omega|, j-|\omega|)_S \rightarrow (i-|\omega|+1, j-|\omega|+1)_S \rightarrow \cdots (i, j)_S$ is a path in DAG_Δ. An occurrence edge from $(i-|\omega|, j-|\omega|)_S$ to $(i, j)_S$, denoted by $(i-|\omega|, j-|\omega|)_S \rightarrow_\omega (i, j)_S$, is augmented to DAG_Δ if ω occurs at (i, j) for some pattern ω in the given pattern set.

In order to augment DAG_Δ with all such occurrence edges, a finite state pattern matching machine, following the scheme of Aho and Corasick (1975), is constructed. It is operated by three functions: a goto function g, a failure function f, and an output function $output$ (see Aho and Corasick, 1975). Figure 3 outlines the algorithm for constructing all occurrence edges.

```
for each Δ-point (i, j) in topological order do
    if (i − 1, j − 1)ₛ → (i, j)ₛ is not a Δ-edge then
    {  state ← 0
       k ← 0
       while (i + k, j + k)ₛ → (i + k + 1, j + k + 1)ₛ is a Δ-edge
       {  if aᵢ₊ₖ₊₁ = bⱼ₊ₖ₊₁ then
          {  while g(state, aᵢ₊ₖ₊₁) = fail do state ← f(state)
             for each pattern ω in output(state) do
                  Construct (i + k + 1 − |ω|, j + k + 1 − |ω|)ₛ →ω (i + k + 1, j + k + 1)ₛ
          }
          else state ← 0
          k ← k + 1
       }
    }
}
```

Figure 3. The algorithm for constructing all occurrence edges.

Let l be the sum of the pattern lengths. Let Num_s be the number of the patterns recognized by state s. The time for constructing an Aho-Corasick pattern matching machine is $O(l|\Sigma| + \sum_s Num_s)$. It can be shown that the total number of state transitions made by the algorithm in Figure 3 is $O(|V_\Delta|)$. If a pattern ω occurs at a Δ-point (i, j), we have to construct an occurrence edge $(i-|\omega|, j-|\omega|)_S \rightarrow_\omega (i, j)_S$. A stack can be used to backtrack the starting location of the occurrences on the same diagonal. The time for constructing all occurrence edges is $O(|V_\Delta| + Occ)$, where Occ is the number of occurrences.

Let $Pat\_Score(u)$ be the maximum pattern score of any path from u to t in DAG_Δ. The following recurrence relation computes $Pat\_Score(u)$.

$$Pat\_Score(u) = \max \{ \max \{ Pat\_Score(v) \mid u \rightarrow v \text{ is a } \Delta\text{-edge.} \},$$

$$\max \{ Pat\_Score(v) + \omega_{score} \mid u \rightarrow_\omega v \text{ is an occurrence edge.} \} \}$$

It can be computed in $O(|V_\Delta| + Occ)$ time for all nodes in DAG_Δ. A simple backtracking method with the tie-breaking rules yields an s-t path P_Δ in DAG_Δ such that $Score(P_\Delta)$ is maximum among all s-t paths in DAG_Δ with the maximum pattern score. It should be noted that $Score(P_\Delta)$ may be worse than Δ.

In particular, when the threshold score Δ is the optimum score, it is easy to see that every s-t path in DAG_Δ is an optimal path. Therefore, the algorithm presented in this section can be used to deliver an optimal alignment with the maximum pattern score. It should be noted that the problem of finding optimal alignments containing patterns has been explored before. For example, Lawerence *et al.* (1986) compute the alignment score as the score of the concatenated optimal local alignments which were extended from homologies exceeding or equal to a specified minimum length.

5. An example

We have implemented the algorithms in Sections 2 and 4. The conducted experiments showed that with threshold score Δ close to the optimum score, $T < 2(M+1)(N+1)$. Surprisingly, in that reasonable range, it even ran faster than the quadratic-time, linear-space *left_right* program (Chao *et al.*, 1993) that locates merely the left and right extents of Δ-paths.

To illustrate the utility of the algorithm developed in Section 3, we aligned the ε-globin gene regions of human and rabbit. Identical matching nucleotides scored 1, mismatches scored -1 and k-symbol gaps were penalized $6 + 0.2k$. Figures 4 and 5 are a portion of two different optimal alignments. Figure 6 is a multiple alignment shown in Hardison *et al.* (1993). The human sequence is given in full, and periods denote a matching nucleotide in the other species.

The alternate optimal alignment in Figure 5 reveals three interesting features that are not in the optimal alignment in Figure 4. Box 1 contains the same gap revealed in the multiple alignment in Figure 6. Box 2 contains a gap followed by a matching block instead of two matching blocks split by a gap. Incidentally, this can be used to improve the multiple alignment in Figure 6. Finally, box 3 contains a matching block GAAGAG, a candidate for the phylogenetic footprint, which is defined as at least six consecutive invariant positions (Tagle *et al*, 1988; Gumucio *et al.*, 1993). Phylogenetic footprints have been demonstrated to be useful as a guide to identifying nuclear protein binding sites. In fact, GAAGAG also appears in the corresponding region of galago.

```
           |              |         |          |        |
19091:  TTTGTCAACTGTCACCACCTTTAAGGCAAATGTTAAATGTGCTTTGGCTGAAACTTTTTT   human
 5544:  .......--------..A.C..G.CC.....A.G...C-.A....A...CT...AC...-   rabbit

          |                 |                |             |
19151:  TCCTATTTTGAGATTTGCTCCTTTATATGAGGCTTTCTTGGAAAAGGAGAATGGGAGAGA   human
 5595:  -..AG.C..AT.C..A...G..C.C.....AT...............G........AT..   rabbit

          |              |      |          |              |
19211:  TGGATATCATTTTGGAAGATGATGA-----------AGAGGGTAAAAAAGGGGACAAATG   human
 5654:  .....GC...C.........T.CATGGAAAAAGAAG....T.A...C.T.ATA.TGT...   rabbit
```

Figure 4. An optimal alignment.

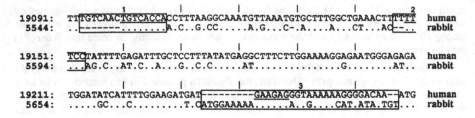

```
         1              |              |              |              |            2
19091:  TT TGTCAACTGTCACCA CCTTTAAGGCAAATGTTAAATGTGCTTTGGCTGAAACTT TTTT   human
5544:   .. --------........ A.C..G.CC.....A.G...C-.A....A...CT...AC --..   rabbit

            |              |              |              |              |
19151:  TCC TATTTTGAGATTTGCTCCTTTATATGAGGCTTTCTTGGAAAAGGAGAATGGGAGAGA     human
5594:  ... AG.C..AT.C..A...G..C.C.....AT...............G........AT..      rabbit

          |              |              |        3      |
19211:  TGGATATCATTTTGGAAGATGAT ---------GAAGAGGGTAAAAAAGGGGACAA-- ATG     human
5654:  .....GC...C.........T.C ATGGAAAAA......A..G....CAT.ATA.TG T...      rabbit
```

Figure 5. An alternate optimal alignment. TGTCACCA, TTTCC and GAAGAG are in the given pattern set.

```
              εNEBP         |              |              |              |
19091:  TTTGTCAAC TGTCACCA CCTTTAAG GCAAAT GTTAAATGTGCTTTGG-CTGAAACTTTTTT   human
3960:   ........T- ......T.T.CA... TC.....T.....CTCAT.....-.....G..G...     galago
5544:   ..------- .......A.C..G.CC.....A.G...CT-A....A.-.CT...AC...         rabbit
76:                                                                 ..     goat
21352:  C..A.T.T .A.AG... TTC..TCCCT...C.G....GT.AT...A.A...G.TT.....         mouse

               |              |              |       YRC    |    YY1
19150:  TTCCTATTTTGA GATTTGCT CCTTTATATGAGGCTTT CTTGGAAAAGG AG-AATGGGAGA   human
4018:   .-....CC.GAG.C..A.T..A.---.....A..--.........G.G....A....            galago
5595:   --..AG.C..AT.C..A...G..C.C.....AT............G.-......AT             rabbit
78:    C-.TAG.C.GAT.C..A...A.AC.CC...GAT.A...G.A.G....-.........             goat
21411:  .-.....AGC.G.C.AAA....C.T.C..T.C....G........A.T.--...A....            mouse

         YY1  GATA1      Ets           |              |              |
19209:  GATGGATATCATTT TGGAAG ATGATGAAGAGGGTAAAAAAGGGGACAAATG             human
4072:   A...TT.G.T.....T...G.....TG.G.AAAG..G.GA.T..GT.GA.                   galago
5652:   .......GC...C......T.CATG..AAAAG..G.GA.T..A..CAT                     rabbit
136:    .......GC...CG....G.....GAGG.GAA.A.C.G.G..T..A..TG.                  goat
21468:  A....GCG.T..AC...G..T.T.CATGTG.A.AGT.G..AAT..A..GA.                  mouse
```

Figure 6. Multiple alignment of the 5' flank of mammalian ε-globin genes.

6. Discussion

It has been shown that Δ-points are useful in speeding up the computation for multiple sequence alignment problem (Carrillo and Lipman, 1988; Altschul and Lipman, 1989). As noted by Kececioglu(1989), the $O(MN)$ space, which is required by a straightforward method for computing all Δ-points, may be the dominant space requirement for inputs consisting of a few long sequences. The linear-space algorithm presented here can be applied in this context.

It is natural to design a model that does some pattern matching to extract more information from the DAG. For instance, it is hoped that the DAG will provide a good estimate of how robust an optimal alignment is. Also, the DAG might be utilized for finding genes in a given sequence. It remains to be investigated what kind of language would be appropriate for these purposes.

A *local alignment* is an alignment where the end-nodes can be arbitrary, i.e., they are not restricted to $(0,0)_S$ and $(M,N)_S$. One can define a grid point to be a *local* Δ-point if at least one of its nodes appears in some local alignment with score at least Δ. The divide-and-conquer approach described in Section 3 yields a $O(MN \log \log \min \{M, N\})$-time, linear-space method for computing all local Δ-points. Can it be done more efficiently?

Acknowledgements

I would like to thank Dr. Webb Miller for valuable guidance and encouragement. His comments resulted in numerous improvements in the presentation. I would also like to thank Dr. John Kececioglu for helpful discussions during his visit at Penn State.

References

Aho, A. V. and Corasick, M. J. (1975) Efficient string matching: an aid to bibliographic search. *Comm. ACM*, **18**, 333-340.

Altschul, S. F. and Lipman, D. J. (1989) Trees, stars, and multiple biological sequence alignment. *SIAM J. Appl. Math.*, **49**, 197-209.

Carrillo, H., and Lipman, D. J. (1988) The multiple sequence alignment problem in biology. *SIAM J. Appl. Math.*, **48**, 1073-1082.

Chao, K.-M., Hardison, R. C. and Miller, W. (1993) Locating well-conserved regions within a pairwise alignment. *CABIOS*, **9**, 387-396.

Gumucio, D. L., Shelton, D. A., Bailey, W. J., Slightom, J. L., and Goodman, M. (1993) Phylogenetic footprinting reveals unexpected complexity in trans factor binding upstream from the ε-globin gene. *Proc. Natl. Acad. Sci. USA*, **90**, 6018-6022.

Hardison, R. C., Chao, K.-M., Adamkiewicz, M., Price, D., Jackson, J., Zeigler, T., Stojanovic, N., and Miller, W. (1993) Positive and negative regulatory elements of the rabbit embryonic ε-globin gene revealed by an improved multiple alignment program and functional analysis. *DNA Sequence*, **4**, 163-176.

Hirschberg, D. S. (1975) A linear space algorithm for computing maximal common subsequences. *Comm. ACM*, **18**, 341-343.

Kececioglu, J. D. (1989) Notes on a multiple sequence alignment cost bound of Carrillo and Lipman. Manuscript.

Lawerence, C. B., Goldman, D. A., and Hood, R. T. (1986) Optimized homology searches of the gene and protein sequence data banks. *Bull. Math. Biol.*, **48**, 569-583.

Myers, E. W. and Miller, W. (1988) Optimal alignments in linear space. *CABIOS*, **4**, 11-17.

Myers, E. W. and Miller, W. (1989) Approximate matching of regular expressions. *Bull. Math. Biol.*, **51**, 5-37.

Naor, D. and Brutlag, D. (1993) On suboptimal alignments of biological sequences. In Proceedings of the 4th Symposium on *Combinatorial Pattern Matching*, Lecture Notes in Computer Science, **684**, 179-196.

Saqi, M. and Sternberg, M. (1991) A simple method to generate non-trivial alternative alignments of protein sequences. *J. Mol. Biol.*, **219**, 727-732.

Tagle, D. A., Koop, B. F., Goodman, M., Slightom, J., Hess, D. L. and Jones, R. T. (1988) Embryonic ε and γ globin genes of a prosimian primate (Galago crassicaudatus): Nucleotide and amino acid sequences, developmental regulation and phylogenetic footprints. *J. Mol. Biol.*, **203**, 7469-7480.

Vingron, M. and Argos, P. (1990) Determination of reliable regions in protein sequence alignment. *Protein Engineering*, **3**, 565-569.

Waterman, M., and Byers, T. (1985) A dynamic programming algorithm to find all solutions in a neighborhood of the optimum. *Math. Biosciences*, **77**, 179-185.

Zuker, M. (1991) Suboptimal sequence alignment in molecular biology: alignment with error analysis. *J. Mol. Biol.*, **221**, 403-420.

Approximation Algorithms for Multiple Sequence Alignment

Vineet Bafna[1]*, Eugene L. Lawler[2]** and Pavel A. Pevzner[1]*

[1] Department of CSE
The Pennsylvania State University
University Park, PA 16802

[2] Computer Science Division
University of California
Berkeley, CA 94720

Abstract. We consider the problem of aligning of k sequences of length n. The cost function is sum of pairs, and satisfies triangle inequality. Earlier results on finding approximation algorithms for this problem are due to Gusfield, 1991, who achieved an approximation ratio of $2 - \frac{2}{k}$, and Pevzner, 1992, who improved it to $2 - \frac{3}{k}$. We generalize this approach to assemble an alignment of k sequences from optimally aligned subsets of $l < k$ sequences to obtain an improved performance guarantee. For arbitrary $l < k$, we devise deterministic and randomized algorithms yielding performance guarantees of $2 - l/k$. For fixed l, the running times of these algorithms are polynomial in n and k.

1 Introduction

Multiple sequence alignment is a fundamental problem in computational molecular biology. Alignments of multiple sequences are commonly computed for the purpose of discovering 'homologous', that is, evolutionarily or functionally related, regions of the sequences. An *optimal* multiple alignment can be computed by dynamic programming. However, the running time of dynamic programming algorithms increases rapidly with k, the number of sequences to be aligned. Accordingly, many heuristics and approximation algorithms have been proposed (Altschul and Lipman, 1989, Lipman et al., 1989, Chan et al. 1992, Kececioglu, 1993).

Many objective functions have been suggested for the multiple sequence alignment problem. One of the most widely used is the 'sum-of-pairs' (SP) criterion. The problem of computing an optimal alignment with respect to the sum-of-pairs criterion is NP-hard (Wang and Jiang,1993). The advanced algorithms (Kececiouglu, 1993) allow one to construct *optimal* alignments of $k \leq 6$ sequences, each of length around 200, the length of an average protein. Many algorithms

---

* The research was supported in part by the National Science Foundation under grant CCR-9308567, the National Institute of Health under grant R01 HG00987 and the DOE grant DE-FG03-90ER60999.

** Research supported in part by the DOE grant DE-FG03-90ER60999.

for k sequences use optimal multiple alignment of $l < k$ sequences with further assembling of these "partial" alignments into an *approximate* alignment of k sequences. This approach requires an efficient "assembly" procedure providing an approximate alignment of k sequences close to the optimal one. However, no 'performance guarantee' algorithms for multiple alignment have been known until recently, although a number of heuristics for *suboptimal* multiple alignment have been developed (see the recent review, Chan et al., 1992).

Gusfield,1991,1993, achieved an approximation ratio of $2-2/k$ by assembling an alignment of k sequences from optimal alignments of pairs of sequences. It is known that models currently employed to align sequences are not quite adequate; thus, for practical sequence alignment it is not always necessary to produce an optimal alignment but only one that is plausible. The Gusfield algorithm produces plausible alignments; a computational experiment with an alignment of 19 sequences gave a suboptimal solution only 2% worse than the optimal one. An obvious direction for improvement is to use optimal alignments of $l > 2$ sequences, and then assemble them to approximately align k sequences. However, devising an efficient "assembling" procedure for an arbitrary l remained an open problem.

Pevzner, 1992 improved the performance guarantee to $2 - 3/k$ by assembling optimal alignments of triples of strings. This suggests the possibility of achieving a further improvement in the performance guarantee to $2 - l/k$ by assembling l-way alignments. We investigate this possibility, and show that for arbitrary $l < k$ it is possible to obtain such a performance guarantee with a running time that is polynomial in n and k.

In sections 2 and 3 we define SP-alignment formally, and outline a heuristic approach to constructing SP-alignments of k sequences by combining alignments of l sequences. In section 4, we show that the problem of constructing SP-alignments within a desired performance ratio reduces to constructing *balanced* sets of *l-stars*. In section 5, we use dynamic programming to get some improvement over the brute-force approach. Sections 6 and 7 deal with constructing small balanced sets to ensure small running time. Finally, in section 8, we show how to obtain an efficient randomized algorithm for SP-alignment.

2 Definitions

Let $\mathcal{A}$ be a finite *alphabet* and $a_1, \ldots a_k$ be k sequences (strings) over $\mathcal{A}$. For convenience, we assume that each of these strings contains n characters. Let $\mathcal{A}'$ denote $\mathcal{A} \bigcup \{-\}$, where $'-'$ denotes 'space'. An *alignment* of strings $a_1, \ldots, a_k$ is specified by a $k \times m$ matrix A, where $m \geq n$. Each element of the matrix is a member of $\mathcal{A}'$, and each row i contains the characters of a_i in order, interspersed by $m - n$ spaces.

Given an alignment A we denote A_{ij} a pairwise alignment formed by the rows i and j of A. The score of an alignment is determined with reference to a symmetric matrix D specifiying the dissimilarity or *distance* between elements of $\mathcal{A}'$. We assume the metric properties for distance d, so that $d(x, x) = 0$ and

$d(x, z) \leq d(x, y) + d(y, z)$, for all x, y, z in $\mathcal{A}'$. For a given alignment $A = [a_{ih}]$, the *score* for sequences a_i, a_j is

$$s(A_{ij}) = \sum_{h=1}^{m} d(a_{ih}, a_{jh}),$$

and the *sum-of-pairs score* (SP-score) for the alignment A is given by $\sum_{i,j} s(A_{ij})$. In this definition the score of alignment A is the sum of the scores of *projections* of A onto all pairs of sequences a_i and a_j. Let $C = [c_{ij}]$ be a $k \times k$ matrix of *weights* where c_{ij} is the 'weight' of the pairwise alignment between a_i and a_j. The *weighted sum-of-pairs score* for the alignment A is

$$\sum_{i,j} c_{ij} s(A_{ij})$$

For notational convenience we use matrix dot product to denote scores of alignments. Thus, letting $S(A) = [s(A_{ij})]$ be the matrix of scores of pairs of sequences, the weighted sum-of-pairs score is $C \cdot S(A)$. Letting E be the unit matrix consisting of all 1's except the main diagonal consisting of all 0's, the (unweighted) sum-of-pairs score of alignment A is $E \cdot S(A)$.

Straightforward dynamic programming, with running time $O((2n)^k)$, solves the weighted sum of pairs alignment problem for k sequences. A number of different variations, and some speedups of the basic algorithm have been devised (Sankoff, 1975, Sankoff, 1985, Waterman et al, 1976). Hereafter, we let $g(k, n)$ denote the running time required to obtain an optimal solution to the weighted sum-of-pairs problem for k sequences of length n.

3   Compatible Alignments

Given an alignment A on sequences $a_1, ..., a_k$ and an alignment A' on some subset of the sequences, we say that A is *compatible* with A' if A aligns the characters of the sequences aligned by A' in the same way that A' aligns them. Feng and Doolittle,1987 observed that given any tree in which each vertex is labeled with a distinct sequence a_i, and pairwise alignments specified for each tree edge, there exists an alignment of the k sequences that is compatible with each of the pairwise alignments. A similar result holds for 'l-stars', defined as follows:

Let V be the set $\{1, 2, \ldots, k\}$ representing the sequences $a_1, a_2, \ldots, a_k$, and suppose $l - 1 | k - 1$. An *l-star* $G = (V, E)$ is defined by $r = (k - 1)/(l - 1)$ cliques of size l whose vertex sets intersect in only one *center* vertex (Fig. 1). Let $A_1 \ldots, A_r$, be alignments for the r cliques, with each A_i aligning l sequences. By a construction similar to Feng and Dolittle, 1987, we have the following lemma:

Lemma 1 *For any l-star and any specified alignments $A_1, ..., A_r$ for its cliques, there is an alignment A for the k sequences that is compatible with each of the alignments $A_1, ..., A_r$.*

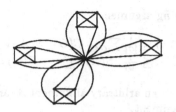

Fig. 1. A 5-star on 17 vertices

Assign weights to the edges of an l-star G as follows.

$$c_{ij} = \begin{cases} k - (l-1) & i = c \text{ or } j = c \\ 1 & i, j \neq c, i \text{ and } j \text{ are contained in the same clique of } G \\ 0 & \text{otherwise} \end{cases}$$

and let $C(G) = [c_{ij}]$ denote the $k \times k$ matrix of weights. Note that

$$C(G) \cdot E = (k - (l-1)) \cdot (k-1) + \left(\frac{k-1}{l-1}\right) \cdot \binom{l-1}{2} = \binom{k}{2} \cdot \left(2 - \frac{l}{k}\right)$$

The pairwise scores of an alignment inherit the triangle inequality property from the distance matrix D. That is, for any alignment A, $s(A_{ij}) \leq s(A_{ik}) + s(A_{kj})$, for all i, j, k. This fact was used by Pevzner, 1992, to prove the following:

Lemma 2 *For any alignment A of the k sequences, and an l-star G, $E \cdot S(A) \leq C(G) \cdot S(A)$.*

Let $C_1, ..., C_r$ denote the submatrices of weights for the r cliques of an l-star G. Let $A_1^*, ..., A_r^*$ be optimal weighted sum-of-pairs alignments for the r cliques. From Lemma 1 and the fact that $d(-, -) = 0$, we obtain the following.

Lemma 3 *Given an l-star G, there is an optimal (weighted with respect to $C(G)$) alignment A_G for the k sequences that is compatible with each of the alignments $A_1^*, ..., A_r^*$. Moreover, $C(G) \cdot S(A_G) = C_1 \cdot S(A_1^*) + ... + C_r \cdot S(A_r^*)$.*

To summarize, for any l-star G we can assemble an alignment A_G, optimal with respect to the weight matrix $C(G)$ specified above, by computing optimal weighted alignments for each clique of G. This can be done in $O(kg(l, n))$ time.

4 Balanced sets of l-stars

Let $\mathcal{G}$ be a collection of l-stars, and let $C(G)$ denote the weight matrix for star G. We say that the collection $\mathcal{G}$ is *balanced* if $\sum_{G \in \mathcal{G}} C(G) = pE$ for some scalar $p > 1$.

Lemma 4 *If $\mathcal{G}$ is a balanced set of l-stars, then*

$$\min_{G \in \mathcal{G}} C(G) \cdot S(A_G) \leq \frac{p}{|\mathcal{G}|} \min_A E \cdot S(A)$$

Proof : We use an averaging argument.

$$\min_{G \in \mathcal{G}} C(G) \cdot S(A_G) \leq \frac{1}{|\mathcal{G}|} \sum_{G \in \mathcal{G}} C(G) \cdot S(A_G)$$
$$\leq \frac{1}{|\mathcal{G}|} \cdot S(A) \cdot \sum_{G \in \mathcal{G}} C(G) = \frac{p}{|\mathcal{G}|} \cdot E \cdot S(A)$$

Here the inequality holds for an arbitrary alignment A, and in particular, it also holds for the optimum alignment. □

Lemmas 2 and 4 motivate the algorithm *Align* (Fig. 2).

Procedure *Align*

1. Construct a balanced set of l-stars, $\mathcal{G}$.
2. For each l-star G in $\mathcal{G}$, assemble an alignment A that is optimal with respect to $C(G)$ from alignments that are optimal for each of its cliques (lemma 3).
3. Choose G with the corresponding alignment A_G such that $C(G) \cdot S(A_G)$ is the minimum over all l-stars in G. Return A_G.

Fig. 2. Deterministic algorithm for multiple alignment

Theorem 1 *Given a balanced collection of l-stars G, Align returns an alignment with a performance guarantee of $2 - l/k$ in $O(k \cdot |\mathcal{G}| \cdot g(l, n))$ time.*

Proof : Note that $\frac{p}{|\mathcal{G}|} = \frac{C(G) \cdot E}{E \cdot E} = 2 - \frac{l}{k}$. Now, *Align* returns the alignment A_G which is optimal for l-star $G \in \mathcal{G}$, and for which the smallest weighted score, $\min_{G \in \mathcal{G}} C(G) \cdot S(A_G)$ is achieved. Lemmas 2 and 4 imply that $E \cdot S(A_G) \leq C(G) \cdot S(A_G) \leq \left(2 - \frac{l}{k}\right) \cdot \min_A E \cdot S(A)$. □

5 Optimizing over all l-stars

We have reduced our approximation problem to that of finding an optimal alignment for each l-star in a balanced set. How hard is it to find a balanced set $\mathcal{G}$? A trivial candidate is simply the set of all l-stars, which is clearly balanced by symmetry. Note that for $l = 2$, there are only k l-stars. This fact was exploited by Gusfield, 1991, to obtain an approximation ratio of $2 - \frac{2}{k}$. This is really a special case, as for $l > 2$, the number of l-stars grows exponentially with k making the algorithm computationally infeasible. Pevzner, 1992, solved the case of $l = 3$, by mapping the problem to weighted matching on graphs.

In this section, we show that it is not necessary to exhaustively compute alignments for all possible l-stars. Dynamic programming provides a shortcut. Later, we shall construct small balanced sets.

For simplicity, consider at first the case when $l - 1 | k - 1$. Fix a center vertex c. For each of the $\binom{k-1}{l-1}$ possible cliques Q which may appear in an l-star with the chosen center vertex c, compute the optimal score $opt(Q)$ of a weighted alignment. Then, solve the following recurrence relations over all $Q \subseteq V \setminus c$, such that $|Q|$ is a multiple of $l - 1$:

$$s(\phi) = 0$$
$$s(Q) = \min_{Q' \subseteq Q, |Q'| = l-1} \{s(Q \setminus Q') + opt(Q')\}$$

Repeating the procedure for each of the k choices of the center vertex, we can compute the optimum score in $O(k^l(2^k + k \cdot g(l, n)))$ time. In the general case, when $l - 1 \nmid k - 1$, we need to consider *hybrid* stars which contain cliques of size l as well as $l + 1$, and compute the recurrence over all $Q \subseteq V \setminus c$. Note that these hybrid l-stars preserve the performance bound. The running time increases to $O(k^{l+1}(2^k + k \cdot g(l + 1, n)))$.

This approach may be computationally tractable for many problem instances. However, in order to obtain a time bound that is polynomial in n and k, for fixed l, we need to construct balanced sets of l-stars of small size.

6 Optimizing Over Small Balanced Sets

Constructing a *small* balanced set of l-stars is not trivial, except for some specific values of l and k.

Let G be an l-star in a complete k-vertex graph. Any permutation of the vertices corresponds to an *isomorph* of G. From symmetry, it is clear that all isomorphs of G form a (large) balanced set. Would a few well chosen isomorphs suffice?

One way of constructing such a set $\mathcal{G}$ is to consider a *sharply doubly transitive* set of permutations. Specifically, a set of permutations $\mathcal{P}$ is sharply doubly transitive if for every two pairs of vertices (x, y) and (x', y') with $x \neq y$ and $x' \neq y'$, there is a unique permutation $p \in \mathcal{P}$, such that $p(x) = x'$ and $p(y) = y'$. Every sharply doubly transitive set of permutations generates a balanced set $\mathcal{G}$, of $k(k - 1)$ isomorphs of the original l-star. With this observation, the problem of searching for a balanced set of isomorphs is reduced to searching for sharply doubly transitive set of permutations. Unfortunately, sharply doubly transitive sets of permutations do not exist for most values of k. One of the few exceptions is described in the following lemma. See Lorimer, 1973, for a complete characterization of sharply doubly transitive sets.

Lemma 5 *Let k be prime. Then the $k(k - 1)$ permutations, $p_{\alpha,\beta}(x) = \alpha x + \beta$ (mod k), for $\alpha \in \{1, \ldots k\}$, $\beta \in \{0, \ldots k\}$, form a sharply doubly transitive set of permutations of $1, \ldots, k$.*

This lemma and theorem 1 imply

Theorem 2 *For any set of k sequences, where k is prime, and l such that $l - 1 | k - 1$, an alignment with a performance ratio of $(2 - l/k)$ can be computed in $O(k^3 g(l, n))$ time.*

The construction described in theorem 2 is the basis of *universal hashing* (Carter and Wegman, 1979). At first glance, construction of *almost* balanced set seems similar to the problem of constructing $(2)_\beta$-*wise independent* hashing functions used in universal hashing (Schmidt and Siegel, 1990). However, there are important differences between universal hashing and our problem, in particular (i) In universal hashing there is a freedom in choosing a prime k; (ii) To derive perfomance guarantee we need $(2)_\beta$-wise independent *permutations*. However, $(2)_\beta$-wise independent hashing *functions* used in universal hashing generally are not permutations.

To overcome this problem we partition the vertices of G into two subsets of prime order or into a single vertex and two subsets of prime order (assuming Goldbach's conjecture, that each even number is the sum of two primes) with further considering the *product of actions* of two sharply doubly transitive groups acting on this subsets. This construction provides an *almost balanced* set $\mathcal{G}$ of isomorphs thus implying a perfomance ratio $2 - \frac{l}{k}$. However, the 'almost' balanced property of the edges decreases the quality of approximation, which means that in order to achieve the desired approximation, we need to expend more computational effort, thereby increasing the running time. The analysis for a general k is rather tedious (omitted) and results in the following theorem:

Theorem 3 *If $\mathcal{G}$ is the set of l-stars defined as above, and either $l \geq 3$, $l-1|k-1$ or $l \geq 4, k \geq 5l$, then Align returns an alignment A, which is within $(2 - \frac{l}{2k})$ of the optimum score.*

Another way of constructing small balanced sets for specific l, k is through block designs (Baranyai, 1975, Bosak, 1990). It is known that balanced sets of size $O(k^2)$ for $l = 4$, $k \equiv 1 \pmod 3$ and $l = 5$, $k \equiv 1 \pmod 4$ exist (Bosak, 1990) .

This implies the following theorem:

Theorem 4 *For $l = 4, 5$, let $k \equiv 1 \pmod{l-1}$. There exists an $O(k^3 g(l, n))$ algorithm for multiple alignment which finds an alignment with a score within $2 - \frac{l}{k}$ of the optimal score.*

In section 7, we get around the difficulty of constructing small balanced sets by constructing a balanced set that is exponentially large, but on which we can quickly find a minimum score l-star by solving matching problems.

7 Balanced Sets of $(2l - 1)$-stars

For simplicity, let us first assume that $2(l-1)|k-1$. For each choice of a center vertex c, let G be an arbitrary l-star with r cliques. Define a *configuration G'* by combining the cliques of G in a pairwise fashion (to form a $(2l-1)$-cliques), and assigning weights as follows:

$$
c_{ij} = \begin{cases} k - (l-1) - 1/2 & i = c \text{ or } j = c \\ 1 & i, j \neq c, \, i \text{ and } j \text{ are contained in the same clique of } G', \\ & \text{but different cliques of } G \\ 0 & \text{otherwise} \end{cases}
$$

Note that, as in the case of l-stars,

$$C(G) \cdot E = (k-1) \cdot [k - (l-1) - \tfrac{1}{2}] + \tfrac{k-1}{2(l-1)} \cdot (l-1)^2 = \binom{k}{2} \cdot (2 - \tfrac{l}{k})$$

Trivially, lemmas 2 and 3 hold for a configuration also.

For an arbitrary l-star G with center c, consider the set of all configurations obtained by pairing up cliques in G. Consider an arbitrary edge (i, j) such that $i, j \neq c$, and i, j do not belong to the same clique of G. By symmetry, each such edge will appear an equal number of times, say x, in the set of all configurations.

Now, for each l-star G in a set of k arbitrary l-stars, each with a different center vertex, consider the set of all configurations obtained by pairing up cliques in G. We assert that this set of configurations, along with x copies of each l-stars, forms a balanced set $\mathcal{G}$. For an arbitrary entry in $C(G)$, $C(G)[i, j] = k - (l-1) - 1/2$ exactly $\frac{2}{k} \cdot |\mathcal{G}|$ times (when i or j is the center vertex of G), and $c(G)[i, j] = 1$ exactly $\frac{k-2}{k} x$ times. Therefore $\sum_{G \in \mathcal{G}} C(G) = p \cdot E$, where $p = \frac{\sum_{G \in \mathcal{G}} C(G)}{\binom{k}{2}}$. Furthermore, $\sum_{i,j} C(G)[i, j] = (2 - l/k) \cdot \binom{k}{2}$ is the same for all $G \in \mathcal{G}$, implying that $p = (2 - l/k) \cdot |\mathcal{G}|$.

For the set of all configurations of an l-star G, form a complete graph of r vertices H_r, with each node corresponding to a clique of the l-star and the weight of an edge being the cost of an optimal weighted alignment on the corresponding $(2l - 1)$-clique. A minimum cost matching on this graph gives the cost of an optimal weighted configuration of G. In order to find the optimal weighted cost configuration in $\mathcal{G}$, we solve k matching problems, and pick one with the minimum cost. Finally compare the optimal configuration with each of the k l-stars, and return one with the minimum cost. From earlier arguments, the corresponding alignment achieves the desired performance ratio.

This method may be generalized for arbitrary l as follows: Consider a hybrid l-star G with an even number of cliques of size l and $l + 1$. As before, define a configuration G' by combining cliques of G arbitrarily in a pairwise fashion to form new cliques of sizes $2l - 1, 2l$ and $2l + 1$. Assign weights exactly as before. Note that lemmas 2 and 3 still hold. Also, from symmetry, if we take the set of all configurations of l-star G, then each edge that does not belong to a clique of G will appear an equal number of times, say x. Combining this with x copies of G, each edge appears exactly x times. By earlier arguments, this set is also balanced. The only thing that remains is to estimate the value of p. Note that,

$$C(G) \cdot E \leq (k-1) \cdot [k - (l-1) - \tfrac{1}{2}] + \tfrac{k-1}{2(l-1)} \cdot l^2 \leq \binom{k}{2} \cdot (2 - \tfrac{l-2}{k})$$

Therefore, $p \leq (2 - \frac{l-2}{k})$. Summarizing,

Theorem 5 *For all k, l, it is possible to compute an alignment with a performance guarantee of $2 - l/k$, in $O(k^3 g(2l + 3, n))$ time.*

As an aside, a smaller balanced set can be explicitly constructed. Let $r = \lfloor \frac{k-1}{2(l-1)} \rfloor$. A *perfect matching* on H_{2r} corresponds to a configuration in the original graph. It is easy to see that a set of configurations corresponding to a 1-*factorization* (edge-disjoint collection of perfect matchings) of H_{2r} (edge-disjoint decomposition of H_{2r} into 1-factors), for each of the k l-stars, along with a single copy of each l-star, forms a balanced set of size $O(k^2)$.

8 Random Sampling of l-stars

What is the performance bound if we choose an l-star at random? Gusfield studied this for 2-stars and gave a bound on the expected score of the alignment (Gusfield, 1993). Consider the sample space of all l-stars and the random variable $C(G) \cdot S(A_G)$. As the set of all l-stars is balanced, for a randomly chosen l-star, $Exp[C(G) \cdot S(A_G)] \le (2 - l/k) \cdot \min_A E \cdot S(A)$. However, it is not clear if we can pick with high probability, an l-star that achieves the $2 - l/k$ performance.

Let $\mathcal{G}_c$ be the set of all l-stars, with a fixed center c. For G in $\mathcal{G}_c$, let $C(G) = C_1(G) + C_2(G)$ be the partition of weight matrix into *Border and Center* weights, with $C_1(G)$ being the same as $C(G)$ except for the c−th row and column which are 0. Define $E = E_1 + E_2$ in an identical manner. Observe the balancing property of $C_1(G)$, that is, $\sum_{G \in \mathcal{G}_c} C_1(G) = p_1 \cdot E_1$, where $p_1 = \frac{l-2}{k-2} \cdot |\mathcal{G}_c|$. We have the following lemma:

Lemma 6 *For G chosen uniformly at random from $\mathcal{G}_c$, and any alignment A,*

$$Prob\left[C_1(G) \cdot S(A) > \frac{3}{2}\frac{p_1}{|\mathcal{G}_c|}E_1 \cdot S(A)\right] < \frac{2}{3}$$

Proof : Let $BAD = \{G \in \mathcal{G}_c | C_1(G) \cdot S(A) > \frac{3}{2}\frac{p_1}{|\mathcal{G}_c|}E_1 \cdot S(A)\}$. Then,

$$\frac{3}{2}\frac{p_1}{|\mathcal{G}_c|}E_1 \cdot S(A) \cdot |BAD| < \sum_{G \in BAD} C_1(G) \cdot S(A) \le \sum_{G \in \mathcal{G}_c} C_1(G) \cdot S(A)$$
$$= p_1 \cdot E_1 \cdot S(A)$$

which implies that $|BAD| \le \frac{2}{3}|\mathcal{G}_c|$. □

Pick m l-stars randomly from $\mathcal{G}_c$. It follows from the proof of lemma 6 that the l-star with the minimum weight alignment(among these m stars) is in BAD with probability less than or equal to $(\frac{2}{3})^m$. *Randomized_Alignment* (Figure 3) uses this fact to construct a set of l-stars which is balanced with arbitrarily high probability.

Theorem 6 *If $l - 1|k - 1$, then for an arbitrary $\epsilon > 0$, Randomized_Alignment runs in time $O(k^2\lceil\lg(\frac{k}{\epsilon})\rceil g(2l, n))$, and returns an alignment that, with probability $1 - \epsilon$, achieves a performance bound of $2 - l/k$.*

Proof : Consider the set of l-stars in $\mathcal{G} = \{G_c : 1 \le c \le k\}$, constructed by the outer loop. To begin with, assume that none of the l-stars in $\mathcal{G}$ is in BAD. In other words, for all $G \in \mathcal{G}$, $C_1(G) \cdot S(A) \le \frac{3}{2}\frac{p_1}{|\mathcal{G}_c|}E_1 \cdot S(A)$. *Randomized_Alignment* returns an l-star G with the minimum weighted score from $\mathcal{G}$. We give a bound on its score by a counting argument. For every alignment A,

$$\min_{G \in \mathcal{G}} C(G) \cdot S(A_G) \le \min_{G \in \mathcal{G}} C(G) \cdot S(A) \le \frac{1}{k}\sum_{G \in \mathcal{G}} C(G) \cdot S(A)$$
$$= \frac{1}{k}\sum_{G \in \mathcal{G}} C_2(G) \cdot S(A) + \frac{1}{k}\sum_{G \in \mathcal{G}} C_1(G) \cdot S(A)$$
$$\le \frac{2}{k}(k - (l - 1))E \cdot S(A) + \frac{k-2}{k} \cdot \frac{3}{2} \cdot \frac{l-2}{k-2} \cdot E \cdot S(A)$$
$$= (2 - \frac{l}{2k}) \cdot E \cdot S(A)$$

Procedure *Randomized_Alignment(l,k,ε)*
   $\mathcal{G} \leftarrow \emptyset$
   for $c \in \{1, \ldots, k\}$
      repeat $2\lceil \lg \frac{k}{\epsilon} \rceil$ times
         choose a random l-star G with center c
         compute an alignment A_G with the minimum weighted score $\min_A C(G) \cdot S(A)$
      $G_c \leftarrow$ an l-star with minimum weighted score among the $2\lceil \lg \frac{k}{\epsilon} \rceil$ l-stars,
      $\mathcal{G} \leftarrow \mathcal{G} \cup \{G_c\}$.
   $G \leftarrow$ an l-star with the minimum weighted score $\min_{G \in \mathcal{G}} C(G) \cdot S(A_G)$
   return A_G

Fig. 3. Randomized Algorithm for Multiple Alignment

Now, recall from lemma 2 that $E \cdot S(A_G) \leq C(G) \cdot S(A_G)$, which implies that if none of the l-stars in $\mathcal{G}$ is in BAD, the algorithm achieves a performance bound of $(2 - \frac{l}{2k})$. In order to achieve a performance ratio of $(2 - \frac{l}{k})$, we need to consider $(2l)$-stars.

Next, we show that none of the l-stars in $\mathcal{G}$ is in BAD with high probability. In each iteration of the inner loop, we consider $2\lceil \lg \frac{k}{\epsilon} \rceil$ random l-stars, and pick a G_c with the minimum weighted score. By definition, this l-star is in BAD only if each l-star picked in that iteration is in BAD. Therefore, for all $1 \leq c \leq k$, the probability that $G_c \in \mathcal{G}$ is in BAD is less than $(\frac{2}{3})^{2\lceil \lg(\frac{k}{\epsilon}) \rceil} < \frac{\epsilon}{k}$. The probabilty that none of the k $G_c \in \mathcal{G}$ are in BAD is greater than or equal to $1 - \epsilon$. Theorem 6 follows. $\square$

References

[AL89] Altschul S.F., Lipman D.J., *Trees, stars, and multiple biological sequence alignment*. SIAM J. Appl. Math., 49, (1989), pp. 197-209.

[B75] Baranyai, Z., *On the factorization of the complete uniform hypergraph*, Infinite and Finite Sets, A. Hajnal, T. Rado, V. T. Sós, eds., North-Holland, Amsterdam, (1975), pp. 91-108.

[B90] Bósak, J., *Decompositions of Graphs*, Kluwer Academic Publishers, (1990).

[CW79] Carter J.L., Wegman M.N.,*Universal classes of hash functions*, Journal of Computer and System Sciences, 18(1979), pp. 143-154.

[CWC92] Chan S.C., Wong A.K.C., Chiu D.K.Y., *A survey of multiple sequence comparison methods*, Bull. Math. Biol., 54(1992), pp. 563-598.

[FD87] Feng D., Doolittle R., *Progressive sequence alignment as a prerequisite to correct phylogenetic trees*, Journal of Molec. Evol., 25(1987), pp. 351-360.

[G91] Gusfield, D., *Efficient methods for multiple sequence alignment with guaranteed error bounds*. Tech. Report, Computer Science Division, Uiversity of California, Davis, CSE-91-4, (1991).

[G93] Gusfield, D., *Efficient methods for multiple sequence alignment with guaranteed error bounds*, Bulletin of Mathematical Biology, 55(1993), pp. 141-154.

[K93] Kececioglu J., *The maximum weight trace alignment problem in multiple sequence alignment*, eds. A. Apostolico, M. Crochemore, Z. Galil, U. Manber, Combinatorial Pattern matching 93, Padova, Italy, June 1993, LNCS 684, 106-119.

[LAK89] Lipman D.J., Altschul S.F., Kececioglu J.D., *A tool for multiple sequence alignment*, Proc. Natl. Acad. Sci. USA, **86**(1989), pp. 4412-4415.

[L73] Lorimer, P., *Finite Projective Planes and Sharply 2-transitive Subsets of Finite Groups*, Proc. Second Internat. Conf. Theory of Groups, Canberra, (1973), pp. 432-436.

[P92] Pevzner, P., *Multiple Alignment, Communication Cost, and Graph Matching*, SIAM J. Applied Math., **52**, (1992), pp. 1763-1779.

[S75] Sankoff D., *Minimum mutation tree of sequences*, SIAM J. Appl. Math., **28**, (1975), pp. 35-42.

[S85] Sankoff D., *Simultaneous solution of the RNA folding, alignment and protosequence problems*, SIAM J. Appl. Math., **45** (1985), pp. 810-825.

[SS90] Schmidt J., Siegel A., *The analysis closed hashing under limited randomness*, Proceedings of the 22nd ACM Symposium on Theory of Computing, (1990), pp. 224-234.

[WJ93] Wang L., Jiang, T., *On the Complexity of Multiple Sequence Alignment*, 1993, J. of Comp. Biol. (to appear).

[WSB76] Waterman M.S., Smith T.F., Beyer W.A., *Some biological sequence metrics*. Adv. in Math., **20**(1976), pp. 367-387.

A Context Dependent Method for Comparing Sequences

Xiaoqiu Huang

Department of Computer Science
Michigan Technological University
Houghton, MI 49931

Abstract. A scoring scheme is presented to measure the similarity score between two biological sequences, where matches are weighted dependent on their context. The scheme generalizes a widely used scoring scheme. A dynamic programming algorithm is developed to compute a largest-scoring alignment of two sequences of lengths m and n in $O(mn)$ time and $O(m+n)$ space. Also developed is an algorithm for computing a largest-scoring local alignment between two sequences in quadratic time and linear space. Both algorithms are implemented as portable C programs. An experiment is conducted to compare protein alingments produced by the new global alignment program with ones by an existing program.

1 Introduction

Sequence comparison deals with the problem of finding a best correspondence between sequences of symbols [9]. A correspondence between two biological sequences is usually represented by an alignment of the sequences, an ordered list of pairs of their symbols. A symbol of the first sequence paired with a symbol of the second sequence is called a substitution pair, a symbol of the first sequence paired with a blank a deletion pair, and a symbol of the second sequence paired with a blank an insertion pair. The quality of an alignment is measured with a numerical value, which is usually defined as the sum of scores of substitutions, deletions and insertions. A best correspondence between the sequences is an alignment with the minimum distance or maximum score.

Sellers [10] introduced a distance metric where substitutions, deletions and insertions are scored independent of their context. Waterman *et al.* [12] generalized the metric of Sellers by giving deletions and insertions context-dependent scores. Wilbur and Lipman [13] presented a general framework for context-dependent evaluation of substitutions. The main reason for introducing context dependence is that biological mutations occur non-uniformly throughout sequences [7, p. 70]. This results in uneven distribution of matches in the alignment of distantly related sequences, where there are more matches in some regions of the alignment and fewer matches in other regions. In contrast, matches are uniformly distributed in the alignment of random or unrelated sequences. This observation may be used to improve the chances of finding the biologically significant alignment of distantly related sequences. For example, in aligning sequences, an occurrence of five close matches is more favorable than an occurrence of five isolated matches and should be given a bonus.

This research was supported in part by NSF Grant DIR-9106510.

In this paper, we introduce a scoring scheme for addressing the arrangement of exact matches within blocks of substitution pairs, and develop a dynamic programming algorithm to compute a largest-scoring alignment in quadratic time and in linear space. The scheme generalizes a widely used scoring scheme of Waterman *et al.* [12]. Although the general algorithm of Wilbur and Lipman [13] can be used to compute the alignment, the algorithm takes cubic time and quadratic space. We also develop an algorithm for computing a largest-scoring local alignment between sequences by following the strategy of Smith and Waterman [11]. Both algorithms are implemented as portable programs in the C programming language. Alignments generated by the new global alignment program on some protein sequences are compared with ones by an existing program.

2 A Scoring Scheme

Let $A = a_1 a_2 \cdots a_m$ and $B = b_1 b_2 \cdots b_n$ be two sequences over an alphabet Σ. The alphabet for protein sequences contains 20 symbols representing 20 different amino acids. Let $-$, a unique symbol not in Σ, denote the sequence of zero symbols. Then $-$ is the identity element with respect to the concatenation of sequences. Use $A_{i,i'}$ to denote $a_i a_{i+1} \cdots a_{i'}$ if $i \le i'$ and $-$ otherwise. An aligned pair consists of two ordered elements in $\Sigma \cup \{-\}$. An alignment of A and B is a finite sequence of aligned pairs, where A is the concatenation of first elements in those pairs and B is that of second elements. Several alignments are shown in Section 5. Since the pair of two $-$'s is the identity element with respect to the concatenation of alignments, we only consider alignments without this null pair. There are three types of non-null aligned pairs: (1) substitution pairs (a, b), (2) deletion pairs $(a, -)$, and (3) insertion pairs $(-, b)$, with a and b in Σ. Substitution pairs are further divided into matches and mismatches depending on whether the two symbols in the pair are the same or not. A substitution block in an alignment is a contiguous subsequence of substitution pairs delimited by aligned pairs of other types or an end of the alignment. A deletion gap in an alignment is a contiguous subsequence of deletion pairs delimited by aligned pairs of other types or an end of the alignment. An insertion gap is similarly defined.

Each deletion or insertion gap of k pairs receives a score of $-(q + r \times k)$, where non-negative numbers q and r are gap-open and gap-extension penalties. Let σ be a function that gives a context-independent score to each substitution pair. A match (a, a) in a substitution block receives a single left bonus of $\delta(a, a)$ if at least one other match occurs within l positions to the left of (a, a) in the same block, where a non-negative integer l is the context length. Similarly, a match (a, a) receives a single right bonus of $\delta(a, a)$ if at least one other match occurs within l positions to the right of (a, a) in the same block. Here use of left and right bonuses, instead of a single bonus, favors a long chain of matches over a few short chains of matches. Use of a symbol-dependent bonus function may improve the sensitivity of the algorithm in aligning protein sequences. The score of a substitution block is the sum of the context-independent scores of each pair in the block plus the sum of the bonuses of each match in the block. Let $c(A_{i-k+1,i}, B_{j-k+1,j})$ denote the score of the substitution block involving two segments $A_{i-k+1,i}$ and $B_{j-k+1,j}$ for some $k \le \min(i, j)$. Assume that $c(-, -) = 0$. For example, for any a and b in $\{C, G, T\}$, assume that $\sigma(a, b) = 2$ if $a = b$ and -2 otherwise, and that $\delta(a, a) = 1$. Let $l = 2$. Then $c(CTGTCTT, CTGTGCT) = 12$. The score of an alignment is the sum of the scores of each substitution block and each gap in the alignment.

The global alignment problem is to compute an alignment of A and B with the maximum score. Such an alignment is called an optimal alignment. Let $S(i,j)$ be the score of an optimal alignment of $A_{1,i}$ and $B_{1,j}$. Let $H(i,j)$ be the maximum score of alignments of $A_{1,i}$ and $B_{1,j}$ ending with aligned pair (a_i, b_j). We obtain the following recurrence for computing the matrix S by generalizing the recurrence of Gotoh [3]:

$$S(i,j) = \begin{cases} \max\{H(i,j), E(i,j), F(i,j)\} & \text{if } i>0 \text{ and } j>0 \\ -(q+r\times i) & \text{if } i>0 \text{ and } j=0 \\ -(q+r\times j) & \text{if } i=0 \text{ and } j>0 \\ 0 & \text{if } i=0 \text{ and } j=0 \end{cases}$$

$$H(i,j) = \max\{S(i-k,j-k) + c(A_{i-k+1,i}, B_{j-k+1,j}) : 1\leq k \leq \min(i,j)\} \quad \text{if } i>0 \text{ and } j>0$$

$$E(i,j) = \begin{cases} \max\{E(i-1,j), S(i-1,j) - q\} - r & \text{if } i>0 \text{ and } j>0 \\ S(0,j) - q & \text{if } i=0 \text{ and } j>0 \end{cases}$$

$$F(i,j) = \begin{cases} \max\{F(i,j-1), S(i,j-1) - q\} - r & \text{if } i>0 \text{ and } j>0 \\ S(i,0) - q & \text{if } i>0 \text{ and } j=0 \end{cases}$$

The recurrence leads to an $O(mn(m+n))$ algorithm since it takes $O(m+n)$ time to compute each $H(i,j)$.

Below we show that the matrix H can be computed efficiently by making use of the property of the c function. For $i \geq 1$ and $j \geq 1$, define $L(i,j)$ to be the smallest $t \geq 1$ such that $a_{i-t+1} = b_{j-t+1}$ if such an t exists and define $L(i,j)$ to be $l+1+\min(i,j)$ otherwise. We introduce a matrix D that can be computed efficiently and use D to express the matrix H. For $i \geq 1$ and $j \geq 1$, if $L(i,j) \leq \min(i,j)$, then define

$$D(i,j) = \max\{S(i-k,j-k) + c(A_{i-k+1,i}, B_{j-k+1,j}) : L(i,j)\leq k \leq \min(i,j)\},$$

otherwise, define

$$D(i,j) = S(i-\min(i,j), j-\min(i,j)) + c(A_{i-\min(i,j)+1,i}, B_{j-\min(i,j)+1,j}).$$

In words, $D(i,j)$ is the maximum score of those alignments of $A_{1,i}$ and $B_{1,j}$ that end with a substitution block containing a match if the match exists; otherwise, $D(i,j)$ is the score of the alignment of $A_{1,i}$ and $B_{1,j}$ that ends with the longest substitution block. The following lemma shows that the matrix D can be used to express the matrix H.

Lemma 1. For any $i>0$ and $j>0$, $H(i,j) = \max\{S(i-1,j-1)+\sigma(a_i,b_j), D(i,j)\}$.

Proof. Assume that $i>0$ and $j>0$. If $a_i = b_j$, then $L(i,j) = 1$. By the definition of D and H, we have $S(i-1,j-1)+\sigma(a_i,b_j) \leq D(i,j)$ and $D(i,j) = H(i,j)$. Thus, the lemma is true in this case. Next consider the case where $a_i \neq b_j$. Let $\Pi_{i,j}$ be an alignment of the maximum score that ends with a substitution block containing (a_i, b_j). It follows from the definition of $H(i,j)$ that $score(\Pi_{i,j}) = H(i,j)$. Let $\Pi_{i-1,j-1}$ be the remaining portion of $\Pi_{i,j}$ with the last substitution pair (a_i, b_j) removed. Then $H(i,j) = score(\Pi_{i-1,j-1}) + \sigma(a_i,b_j)$. Since $\Pi_{i,j}$ is of the maximum score, we have $score(\Pi_{i-1,j-1}) = S(i-1,j-1)$. Thus, we obtain that $H(i,j) = S(i-1,j-1) + \sigma(a_i,b_j)$. The definition of D and H implies that $D(i,j) \leq H(i,j)$. Therefore, $H(i,j) = \max\{S(i-1,j-1)+\sigma(a_i,b_j), D(i,j)\}$.

In the following lemma, we show that each entry of the matrix D can be computed in constant time. For convenience, assume that $D(i,j) = S(i,j)$ if $i = 0$ or $j = 0$.

Lemma 2. The following recurrence correctly computes the matrix D.

$$
D(i,j) = \begin{cases}
\begin{aligned}
&\max\{S(i-1,j-1), D(i-1,j-1) \\
&\quad + \delta(a_{i-L(i-1,j-1)}, b_{j-L(i-1,j-1)}) \\
&\quad + \delta(a_i, b_j)\} + \sigma(a_i, b_j)
\end{aligned} & \text{if } i > 0,\ j > 0,\ a_i = b_j \text{ and } L(i-1,j-1) \le l \\[1em]
S(i-1,j-1) + \sigma(a_i, b_j) & \text{if } i > 0,\ j > 0,\ a_i = b_j \text{ and } L(i-1,j-1) > l \\[0.5em]
D(i-1,j-1) + \sigma(a_i, b_j) & \text{if } i > 0,\ j > 0 \text{ and } a_i \ne b_j \\[0.5em]
S(i,j) & \text{if } i = 0 \text{ or } j = 0
\end{cases}
$$

Proof. We just show that the recurrence is correct for the case where $i > 0$, $j > 0$, $a_i = b_j$ and $L(i-1,j-1) \le l$. In this case, $D(i,j)$ is the maximum score of those alignments of $A_{1,i}$ and $B_{1,j}$ that end with the match (a_i, b_j). Let $\Pi_{i,j}$ be an alignment of score $D(i,j)$ that ends with the longest substitution block containing (a_i, b_j). Let $\Pi_{i-1,j-1}$ be the remaining portion of $\Pi_{i,j}$ with the last pair (a_i, b_j) removed. Let $k = L(i-1,j-1)$. If the last substitution block of $\Pi_{i,j}$ does not contain the match (a_{i-k}, b_{j-k}), then we have $D(i,j) = score(\Pi_{i-1,j-1}) + \sigma(a_i, b_j)$. Then we have $score(\Pi_{i-1,j-1}) = S(i-1,j-1)$. Otherwise, $score(\Pi_{i-1,j-1}) < S(i-1,j-1)$. We could construct an alignment of $A_{1,i}$ and $B_{1,j}$ with a score greater than $D(i,j)$ by appending (a_i, b_j) to an alignment of score $S(i-1,j-1)$, a contradiction to the fact that $\Pi_{i,j}$ is a largest-scoring alignment ending with (a_i, b_j).

If the last substitution block of $\Pi_{i,j}$ contains the match (a_{i-k}, b_{j-k}), then (a_{i-k}, b_{j-k}) is the second match from the right end of the block. This means that (a_{i-k}, b_{j-k}) is the last match in the last substitution block of $\Pi_{i-1,j-1}$. Since $k = L(i-1,j-1) \le l$, we have

$$D(i,j) = score(\Pi_{i-1,j-1}) + \delta(a_{i-L(i-1,j-1)}, b_{j-L(i-1,j-1)}) + \delta(a_i, b_j) + \sigma(a_i, b_j).$$

Since $\Pi_{i,j}$ is of the maximum score, we conclude that $score(\Pi_{i-1,j-1}) = D(i-1,j-1)$. Therefore, $D(i,j)$ is

$$\max\{S(i-1,j-1), D(i-1,j-1) + \delta(a_{i-L(i-1,j-1)}, b_{j-L(i-1,j-1)}) + \delta(a_i, b_j)\} + \sigma(a_i, b_j).$$

$\square$

Combining the results of the two lemmas, we have the main theorem.

Theorem. The matrix S is correctly computed by the following recurrence:

$$
S(i,j) = \begin{cases}
\max\{S(i-1,j-1) + \sigma(a_i, b_j), D(i,j), E(i,j), F(i,j)\} & \text{if } i > 0 \text{ and } j > 0 \\[0.5em]
-(q + r \times i) & \text{if } i > 0 \text{ and } j = 0 \\[0.5em]
-(q + r \times j) & \text{if } i = 0 \text{ and } j > 0 \\[0.5em]
0 & \text{if } i = 0 \text{ and } j = 0
\end{cases}
$$

$$
D(i,j) = \begin{cases}
\max\{S(i-1,j-1),\ D(i-1,j-1) & \text{if } i>0,\ j>0,\ a_i=b_j \text{ and } L(i-1,j-1)\le l \\
\quad + \delta(a_{i-L(i-1,j-1)},\ b_{j-L(i-1,j-1)}) & \\
\quad + \delta(a_i,b_j)\} + \sigma(a_i,b_j)\} & \\
S(i-1,j-1) + \sigma(a_i,b_j) & \text{if } i>0,\ j>0,\ a_i=b_j \text{ and } L(i-1,j-1)>l \\
D(i-1,j-1) + \sigma(a_i,b_j) & \text{if } i>0,\ j>0 \text{ and } a_i \ne b_j \\
S(i,j) & \text{if } i=0 \text{ or } j=0
\end{cases}
$$

$$
E(i,j) = \begin{cases}
\max\{E(i-1,j),\ S(i-1,j) - q\} - r & \text{if } i>0 \text{ and } j>0 \\
S(0,j) - q & \text{if } i=0 \text{ and } j>0
\end{cases}
$$

$$
F(i,j) = \begin{cases}
\max\{F(i,j-1),\ S(i,j-1) - q\} - r & \text{if } i>0 \text{ and } j>0 \\
S(i,0) - q & \text{if } i>0 \text{ and } j=0
\end{cases}
$$

$$
L(i,j) = \begin{cases}
L(i-1,j-1) + 1 & \text{if } i>0,\ j>0 \text{ and } a_i \ne b_j \\
1 & \text{if } i>0,\ j>0 \text{ and } a_i = b_j \\
l + 1 & \text{if } i=0 \text{ or } j=0
\end{cases}
$$

The matrices are computed in order of rows (or columns). For $i=0$ or $j=0$, compute $S(i,j)$ directly and initialize the other matrices at (i,j). Otherwise, compute the other matrices at (i,j) first and then compute $S(i,j)$. If only $S(m\ n)$, the score of an optimal alignment, is needed, then the computation of the matrices can be done in linear space by saving only newly-computed rows of the matrices. If an optimal alignment is desired, then a simple traceback procedure requires that the entire matrices be saved. Saving the matrices takes $O(mn)$ space.

3 Optimal Global Alignment

We consider extending the algorithm of Myers and Miller [8] to obtain an optimal alignment in $O(m+n)$ space. Myers and Miller [8] generalized the divide-conquer technique of Hirschberg [4] to accommodate linear gap penalties. The idea of the divide-conquer technique is to find the 'middle' position pair of an optimal alignment and then to construct recursively two smaller optimal alignments before and after the middle pair. The optimal alignment is constructed by concatenating the smaller alignments in the proper order. The middle pair can be found without saving the whole matrices.

The matrices introduced previously are defined with respect to the left end of the sequences and computed in increasing order of indices. By symmetry, another group of matrices can be defined with respect to the right end of the sequences and computed in decreasing order of indices. Let R, C, U, V, K be the group of matrices, where R corresponds to S, C to D, U to E, V to F and K to L. For example, $R(i,j)$ is the score of an optimal alignment of $A_{i+1,m}$ and $B_{j+1,n}$, and $R(0,0)$ is the score of an optimal alignment of A and B. For any i and j, $0\le i \le m$ and $0\le j \le n$, the term $S(i,j)+R(i,j)$ is the score of the alignment obtained by concatenating an optimal alignment of $A_{1,i}$ and $B_{1,j}$ with an optimal alignment of $A_{i+1,m}$ and $B_{j+1,n}$. For any i and j, $1\le i \le m-1$ and $0\le j \le n$, the term $E(i,j)+U(i,j)+q$ is the score of the concatenation of an optimal alignment of $A_{1,i}$ and $B_{1,j}$ that ends with a delete and an optimal alignment of $A_{i+1,m}$ and $B_{j+1,n}$ that begins with a delete, where the two deletes are combined into a single delete and hence q is added to ensure that the open penalty is charged only once for the delete [8]. For any i and j, $1 \le i \le m-1$ and $1 \le j \le n-1$, the term $D(i,j)+C(i,j)+w(i,j)$ is the score of the concatenation of an optimal alignment of $A_{1,i}$ and $B_{1,j}$ that ends with a substitution block and an optimal alignment of $A_{i+1,m}$ and $B_{j+1,n}$ that begins with a substitution block, where the two substitution blocks are joined into a single one and $w(i,j)$ is the right bonus of the match before the joint plus the left bonus of the match after the joint if the matches are at most l positions away. In symbols, if $L(i,j)+K(i,j)-1 \le l$, we have

$$w(i,j) = \delta(a_{i-L(i,j)+1}, b_{j-L(i,j)+1}) + \delta(a_{i+K(i,j)}, b_{j+K(i,j)}).$$

Otherwise, $w(i,j) = 0$. Note that for any i and j, $0 \le i \le m-1$ and $0 \le j \le n-1$, $K(i,j)$ is the smallest $t \ge 1$ such that $a_{i+t} = b_{j+t}$ if such an t exists and $l+1+\min(m-i, n-j)$ otherwise. Assume that $K(i,j) = l+1$ if $i=m$ or $j=n$. We use the following observation to find a middle pair. Let $i^* = \lfloor m/2 \rfloor$. The score of an optimal alignment of A and B is

$$\max\{S(i^*,j)+R(i^*,j), E(i^*,j)+U(i^*,j)+q, D(i^*,j)+C(i^*,j)+w(i^*,j) : 0 \le j \le n\},$$

where assume that $D(i^*,n)+C(i^*,n)+w(i^*,n) = -\infty$. If the maximum is obtained at j^*, then (i^*,j^*) is a middle pair. So a middle pair can be found in linear space by computing the two groups of matrices and saving only row i^* of each matrix.

The pair (i^*,j^*) is of type 1 if $S(i^*,j^*)+R(i^*,j^*)$ is the maximum. Otherwise, it is of type 2 if $E(i^*,j^*)+U(i^*,j^*)+q$ is the maximum and of type 3 if not. If a type 2 middle pair is found, recursive calls are constrained to optimal alignments that begin or end with a delete. A method for constraining recursive calls was given by Myers and Miller [8]. We use a similar strategy to handle type 3 pairs. If a type 3 middle pair is found, we require recursive calls to find optimal alignments that begin or end with a substitution block. An optimal alignment of A and B with a type 3 middle pair (i^*,j^*) is divided into three parts: (i) an optimal alignment of $A_{1,i^*-L(i^*,j^*)+1}$ and $B_{1,j^*-L(i^*,j^*)+1}$, where the ending match is given an extra bonus of $w(i^*,j^*)$, (ii) the alignment without any gap of $A_{i^*-L(i^*,j^*)+2, i^*+K(i^*,j^*)-1}$ and $B_{j^*-L(i^*,j^*)+2, j^*+K(i^*,j^*)-1}$, and (iii) an optimal alignment of $A_{i^*+K(i^*,j^*),m}$ and $B_{j^*+K(i^*,j^*),n}$, where the initial match is given an extra bonus of $w(i^*,j^*)$. A recursive call is made to compute an optimal alignment in cases (i) and (iii). The call always produces an optimal alignment ending with a match in case (i) or an optimal alignment beginning with a match in case (iii). This is because the extra bonus makes the optimal alignment ending (or beginning) with a match have a larger score than any alignment ending (or beginning) with a delete or insert. To implement this strategy, we introduce two more parameters in the recursive procedure diff of Myers and Miller [8], one for the initial match bonus and the other for the ending match bonus. For other types of middle pairs, the parameters are set to 0. Using the time analysis given by Myers and Miller [8], we can easily show that the time requirement of the new algorithm is $O(mn)$.

4 Optimal Local Alignment

A local alignment between two sequences is simply an alignment of two substrings from the sequences, respectively. An optimal local alignment between two sequences is one with the maximum score. An optimal local alignment can be computed by generalizing the algorithm of Smith and Waterman [11]. Let $T(i,j)$ be the score of an optimal local alignment ending at positions i of A and j of B. Then the recurrence for computing matrix T is given below:

$$T(i,j) = \begin{cases} \max\{0, T(i-1,j-1) + \sigma(a_i,b_j), D(i,j), E(i,j), F(i,j)\} & \text{if } i > 0 \text{ and } j > 0 \\ 0 & \text{if } i = 0 \text{ or } j = 0 \end{cases}$$

The recurrences for the other matrices are the same as those given in Section 2. An optimal local alignment between A and B ends at an entry with the maximum T value.

To construct an optimal alignment in linear space, we use the technique of Huang *et al.* [5]. In a forward phase, compute the matrix T and its auxiliary matrices in linear space and save an entry (i_2, j_2) with the maximum T value. In a backward phase, compute the matrix R and its auxiliary matrices in linear space with respect to sequences A_{1,i_2} and B_{1,j_2} and save an entry (i_1, j_1) with the maximum R value. (Recall that R is defined in the previous section.) An optimal local alignment between A and B is an optimal global alignment of A_{i_1+1, i_2} and B_{j_1+1, j_2}, which is computed in linear space by the algorithm in the previous section. So an optimal local alignment between A and B can be computed in $O(mn)$ time and in $O(m+n)$ space. To compute several best local alignments in linear space, we can generalize the algorithm of Huang and Miller [6].

5 Experimental Results

The global and local alignment algorithms introduced in this paper were implemented as portable computer programs in the C programming language. Both programs are able to align very long sequences. The global alignment program is called CDA (Context Dependent Alignment). To get a feel for the performance of CDA, we conducted an experiment of comparing protein alignments of the CDA program with those of the LINEAR program based on the algorithm of Myers and Miller [8]. The algorithm of Myers and Miller [8] is a space-efficient version of the algorithm of Gotoh [3], which in turn is a time-efficient version of the algorithm of Waterman *et al.* [12].

For this experiment, we selected two large protein sequence families with different levels of similarity among their members. The data are the globin and immunoglobulin variable region superfamilies from the Protein Identification Resource (PIR) protein database in release 34, September 1992 [1]. Globin proteins carry and store oxygen in the blood. Immunoglobulin proteins are antibodies that are composed of a variable region and a constant region, where different immunoglobulins have different variable regions. The globin and immunoglobulin variable region superfamilies are two large superfamilies in the database with 626 protein sequences in the globin superfamily and 468 ones in the immunoglobulin superfamily. The sequences in the globin superfamily are in general more similar to each other than the ones in the immunoglobulin superfamily. In the globin superfamily, the longest sequence is of 162 residues, shortest sequence is of 8 residues, and the average sequence length is 143, while in the immunoglobulin superfamily, the longest one is of 773 residues, the shortest one is of 15 residues, and the average length is 126.

In the experiment, the PAM250 substitution matrix [2] was selected as σ, and a gap-open penalty of 10 and a gap-extension penalty of 2 were used. The context-dependent substitution matrix δ was set as follows: $\delta(a,b) = 1$ if $a = b$ and 0 otherwise. The context length l was set to 2. Note that only the PAM250 matrix was used by LINEAR. Two sequences were randomly selected from a superfamily with replacement and were aligned by CDA and LINEAR. The alignment produced by CDA was compared with the alignment produced by LINEAR by calculating the number of differences between the two alignments. The process was repeated 10,000 times for each superfamily. The number of differences between two alignments with respect to one of the two sequences is defined as the number of positions of the sequence that are aligned differently in the two alignments. The rate of difference is the number of differences divided by the length of the sequence.

The average rate of difference for 10,000 pairs of CDA and LINEAR alignments is 2.0% for the globin superfamily and 5.8% for the immunoglobulin superfamily. The minimum and maximum rates of difference are 0% and 100% for each superfamily. Table I shows rates of difference between the CDA and LINEAR alignments of each of 10,000 pairs of sequences for each superfamily. For each superfamily, 10,000 pairs of sequences were partitioned into six groups according to the match percentage of their LINEAR alignment. The number of sequence pairs and the average rate of difference between CDA and LINEAR alignments in each group were calculated. Several observations can be made from the statistics in Table I. CDA and LINEAR alignments are, in general, very similar. The more similar the sequences are, the more similar the CDA and LINEAR alignments are. The average rate of difference between CDA and LINEAR alignments of immunoglobulin sequences is larger than that of globin sequences with the same level of similarity.

Table I. Rates of difference between CDA and LINEAR alingments

Similarity interval	Globin		Immunoglobulin	
	Number of pairs	Average rate of difference	Number of pairs	Average rate of difference
0–9%	257	10.06%	640	22.09%
10–19	1775	6.03	2433	10.40
20–29	2378	1.96	4044	4.16
30–39	1891	0.83	1054	1.17
40–49	1109	0.37	965	0.79
50–100	2590	0.03	864	0.27

Shown in Figure 1 are four pairs of alignments generated by LINEAR and CDA on globin sequences and immunoglobulin sequences, respectively. The match percentage of each alignment is at least 25%. As shown, the CDA alignments contain more close matches than the corresponding LINEAR alignments. As an indication on the running time of the new algorithm, we found that the CDA program was just slower by a few percentage points than the LINEAR program in those tests.

We have introduced a scoring scheme for addressing context dependence in evaluating matches and developed an algorithm for computing a largest-scoring alignment. More experience with the CDA program needs to be gained to see if it produces more biologically significant alingments. Also a search needs to be performed to see if a more suitable context dependent scoring matrix exists. The programs are freely available by e-mail from the author at huang@cs.mtu.edu.

```
Alignment pair 1

V-LSSKDKTNVKTAFGKIGGHAAEYGAEALERMFLGFPTTKTYFPHF-DLSH-----GSAQVKAHGKKVGDALTKAADHLDDLPSALSALSDLHAHKLRV
| |    | | | |  ||       | | |||  |  | |   | |||       |||||||| |    ||| |     || ||  || |
VHLTGEEKSAVTTLWGKV--NVEEVGGEALGRLLVVYPWTQRFFDSFGDLSSPDAVMNNPKVKAHGKKVLGAFSDGLAHLDNLKGTFAQLSELHCDKLHV

DPVNFKLLSHCLLVTVAAHHPGDFTPSVHASLDKFLANVSTVLTSKYR          Arabian camel hemoglobin alpha (PIR1:HACMA)
|| || ||  | | | |   ||| |||   | |    |  |  | |                               LINEAR alignment
DPENFRLLGNVLVCVLAHHFGKEFTPQVQAAYQKVVAGVANALAHKYH          Moustached tamarin hemoglobin beta (PIR1:HBMKM)

V-LSSKDKTNVKTAFGKIGGHAAEYGAEALERMFLGFPTTKTYFPHF-DLSH-----GSAQVKAHGKKVGDALTKAADHLDDLPSALSALSDLHAHKLRV
| |    | | | |  ||       | | |||  |  | |   | |||       |||||||| |    ||| |     || ||  || |
VHLTGEEKSAVTTLWGKV--NVEEVGGEALGRLLVVYPWTQRFFDSFGDLSSPDAVMNNPKVKAHGKKVLGAFSDGLAHLDNLKGTFAQLSELHCDKLHV
          **********
DPVNFKLLSHCLLVTVAAHHPG-DFTPSVHASLDKFLANVSTVLTSKYR         Arabian camel hemoglobin alpha (PIR1:HACMA)
|| || ||  | | || |  | | |||   | ||    |  |  | ||                             CDA alignment
DPENFRLLGN-VLVCVLAHHFGKEFTPQVQAAYQKVVAGVANALAHKYH         Moustached tamarin hemoglobin beta (PIR1:HBMKM)


Alignment pair 2

MVHLTDAEKATVSGLWGKVNPDNVGAEALGRLLVVYPWTQRYFSKFGDLSSASAIMGNPQVKAHGKKVINAFNDGLKHLDNLKGTFAHLSELHCDKLHVD
 |      || ||   |   | |||    ||| ||  |||  | |   |        |       |||||  ||||| || |||||| ||  | |  |||
-SPFSAHEEKLIVDLWAKVDVASCGGDALSRMLIIYPWKRRYFEHFGKLSTDQDVLHNEKIREHGKKVLASFGEAVKHLDNIKGHFAHLSKLHFEKFHVD

PENFRLLGNMIVIVLGHHLGKEFTPSAQAAFQKVVAGVASALAHKYH           Rat hemoglobin beta (PIR2:S06749)
||| ||  | | ||    | |  |||| |  |  ||| || |                             LINEAR alignment
CENFKLLGDIIIVVLGMHHPKDFTLQTHAAFQKLVRHVAAALSAEYH           Spectacled caiman hemoglobin beta (PIR1:HBCQ)
************
MVHLTDAEKATVSGLWGKVNPDNVGAEALGRLLVVYPWTQRYFSKFGDLSSASAIMGNPQVKAHGKKVINAFNDGLKHLDNLKGTFAHLSELHCDKLHVD
 |      || ||   |   | |||    ||| ||  |||  | |   |        |       |||||  ||||| || |||||| ||  | |  |||
SPFSAHEEKLIV-DLWAKVDVASCGGDALSRMLIIYPWKRRYFEHFGKLSTDQDVLHNEKIREHGKKVLASFGEAVKHLDNIKGHFAHLSKLHFEKFHVD

PENFRLLGNMIVIVLGHHLGKEFTPSAQAAFQKVVAGVASALAHKYH           Rat hemoglobin beta (PIR2:S06749)
||| |||  | | ||    | |  |||| |  |||| || ||  |                          CDA alignment
CENFKLLGDIIIVVLGMHHPKDFTLQTHAAFQKLVRHVAAALSAEYH           Spectacled caiman hemoglobin beta (PIR1:HBCQ)

Alignment pair 3

MDMRVPALLLGLLLLWLPGAKCDIQMTQSPSTLSASVGDSITITCRASQSIGNWLAWFQQKPGKAPNVLIYKASNLKNGIPSRFSGSGSGTEFTLTVINL
                                         | | || ||    | | ||| |   ||| ||  ||||| |||||||||| ||| ||
YELT----------------------QPPSVSVSPGQTATISCSGDKLGESYYDWYQQSPGQSPLLVIYEGDKRPSGIPZRFSGSNSGNTATLTISGT

QSDDFATYYCHHNDTFSWTFGQGTKVLIKR            Human Ig kappa chain precursor V-I region (PIR2:PL0113)
| | ||| |   | || |||  |                                  LINEAR alignment
ESMDEADYYCQAWNSSSVLFGGGTKLTVLG            Human Ig lambda chain V-IV region (PIR1:L4HUML)

    *              ******
MDMRVPALLLGLLLLWLPGAKCDIQMTQSPSTLSASVGDSITITCRASQSIGNWLAWFQQKPGKAPNVLIYKASNLKNGIPSRFSGSGSGTEFTLTVINL
                                         || ||  | |   |  | | ||  |   ||| |||  || | |||||||| ||  |||
YEL--------------------TQPPS-VSVSPGQTATISCSGDKLGESYYDWYQQSPGQSPLLVIYEGDKRPSGIPZRFSGSNSGNTATLTISGT

QSDDFATYYCHHNDTFSWTFGQGTKVLIKR            Human Ig kappa chain precursor V-I region (PIR2:PL0113)
| | ||| |   | || |||  |                                  CDA alignment
ESMDEADYYCQAWNSSSVLFGGGTKLTVLG            Human Ig lambda chain V-IV region (PIR1:L4HUML)


Alignment pair 4

DIQMTQSPSTLSASVGDRVTITCRPSQ-AFGSW-LAWYQQKPGKAPELL--IYRVSSL---QSGVPSRFSGS--GSGTEFTLTISSLQPDDVATYYCQQA
 |    |    | |    | | |  |     |   ||| | | ||| ||     |  |     | | |||  |   ||| ||| |  | | || ||| ||
EVQLVESGGGLVQP-GGSLRLSCAASGFTFSTSAVYWVRQAPGKGLEWVGWRYEGSSLTHYAVSVQGRFTISRNDSKNTLYLQMLSLEPZBTAVYYCARV

NRY-----FGQG-TKLEIKR              Human Ig kappa chain V-I region (PIR2:A27585)
|                                                    LINEAR alignment
TPAAASLTFSAVWGQGTLVT              Human Ig heavy chain V-III region (PIR1:G1HUTE)

            **********
DIQMTQSPSTLSASVGDRVTITCRPSQ-AFGSW-LAWYQQKPGKAPELL--IYRVSSL---QSGVPSRFSGSGSGTEFTLTIS--SLQPDDVATYYCQQA
 |    |    | |    | | |  |     |   ||| | | ||| ||     | |||     | | |||  | ||  | ||     || || ||| |||
EVQLVESGGGLVQP-GGSLRLSCAASGFTFSTSAVYWVRQAPGKGLEWVGWRYEGSSLTHYAVSVQGRFTISRNDSKNTLYLQMLSLEPZBTAVYYCARV
            **********
NR---------YFGQGTKLEIKR           Human Ig kappa chain V-I region (PIR2:A27585)
           ||||                                       CDA alignment
TPAAASLTFSAVWGQGTLVT---           Human Ig heavy chain V-III region (PIR1:G1HUTE)
```

Figure 1. Four pairs of LINEAR and CDA alignments. The positions of the upper sequence in each CDA alignment, which are aligned differently by LINEAR, are starred.

References

1. Barker, W. C., D. G. George, L. T. Hunt and J. S. Garavelli. The PIR protein sequence database. Nucleic Acids Res. **19** (1991) 2231-2236.
2. Dayhoff, M. O., R. M. Schwartz and B. C. Orcutt. A model of evolutionary change in proteins. In Atlas of Protein Sequence and Structure (Dayhoff, M. O. ed.), Vol. 5, Suppl. 3, pp. 345-358. National Biomedical Research Foundation, Washington, DC (1978).
3. Gotoh, O. An improved algorithm for matching biological sequences. J. Mol. Biol. **162** (1982) 705-708.
4. Hirschberg, D. S. A linear space algorithm for computing maximal common subsequences. Comm. ACM **18** (1975) 341-343.
5. Huang, X., R. C. Hardison and W. Miller. A space-efficient algorithm for local similarities. Comput. Applic. Biosci. **6** (1990) 373-381.
6. Huang, X. and W. Miller. A time-efficient, linear-space local similarity algorithm. Adv. Appl. Math. **12** (1991) 337-357.
7. Lewin, B. Genes IV. Cell Press, Cambridge, MA (1990).
8. Myers, E. W. and W. Miller. Optimal alignments in linear space. Comput. Applic. Biosci. **4** (1988) 11-17.
9. Sankoff, D. and J. B. Kruskal. Time Warps, String Edits, and Macromolecules: The Theory and Practice of Sequence Comparisons. Addison-Wesley, Reading, MA (1983).
10. Sellers, P. H. On the theory and computation of evolutionary distances. SIAM J. Appl. Math. **26** (1974) 787-793.
11. Smith, T. F. and M. S. Waterman. Identification of common molecular subsequences. J. Mol. Biol. **147** (1981) 195-197.
12. Waterman, M. S., T. F. Smith and W. A. Beyer. Some biological sequence metrics. Adv. Math. **20** (1976) 367-387.
13. Wilbur, W. J. and D. J. Lipman. The context dependent comparison of biological sequences. SIAM J. Appl. Math. **44** (1984) 557-567.

Fast Identification of Approximately Matching Substrings

Archie L. Cobbs*

Computer Science Division, University of California Berkeley, Berkeley, CA 94720

Abstract. Let two strings S, T over a finite alphabet Σ be given, and let M be an arbitrary relation on $\Sigma \times \Sigma$. Define an *approximate match* (x, y) of two length m subwords (substrings) $x \subseteq S$, $y \subseteq T$ when $M(x_i, y_i)$ for all $1 \leq i \leq m$. A match implies all the *local alignments* (without insertions and deletions) which are pairings of specific occurrences of x and y. A match (x, y) is *maximal* if there exists no longer match (u, v) such that all of the local alignments implied by (x, y) are contained in a local alignment implied by (u, v). We give an efficient algorithm for finding all maximal matches between S and T. The algorithm runs in time bounded by the sum of the lengths of the maximal matches, at worst $O(|\Sigma|^2 n^2)$. The main application is identifying homologous regions of protein sequences.

1 Introduction

Protein and DNA sequence analysis is a common technique in molecular biology for determining structure, function, and evolutionary relationships. Sequences of related proteins are often characterized by short, highly conserved homolgous regions which translate into portions of the molecule structure critical to its function, separated by less well conserved regions varying in length and residue composition. Much attention has been paid to the "local alignment" problem, which attempts to identify related regions in different protein sequences. Here we give an efficient algorithm useful for finding related regions in protein sequences.

A binary relation on sequence letters provides the basis for approximate matching. A match is a pair of equal length subwords all of whose induced letter pairings satisfy this relation. The relation can be generated using a real-valued sequence element similiarity matrix, such as the PAM matrices described by Dayhoff, et.al. [4], and a match threshold. Sequence elements then satisfy the relation if their similarity score is above the threshold. The threshold can be varied for subsequent runs of the algorithm to adjust the sensitivity and hence the number of matches reported.

The number of matches between S and T depends on the relation M. If M is the identity relation, we find all maximal exact matches between subwords of S and T. This is essentially the case treated by Clift, et.al. [3], in which two-dimensional "sequence landscapes" are created. Another special case is when M

* Supported by DOE Grant 442427-22446

is an equivalence relation, which is equivalent to the exact case over a smaller alphabet. Here M is arbitrary, though usually symmetric.

The algorithm has some properties desirable when dealing with large sequences. First, it only reports matches of maximal length, eliminating redundancy. Second, the total running time is proportional to the sum of the lengths of the maximal matches, in the best case linear and at worst $O(n^2)$ (for a fixed alphabet). In common practice quadratic behavior is likely to be the case; however, if very long strings are analyzed for just a few close matches, the running time will be reduced accordingly. In any case, $O(n)$ space is required.

2 Notation and Definitions

Let Σ be a fixed finite alphabet, and let M be a relation on $\Sigma \times \Sigma$. If $M(a, b)$ we say a and b *approximately match*, denoted $a \sim b$. Let S and T be finite strings over Σ. Let $x \subseteq S$ indicate that x is a subword (substring) of S. The length of x is $|x|$. ϵ denotes the empty word. An *occurrance* of x in S identifies a particular location where x appears. The concatenation of x and a is written xa. The ith letter of x is written x_i, for $1 \le i \le |x|$.

The subwords of S are partitioned into *right equivalence classes*. These are formed by the equivalence relation $u \equiv_R v$ when one of u, v is a suffix of the other (say, u is a suffix of v) and every occurrence of u in S is as a suffix of an occurrance of v. Then u and v have the same *right context* in S. Similarly, $u \equiv_L v$ when u and v have the same *left context* in S.

If $x \subseteq S$, $y \subseteq T$, and $|x| = |y| = m$, a pairing of an occurrance of x in S with an occurrence of y in T is a *local alignment*. This usage of the term is different from the common meaning which includes insertions and deletions. A local alignment *induces* those local alignments which it contains as matched substrings: if $|x| = |y| = m$, there are $m(m + 1)/2$ induced local alignments.

A *match* is defined as an ordered pair (x, y), for $x \subseteq S$, $y \subseteq T$, which satisfies:

1. $|x| = |y| = m$, for some $m \ge 0$.
2. $M(x_i, y_i)$ for all $1 \le i \le m$.

The *length* of the match (x, y) is m. Note that a match is independent of the actual number of occurrences of x or y. A match *implies* all of the local alignments obtained by pairing occurrences.

If $(wxz, w'yz')$ is a match, then so is (x, y). Suppose further every occurrence of x is in the context wxz, and every occurrence of y is in the context $w'yz'$, implying $x \equiv_{L,R} wxz$ and $y \equiv_{L,R} w'yz'$. Then every local alignment implied by (x, y) is induced by a local alignment implied by $(wxz, w'yz')$. If $|w| > 0$ ($|z| > 0$), (x, y) is *left-extendable* (*right-extendable*). Alternately, (x, y) is not left (right) extendable iff one or both of x, y is the longest subword in its respective right (left) equivalence class.

Suppose (x, y) is a match of length m, and let k be such that $0 \le k \le m$. If $(x_1 \cdots x_k, y_1 \cdots y_k)$ is left-extendable, so is (x, y); if $(x_k \cdots x_m, y_k \cdots y_m)$ is right-extendable, so is (x, y). A match not extendable in either direction is *maximal*.

3 Directed Acyclic Word Graphs

As in [3], our algorithm uses the *directed acyclic word graph* (DAWG) data structure of Blumer, et.al. [1] (Figure 1). The DAWG D_S for S is a graph having a node for each right equivalence class of subwords of S and whose edges are labeled with single letters from Σ. R_p^S denotes the right equivalence class associated with node $p \in D_S$. Given two nodes $p, p' \in D_S$, there is an edge labeled a from $p \to p'$ iff there exist subwords $w \in R_p^S$, $w' \in R_{p'}^S$, such that $w' = wa$. The definition of right equivalence insures this is well-defined, and guarantees one (unique) letter per out-going edge of each node.

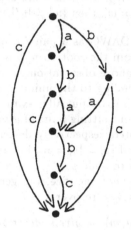

Fig. 1. Directed acyclic word graph (DAWG) on the string *ababc*.

D_S has a single source node, or root, corresponding to $\{\epsilon\}$, and a single sink node corresponding to $\{S\}$. The in-degree of any node p is bounded by $|R_p^S|$; the out-degree of any node is bounded by $|\Sigma|$. Each subword $w \subseteq S$ traces a unique path in D_S of length $|w|$ from the root to the node p for which $w \in R_p^S$. Each path from p to the sink represents a particular occurrence of all the subwords in R_p^S. A path's concatenated labels form the suffix of S that begins with the first character after the corresponding occurrance.

D_S can be thought of as the uncompacted suffix tree for S with edge-isomorphic subtrees identified and the terminator structure removed (see also [2]). A suffix tree's edges of out-degree one are compacted together, resulting in subwords labeling the edges. This compaction enables it to remain linear in size and construction ([6], [5]). The DAWG for S has no edge compaction; yet, as reported in [1], if $|S| = n$, then D_S contains no more than $2n - 1$ nodes, $3n - 3$ edges, and can be constructed on-line in linear time. In [1] it is shown how DAWG edges can also be compacted in linear time, if so desired.

4   Describing Matches

From [1] the following observation is clear:

Observation 1. *There is a one to one correspondence between paths from the root to a node $p \in D_S$ and subwords $w \in R_p^S$. Moreover, the length of the path describing w is $|w|$.*

From this it follows:

Observation 2. *No two words $w, w' \in R_p^S$ have the same length. Consequently, no two paths from the root to p have the same length.*

Proof. Suppose $w, w' \in R_p^S$. Then one is a suffix of the other, and if $|w| = |w'|$, then $w = w'$. Therefore, if $w \neq w'$, then $|w| \neq |w'|$. □

Define a *d-pointer* into a DAWG as a pair $\langle p, m \rangle$, where p is a node and m is the length of some path from the root to p. p is called the *target*, and m is called the *length*. Using the above observations, a d-pointer uniquely *describes* the subword $w \in R_p^S$ corresponding to the unique path of length m to p.

The algorithm's objective is to output all maximal matches between subwords of S and T. It will do this by outputting pairs of d-pointers $\langle p, m \rangle$, $\langle q, m \rangle$, where $p \in D_S$, $q \in D_T$. The d-pointers respectively describe subwords $x \subseteq S$, $y \subseteq T$ such that $x \in R_p^S$, $y \in R_q^T$, and $|x| = |y| = m$. In turn, this describes the match (x, y), which is said to *belong* to node p.

This form of output does not sacrifice any generality, because the actual occurrences of x and y are easy to recover:

Lemma 3. *Let $\langle p, m \rangle$ and $\langle q, m \rangle$ be given, describing a match (x, y). We can find the occurrences of the corresponding subwords $x \subseteq S$ and $y \subseteq T$ in time proportional to the number of such occurrences.*

Proof (Sketch). [1] gives a way to do this: after construction the edges of the DAWG are compacted (in linear time). Then, starting from node p for which $x \in R_p^S$, all of the occurrences of x are found with a depth-first search taking time proportional to the number of occurrences. □

Conversely, given a match (x, y) of length m, there is a unique pair of d-pointers $\langle p, m \rangle$, $\langle q, m \rangle$, where $p \in D_S$, $q \in D_T$, $x \in R_p^S$, $y \in R_q^T$, and $m = |x| = |y|$.

The next lemma describes the connection between extensions of a match and the corresponding d-pointer pairs:

Lemma 4. *Let (x, y) be a match between sequences S and T described by the pair $\langle p, m \rangle$, $\langle q, m \rangle$. Let $(wxz, w'yz')$ be a match. Then $(wx, w'y)$ and (xz, yz') are matches. Moreover,*

(a) *$(wx, w'y)$ left-extends (x, y) if and only if $(wx, w'y)$ is described by the pair $\langle p, t \rangle$, $\langle q, t \rangle$, where $t = m + |w|$.*

(b) (xz, yz') *right-extends* (x, y) *if and only if* (xz, yz') *is described by the pair* $\langle p', t \rangle$, $\langle q', t \rangle$, *where* $xz \in R_{p'}^S$, $yz' \in R_{q'}^T$, $t = m + |z|$, *and there is exactly one path from* p *to* p' *and from* q *to* q', *tracing out* z *and* z', *respectively.*

Proof. That $(wx, w'y)$ and (xz, yz') are matches is clear.

(a) If $(wx, w'y)$ left-extends (x, y) then $wx \equiv_R x$ and $w'y \equiv_R y$, and so $wx \in R_p^S$ and $w'y \in R_q^T$. Therefore $(wx, w'y)$ is described by the pair $\langle p, t \rangle$, $\langle q, t \rangle$ for some t; clearly t must equal $m + |w|$.

On the other hand, suppose the pair $\langle p, t \rangle$, $\langle q, t \rangle$ describes $(wx, w'y)$, where $t = m + |w|$. Then wx traces a path of length t from the root of D_S to p, and so $wx \in R_p^S$ and $wx \equiv_R x$. Similarly, $w'y \equiv_R y$. Since $(wx, w'y)$ is a match, $(wx, w'y)$ left-extends (x, y).

(b) If (xz, yz') right-extends (x, y), then $xz \equiv_L x$, so every occurrence of x is followed by z. Therefore only one path exists from p to p', tracing out z and so $xz \in R_{p'}^S$. Similarly for yz' and q'. Clearly then $\langle p', t \rangle$, $\langle q', t \rangle$ describes (xz, yz').

On the other hand, if there is exactly one path from p to p' tracing out z, then it must be that every occurrence of $x \in R_p^S$ is followed by z; similarly for y and z'. Since (xz, yz') is a match, then (xz, yz') right-extends (x, y). $\square$

5 The Algorithm

5.1 Overview

Let S and T be given. First we construct DAWG's D_S and D_T. The algorithm will maintain for each node $p \in D_S$ a list $L(p)$ of d-pointers, all of whose targets are nodes of D_T. $L(p)$ is called p's *match list* because each d-pointer represents a match: if $\langle q, m \rangle \in L(p)$, then the pair $\langle p, m \rangle$, $\langle q, m \rangle$ represent a match (x, y) as described above. When the algorithm finishes, these lists will contain exactly the maximal matches.

The algorithm *processes* one node at a time. It only processes a node after all of its predecessor nodes have been processed. A linear time topological sort insures proper ordering of the nodes. Initially $L(p)$ is empty. When p is processed, d-pointers *propagate* (as described below) from the match lists of p's predecessors into $L(p)$. Once all of the predecessors' match lists have been propagated, p has been processed and the next node is chosen. At this point, no more d-pointers will be added to $L(p)$. Once all nodes have been processed we make a *final scan* of all the node lists, removing some d-pointers. Then we output all the maximal matches using the match lists.

5.2 Invariants

The following invariants are maintained:

1. The d-pointers in $L(p)$ always describe matches.
2. Immediately after node p has been processed, the d-pointers in $L(p)$ describe exactly those matches belonging to p which are not left-extendable.

Moreover, the final scan removes all right-extendable matches from $L(p)$. Therefore at the end of the algorithm all of the $L(p)$ together describe exactly the maximal matches between subwords of S and T. The output of the algorithm is this combined list.

5.3 Description

Initially all edge lists are empty. The root node $r \in D_S$ is given an edge list containing only $\langle r', 0 \rangle$, where r' is the root of D_T. This operation takes constant time, and at this point r has been processed.

Processing a Node. In order to process the next node, the algorithm chooses a node $p \in D_S$ all of whose predecessor nodes $p_1, \ldots, p_k$ have been processed. For each p_i, we propagate the list associated with p_i to p, one d-pointer at a time, as follows.

Let a_i be the label on the edge from p_i to p. For each d-pointer $\langle q, m \rangle \in L(p_i)$, we try to extend the match: Let $q_1, \ldots, q_l$ be the successor nodes of q in D_T, and let b_j be the label on the edge from q to q_j. For each b_j such that $a_i \sim b_j$, add the d-pointer $\langle q_j, m+1 \rangle$ to $L(p)$. However, if $L(p)$ already contains another d-pointer with the same target q, we throw out the d-pointer having the smaller length. The lengths will never be equal by Observation 2. This implies $L(p)$ contains at most one d-pointer with target q, for any $q \in D_T$.

Final Scan. After all nodes have been processed, we make a final scan of the nodes of D_S: for each $p \in D_S$, if p has out-degree more than one, we do nothing. Otherwise, let p' be p's successor. For each entry $\langle q, m \rangle$ in p's list, if q has out-degree more than one, we go to the next entry. Otherwise, let q' be q's successor. If the labels on the edges $p \to p'$ and $q \to q'$ approximately match, then we remove $\langle q, m \rangle$ from $L(p)$.

Output. After the final scan, the algorithm outputs pairs $\langle p, m \rangle$, $\langle q, m \rangle$ for each d-pointer $\langle q, m \rangle \in L(p)$, for all $p \in D_S$.

5.4 Proofs of the Invariants

First we show that all d-pointers describe matches:

Lemma 5. Let p be a node of D_S, and let $\langle q, m \rangle$ be a d-pointer in p's list. The words described by $\langle p, m \rangle$ and $\langle q, m \rangle$ form a match.

Proof. We prove by induction. In the case of the root node's list $\langle q, m \rangle = \langle r', 0 \rangle$, describing the match (ϵ, ϵ). In the general case, assume the lemma is true for all of p's predecessors. $\langle q, m \rangle$ was propagated to p from the list of some predecessor p_i, where it came from a d-pointer of the form $\langle q_j, m-1 \rangle$, where q_j is a predecessor of q in D_T. Let a and b be the labels on the edges $p_i \to p$ and $q_j \to q$, respectively.

By induction, the pair $\langle p_i, m-1 \rangle$, $\langle q_j, m-1 \rangle$ describe a match (x, y). Consider xa and yb. It must be the case that $a \sim b$, otherwise $\langle q, m \rangle$ could not have been added to p's list. Therefore (xa, yb) is be a match. But this is the match described by $\langle p, m \rangle$ and $\langle q, m \rangle$. □

Lemma 3 proves the second invariant:

Lemma 6. *Let $p \in D_S$ be a processed node before the final scan, and let (x, y) be a match of length m such that $x \in R_p^S$. Let $q \in D_T$ be such that $y \in R_q^T$. Then $\langle q, m \rangle \in L(p)$ if and only if (x, y) is not left-extendable.*

Proof. For the root node r, (ϵ, ϵ) is the only match described, and it is not left-extendable. Assume $p \neq r$.

Suppose (x, y) is not left-extendable, yet $\langle q, m \rangle \notin L(p)$. Let the nodes in D_S on the path from the root to p traced by x be $p_0 = r, p_1, \ldots, p_m = p$; similarly for q_i and y. There must exist some some $0 < k \leq m$ such that $\langle q_{k-1}, k-1 \rangle \in L(p_{k-1})$, yet $\langle q_k, k \rangle \notin L(p_k)$. Since $x_k \sim y_k$, this can only happen if $\langle q_k, k \rangle$ was overriden by another d-pointer $\langle q_k, t \rangle \in L(p_k)$ for which $t > k$. This implies $(x_1 \cdots x_k, y_1 \cdots y_k)$ is left-extendable. But if (x, y) is not left-extendable, then neither can be $(x_1 \cdots x_k, y_1 \cdots y_k)$. This is a contradiction, so it must be that $\langle q, m \rangle \in L(p)$.

For the reverse direction, we prove by induction; assume the lemma is true for all of p's predecessors. Suppose $\langle q, m \rangle \in L(p)$, yet (x, y) is left-extendable. Then there exists a longer match (ux, vy) belonging to p represented by $\langle q, t \rangle \in L(p)$ for some $t > m$. Without loss of generality, assume (ux, vy) is not left-extendable. Then $(ux_1 \cdots x_{m-1}, vy_1 \cdots y_{m-1})$ is not left-extendable. By induction, $(ux_1 \cdots x_{m-1}, vy_1 \cdots y_{m-1})$ is represented by $\langle q_j, m-1 \rangle \in L(p_i)$ for some predecessors p_i of p and q_j of q. As $x_m \sim y_m$, during processing of p, $\langle q, t \rangle$ must have overriden $\langle q, m \rangle$, which contradicts $\langle q, m \rangle \in L(p)$. So it must be that (x, y) is not left-extendable. □

Lemma 7. *The final scan removes exactly those d-pointers from $L(p)$ which represent right-extendable matches.*

Proof. This follows easily from Lemma 4(b). □

Together, these yield:

Lemma 8. *When the algorithm finishes, the algorithm has identified all maximal matches between subwords of S and T. In particular, there is a one to one correspondence between entries in the match lists of all the nodes of D_S and the maximal matches between S and T.*

Proof. Let (x, y) be a maximal match. Let $\langle p, m \rangle$, $\langle q, m \rangle$ be the unique d-pointer pair describing (x, y). Clearly (x, y) can only be described by an entry in $L(p)$; moreover, there is at most one such entry with target q. In fact, there is exactly one: since (x, y) is not left-extendable, by Lemma 6 $\langle q, m \rangle$ appears in $L(p)$ after processing. Since (x, y) is not right-extendable, it is not removed in the final scan.

Conversely, suppose $\langle q, m \rangle$ appears in $L(p)$ for some node p. By Lemma 5, $\langle q, m \rangle$ represents some match (x, y). By Lemma 7, (x, y) is not right-extendable or else $\langle q, m \rangle$ would have been removed in the final scan. By Lemma 6, (x, y) is not left-extendable or else $\langle q, m \rangle \notin L(p)$ after processing. Therefore, (x, y) is a maximal match. (x, y) can not be described by any other d-pointer, so each d-pointer uniquely represents a maximal match. $\qquad\square$

By outputting all pairs $\langle p, m \rangle$, $\langle q, m \rangle$ for $p \in D_S$ and $\langle q, m \rangle \in L(p)$, the algorithm has describes all maximal matches between subwords of S and T.

6 Analysis

6.1 List Operations

The algorithm works by processing each node $p \in D$, propagating d-pointer lists down all the edges coming into p. When a single d-pointer is propagated from $L(v_i)$ to $L(p)$ for some predecessor p_i, it can result in more than one but at most $|\Sigma|$ new d-pointers in p's list. During the final scan, some d-pointers are removed. Call the action of adding, replacing, or deleting a d-pointer in a node's match list a *list operation*.

Lemma 9. *Each list operation can be done in constant time, given a single linear time operation at the start of the algorithm.*

Proof. We only add or replace nodes during processing, and we only delete nodes during the final scan. For the list associated with p we use a doubly linked list. Adding, replacing, and deleting entries can clearly be done in constant time. The only other task is searching the list during propagation, when we must check whether $L(p)$ already contains a d-pointer with target q.

To facilitate this, at the start of the algorithm we create an array A. The array is indexed by the $q \in D_T$, and so is linear size. Each entry contains either NIL, or else a pointer to an entry in $L(p)$ for some $p \in D_S$. $L(p)$ contains at most one d-pointer with target q; we use the array to locate it (if it exists) in constant time. Upon creation all entries are initialized to NIL.

As we process node p, for each $q \in D_T$ the array keeps a pointer to the (unique) d-pointer in p's list with target q. If there is no such d-pointer, the entry is NIL. Before adding a d-pointer $\langle q, m \rangle$ to p's list, we check the entry corresponding to q in order to determine if another d-pointer with target q already exists in $L(p)$. If the entry is NIL, the answer is no, so we insert the d-pointer into $L(p)$ and change the NIL pointer to point to it. If the entry is not NIL, we compare the lengths of the two d-pointers. If the old one is longer, we do nothing. Otherwise, we replace the old d-pointer with the new one and update the array pointer. This all can be done in constant time per operation.

We need to reuse the array for each node treated. However, directly resetting the array to all NIL pointers before processing each node would require linear work per node. To get around this, we store a pointer to p along with each pointer in the array. Then whenever we access the array if the stored p is not the one we're currently treating, we set the pointer back to NIL. This adds only constant time per list operation. □

6.2 An $O(n^2)$ Bound

Lemma 10. *Propagating a d-pointer down an edge requires $O(|\Sigma|)$ work.*

Proof. Let $p_i \rightarrow p$ be an edge. Let $\langle q, m \rangle$ be a d-pointer in p_i's list. Propagating $\langle q, m \rangle$ requires the following for each outgoing edge from q: comparing its edge label with the edge label on $p_i \rightarrow p$, and doing one list operation (if they approximately match). Both are constant time operations. Since node $q \in D_T$ has out-degree at most $|\Sigma|$, the result follows. □

For simplicity in analysis, assume $|S| = |T| = n$. Each match list has size $O(n)$, since it contains no more entries than nodes in D_T. Therefore propagating a match list down an edge requires at most $O(n|\Sigma|)$ work. There are $O(n)$ edges in D_S, so processing all of the nodes takes time in $O(n^2|\Sigma|)$.

Once this is done, we need to do a final scan (and output all of the maximal matches in the form of d-pointer pairs). Since there are a linear number of nodes in D_S, the total size of all the completed lists is at most $O(n^2)$. We can scan each linked list in time proportional to its length, so an upper bound on time required during the final scan is $O(n^2)$, giving a total time bound of $O(n^2|\Sigma|)$.

6.3 A Better Bound

Clearly the algorithm requires at least $O(n)$ time and space. If there are no matches at all, it will in fact run in linear time. So a more precise bound than $O(n^2)$ is possible.

Let $|L(p)|$ be the length of $L(p)$ after p has been processed but before the final scan.

Lemma 11. *The total time required to process all of the nodes of D_S is*

$$O\left(|\Sigma|^2 \sum_{p \in D_S} |L(p)|\right).$$

Proof. By Lemma 10 processing node p takes time proportional to $|\Sigma|$ times the total number of d-pointers in the match lists of p's predecesors. So the total time required is in the order of

$$\sum_{p \in D_S} \sum_{p_i \rightarrow p} |\Sigma| |L(p_i)|.$$

In other words, we do one $O(|\Sigma|)$ operation for each pair $(\langle q, m \rangle \in R_p^\varsigma$, edge $p \rightarrow p_i)$. A node p can have at most $|\Sigma|$ successors. By summing up from the point of view of predecessor nodes instead of successor nodes, the total processing time is bounded by

$$\sum_{p \in D_S} \sum_{p \to p_1} |\Sigma||L(p)| \le \sum_{p \in D_S} |\Sigma|^2 |L(p)| = |\Sigma|^2 \sum_{p \in D_S} |L(p)| \ .$$

□

This immediately yeilds

Lemma 12. *The total time required for the algorithm is*

$$\max \left\{ O(n), O\left(|\Sigma|^2 \sum_{p \in D_S} |L(p)| \right) \right\} \ .$$

Proof. DAWG creation and initialization take linear time. The final scan only takes time

$$O\left(\sum_{p \in D_S} |L(p)| \right) \ .$$

Lemma 11 completes the result. □

For a fixed alphabet, the running time of the algorithm is bounded by the total number of non left-extendable matches, since by Lemma 6 this is equal to $\sum_{p \in D_S} |L(p)|$. This number can be thought of differently: suppose (x, y) is a maximal match. Then each pair of equal length prefixes of x and y form a non left-extendable match. Conversely, every non left-extendable match is a prefix of some maximal match. Therefore, the total number of non left-extendable matches is bounded by the sum of the lengths of the maximal matches. This proves:

Theorem 13. *Let Σ be a fixed finite alphabet. Given M, an arbitrary relation on $\Sigma \times \Sigma$, and $S, T \in \Sigma^*$, $|S| = |T| = n$, we can find all maximal matches between S and T in time*

$$\max \left\{ O(n), O\left(\sum_{(x,y)} |x| \right) \right\} \ ,$$

where (x, y) ranges over all maximal matches.

7 Discussion

There are a few possible modifications and extensions to the above algorithm which might prove useful. For one, very short maximal matches are not interesting; the algorithm could easily be modified to only output matches longer than a certain length. A more general matching function which operates in any constant sized fixed window (rather than the current window size of one) would be easy to implement, and would not change the asymptotic running time. One way to do this is to replace the alphabet Σ with $\Sigma \times \cdots \times \Sigma$, where the number of terms is the width of the window.

Running the algorithm on k strings instead of two is a possible generalization. The matching relation M would be a k-ary relation, and d-pointers would be generalized to point into the $k-1$ other graphs. However, the logic of Section 6.2 yeilds only an $O(|\Sigma|^k n^k)$ upper bound running time. Still, in cases with sparse matching the algorithm would run acceptably fast. One approximation approach to the problem of multiple string alignment is to find similar regions in all the strings, align these, and subdivide the problem. Though any local alignment algorithm suffices for this step, our algorithm's ability to use a varying threshold might make it more suitable for this approach.

A trickier issue is expanding the definition of match to include insertions and deletions. With appropriate bounds on the error rate (e.g., at most a logarithmic number of insertions or deletions) the lists associated with each node could be suitably broadened to include such matches and still keep the total running time polynomial

8 Acknowledgements

Thanks to Eugene Lawler and Z Sweedyk for helpful discussions, and to the reviewers for their comments.

References

1. A. Blumer, J. Blumer, A. Ehrenfeucht, D. Haussler, and R. McConnell. Building a complete inverted file for a set of text files in linear time. In *FOCS*, pages 349–358. ACM, January 1984.
2. M. T. Chen and Joel Seiferas. *Efficient and Elegant Subword Tree Construction*, pages 97–107. Springer-Verlag, Berlin, 1985.
3. B. Clift, D. Haussler, R. McConnell, T. D. Schneider, and G. D. Stormo. Sequence landscapes. *Nucleic Acids Res.*, 14(1):141–158, January 1986.
4. M. O. Dayhoff, R. M. Schwartz, and B. C. Orcutt. *Atlas of Protein Structure*, volume 5. National Biomedical Research Foundation, Washington, DC, 1978. suppl. 3.
5. Edward M. McCreight. A space-economical suffix tree construction algorithm. *JACM*, 23(2):262–272, April 1976.
6. P. Weiner. Linear pattern matching algorithms. In *14th Annual Symposium on Switching and Automata Theory*, pages 1–11. IEEE, 1973.

Alignment of Trees - An Alternative to Tree Edit

Tao, Jiang[1*], Lusheng Wang[2*], Kaizhong Zhang[3**]

[1] Department of Computer Science, McMaster University, Hamilton, Ont. L8S 4K1, Canada.
[2] Department of Electrical and Computer Engineering, McMaster University, Hamilton, Ont. L8S 4K1, Canada.
[3] Department of Computer Science, University of Western Ontario, London, Ont. N6A 5B7, Canada.

Abstract. In this paper, we propose the *alignment of trees* as a measure of the similarity between two *labeled* trees. Both *ordered* and *unordered* trees are considered. An algorithm is designed for ordered trees. The time complexity of this algorithm is $O(|T_1| \cdot |T_2| \cdot (\deg(T_1) + \deg(T_2))^2)$, where $|T_i|$ is the number of nodes in T_i and $\deg(T_i)$ is the degree of T_i, $i = 1, 2$. The algorithm is faster than the best known algorithm for tree edit when $\deg(T_1)$ and $\deg(T_2)$ are smaller than the depths of T_1 and T_2. For unordered trees, we show that the alignment problem can be solved in polynomial time if the trees have a bounded degree and becomes NP-hard if one of the trees is allowed to have an arbitrary degree. In contrast, the edit problem for unordered trees is NP-hard even if both trees have a bounded degree [17]. Finally, multiple alignment of trees is discussed.

1 Introduction

In many fields such as RNA secondary structures comparison, syntactic pattern recognition, image clustering, genetics, and chemical structure analysis, one often faces the problem of finding the similarity of two *labeled trees* [4, 5, 6, 7, 8, 9, 10, 12]. For instance, the comparison of *ordered* trees is very useful in the study of RNA secondary structures. In general, we can decompose an RNA secondary structure into components of fives types: stem (S), hairpin (H), bulge (B), interior loop (I), and multi-branch loop (M). The secondary structure can be conveniently expressed as a tree in which each node is labeled by a letters S, H, B, I, or M, and the left to right order among siblings is significant [5, 8, 11, 16]. For example, the RNA secondary structure in Figure 1(a) can be represented as an ordered labeled tree [11] as shown in Figure 1(b). The comparison of RNA secondary structure trees can help identify conserved structural motifs in an RNA folding process [4] and construct taxonomy trees [11]. The comparison of *unordered trees* has applications to the morphological problems arising in genetics

* Supported in part by NSERC Research Grant OGP0046613.
** Supported in part by NSERC Research Grant OGP0046373.

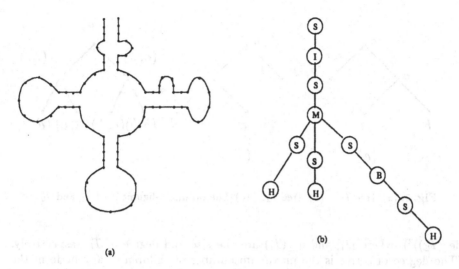

Fig. 1. An RNA secondary structure and its tree representation.

(*e.g.*, determining genetic diseases based on ancestry tree patterns) and other fields [10, 9, 12].

As in the case of sequence comparisons, there are many ways to measure the similarity between two trees. For instance, one could use the *largest common sub-tree*, the *smallest common super-tree*, *tree edit distance*, and the *transferable ratio* between two trees to describe the degree of similarity [3, 5, 7, 11, 13, 15]. Although edit distance and transferable ratio are both sensible measures of the distance between RNA secondary structures [7, 5], each of them only represents a certain approximation of the true functional similarity. Thus, more realistic and feasible measures would be of interest. Here, we introduce the notion of *alignment of trees* as another measure of similarity of labeled trees. The notion is a natural extension of sequence alignment to trees.

First, let's recall the definition of an insertion operation in tree edit [13, 16]. Let T be an ordered (or unordered) tree. Inserting a node u into T means that for some node v in T, we make u the parent of a consecutive subsequence (or a subset, respectively) of the children of v and then v the parent of u. Let T_1 and T_2 be two labeled trees. A deletion is just the complement of an insertion. An alignment A of T_1 and T_2 is obtained by first inserting nodes labeled with *spaces* into T_1 and T_2 such that the two resulting trees T_1' and T_2' have the same structure, *i.e.*, they are identical if the labels are ignored, and then *overlaying* T_1' on T_2'. An example alignment is shown in Figure 2. A score is defined for each pair of labels. The *value* of alignment A is the sum of the scores of all pairs of opposing labels. An *optimal* alignment is one that minimizes the value over all possible alignments. The *alignment distance* between T_1 and T_2 is the value of an optimal alignment of T_1 and T_2.

We present an algorithm for computing the alignment distance between ordered trees. The time complexity of this algorithm is $O(|T_1| \cdot |T_2| \cdot (\deg(T_1) +$

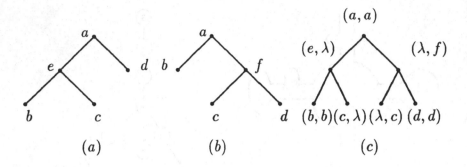

Fig. 2. (a) Tree T_1. (b) Tree T_2. (c) The optimal alignment of T_1 and T_2.

$\deg(T_2))^2)$, where $|T_i|$ and $\deg(T_i)$ are the size and degree of T_i, respectively. (The degree of a tree is the maximum number of children of any node in the tree.) We also show that the alignment distance between two unordered trees can be computed in polynomial time if the trees have bounded degrees and becomes NP-hard if one of the trees is allowed to have an arbitrary degree.

- **Alignment of trees vs tree edit.**

It is well known that edit and alignment are two equivalent notions for sequences. In particular, for any two sequences x_1 and x_2, the edit distance between x_1 and x_2 equals the value of an optimal alignment of x_1 and x_2. However, edit and alignment turn out to be very different for trees. The following are some interesting comparisons between alignment of trees and tree edit.

1. The edit distance and alignment distance between two trees can be different. For example, assume that each edit operation (*i.e.*, insertion, deletion, or replacement) costs 1 and consequently each pair of distinct letters has a score 1. Consider the two ordered trees shown in Figure 2. To optimally edit T_1 into T_2, we simply delete e from T_1 and insert f into the new tree. Thus, the edit distance between T_1 and T_2 is 2. The optimal alignment of the two trees is unique and is shown in 2(c), with a value 4. The difference between edit distance and alignment distance can be made arbitrarily large by adding subtrees below nodes b, c, d in both trees. It is easy to see that in general the edit distance is smaller than the alignment distance for trees. This is because each alignment of trees actually corresponds to a restricted tree edit in which all the insertion precede all the deletions. Note that, the order of edit operations is not important for sequences. Also, it seems that alignment charges more for the structural dissimilarity at the top levels of the trees than at the lower levels, whereas edit treats all the levels the same.

2. The best algorithm computing the edit distance between ordered trees runs in time $O(|T_1| \cdot |T_2| \cdot \min\{\text{depth}(T_1), \text{leaves}(T_1)\} \cdot \min\{\text{depth}(T_2), \text{leaves}(T_2)\})$, where $\text{depth}(T_i)$ and $\text{leaves}(T_i)$ are the depth and number of leaves of tree T_i,

$i = 1, 2$ [16]. Clearly, $\deg(T_i) \leq \text{leaves}(T_i)$. In practice (*e.g.*, RNA secondary structures), $\deg(T_i) \ll \text{leaves}(T_i)$ and $\deg(T_i) \ll \text{depth}(T_i)$. Hence, our above result shows that it is easier (faster) to align ordered trees than to edit. In particular, we can align trees with bounded degrees in time $O(|T_1| \cdot |T_2|)$.

3. The difference in time complexity is even bigger for unordered trees. As mentioned earlier, unordered trees with bounded degrees can be aligned in polynomial time (in fact, in time $O(|T_1| \cdot |T_2|)$). On the other hand, editing unordered trees with bounded degrees is NP-hard [17]. (In fact, it is MAX SNP-hard [15].)

4. The alignment of trees can be easily generalized to more than two trees as in the case of sequences. This provides a way to compare multiple trees simultaneously. Although it is also possible to compare multiple trees based on tree edit distance, as reported in [11], the method seems to be only applicable to situations where clustering is required (*e.g.*, in the construction of a taxonomy tree).

In section 2, we present the algorithm for aligning ordered trees. Section 3 contains some results on unordered trees. We briefly discuss multiple alignment of trees with SP-score in section 4.

2 An efficient algorithm for aligning ordered trees

We need some definitions. The notion of alignment can be easily extended to ordered forests. The only change is that it is now possible to insert a node (as the root) to join a consecutive subsequence of trees in the forest. Denote the alignment distance between forests F_1 and F_2 as $D(F_1, F_2)$. Let θ denote the empty tree, λ denote space, and $\mu(a, b)$ denote the score of the opposing letters a and b. A standard assumption is that the score scheme μ satisfies *triangle inequality*, *i.e.*, for any three letters a, b, and c, $\mu(a, c) \leq \mu(a, b) + \mu(b, c)$. The nodes in an ordered tree of size n are numbered 1 through n according to the *postorder*. Let T_1 and T_2 be two fixed ordered labeled trees throughout this section. Denote the label of node j in tree T_i as $l_i[j]$ and the subtree of T_i rooted at node j as $T_i[j]$.

In the following, let i be a node of T_1 and j a node of T_2. Suppose that the degrees of i and j are m_i and n_j, respectively. Denote the children of i as $i_1, \ldots, i_{m_i}$ and the children of j as $j_1, \ldots, j_{n_j}$. For any $s, t, 1 \leq s \leq t \leq m_i$, let $F_1[i_s, i_t]$ represent the forest consisting of the subtrees $T_1[i_s], \ldots, T_1[i_t]$. For convenience, $F_1[i_1, i_{m_i}]$ is also denoted $F_1[i]$. Note that $F_1[i] \neq F_1[i, i]$. $F_2[j_s, j_t]$ and $F_2[j]$ are defined similarly.

2.1 Properties of the alignment distance

The following lemma below form the basis of our algorithm. The first lemma is trivial.

Lemma 1. $D(\theta, \theta) = 0$;

$$D(F_1[i], \theta) = \sum_{k=1}^{m_i} D(T_1[i_k], \theta); \quad D(T_1[i], \theta) = D(F_1[i], \theta) + \mu(l_1[i], \lambda);$$

$$D(\theta, F_2[j]) = \sum_{k=1}^{n_j} D(\theta, T_2[j_k]); \quad D(\theta, T_2[j]) = D(\theta, F_2[j]) + \mu(\lambda, l_2[j]).$$

Lemma 2.

$$D(T_1[i], T_2[j]) = \min \begin{cases} D(\theta, T_2[j]) + \min_{1 \le r \le n_j} \{D(T_1[i], T_2[j_r]) - D(\theta, T_2[j_r])\} \\ D(T_1[i], \theta) + \min_{1 \le r \le m_i} \{D(T_1[i_r], T_2[j]) - D(T_1[i_r], \theta)\} \\ D(F_1[i], F_2[j]) + \mu(l_1[i], l_2[j]) \end{cases}$$

Proof: Consider an optimal alignment (tree) $\mathcal{A}$ of $T_1[i]$ and $T_2[j]$. There are four cases: (1) $(l_1[i], l_2[j])$ is a label in $\mathcal{A}$, (2) $(l_1[i], \lambda)$ and $(l_1[k], l_2[j])$ are labels in $\mathcal{A}$ for some k, (3) $(l_1[i], l_2[k])$ and $(\lambda, l_2[j])$ are labels in $\mathcal{A}$ for some k, (4) $(l_1[i], \lambda)$ and $(\lambda, , l_2[j])$ are labels in $\mathcal{A}$. We actually need not consider Case 4 since in this case we can delete the two nodes and then add $(l_1[i], l_2[j])$ as the new root, resulting in a better alignment.

Case 1. The root of $\mathcal{A}$ must be labeled as $(l_1[i], l_2[j])$. Clearly, $D(T_1[i], T_2[j]) = D(F_1[i], F_2[j]) + \mu(l_1[i], l_2[j])$.

Case 2. The root of $\mathcal{A}$ must be labeled as $(l_1[i], \lambda)$. In this case k must be a node in $T_1[i_r]$ for some $1 \le r \le m_i$. Therefore,

$$D(T_1[i], T_2[j]) = D(T_1[i], \theta) + \min_{1 \le r \le m_i} \{D(T_1[i_r], T_2[j]) - D(T_1[i_r], \theta)\}$$

Case 3. Similar to Case 2. $\square$

Note, the above lemma implies that $D(F_1[i], F_2[j])$ is required for computing $D(T_1[i], T_2[j])$. We have to be careful here when designing the recurrence relation so that we don't align an arbitrary forest of T_1 with an arbitrary forest of T_2. Otherwise the time complexity would be higher.

Lemma 3. *For any s, t such that $1 \le s \le m_i$ and $1 \le t \le n_j$,*

$$D(F_1[i_1, i_s], F_2[j_1, j_t]) = \min \begin{cases} D(F_1[i_1, i_{s-1}], F_2[j_1, j_t]) + D(T_1[i_s], \theta) \\ D(F_1[i_1, i_s], F_2[j_1, j_{t-1}]) + D(\theta, T_2[j_t]) \\ D(F_1[i_1, i_{s-1}], F_2[j_1, j_{t-1}]) + D(T_1[i_s], T_2[j_t]) \\ \mu(\lambda, l_2[j_t]) + \min_{1 \le k < s} \{D(F_1[i_1, i_{k-1}], F_2[j_1, j_{t-1}]) \\ \qquad\qquad + D(F_1[i_k, i_s], F_2[j_t])\} \\ \mu(l_1[i_s], \lambda) + \min_{1 \le k < t} \{D(F_1[i_1, i_{s-1}], F_2[j_1, j_{k-1}]) \\ \qquad\qquad + D(F_1[i_s], F_2[j_k, j_t])\} \end{cases}$$

Proof: Consider an optimal alignment (forest) $\mathcal{A}$ of $F_1[i_1, i_s]$ and $F_2[j_1, j_t]$). The root of the rightmost tree in $\mathcal{A}$ is labeled by either $(l_1[i_s], l_2[j_t])$, $(l_1[i_s], \lambda)$, or $(\lambda, l_2[j_t])$.

Case 1: the label is $(l_1[i_s], l_2[j_t])$. In this case, the rightmost tree must be an optimal alignment of $T_1[i_s]$ and $T_2[j_t]$. Therefore

$$D(F_1[i_1, i_s], F_2[j_1, j_t]) = D(F_1[i_1, i_{s-1}], F_2[j_1, j_{t-1}]) + D(T_1[i_s], T_2[j_t])$$

Case 2: the label is $(l_1[i_s], \lambda)$. In this case, there is a k, $0 \le k \le t$, such that $T_1[i_s]$ is aligned with the subforest $F_2[j_{t-k+1}, j_t]$. A key observation here is the fact that the subtree $T_2[j_{t-k+1}]$ is not split by the alignment with $T_1[i_s]$. There are three subcases.

2.1 ($k = 0$) i.e., $F_2[j_{t-k+1}, j_t] = \theta$. Therefore,

$$D(F_1[i_1, i_s], F_2[j_1, j_t]) = D(F_1[i_1, i_{s-1}], F_2[j_1, j_t]) + D(T_1[i_s], \theta)$$

2.2 ($k = 1$) i.e., $F_2[j_{t-k+1}, j_t] = T_2[j_t]$. This is the same as in Case 1.

2.3 ($k \ge 2$) This is the most general case. It is easy to see that

$$D(F_1[i_1, i_s], F_2[j_1, j_t]) = \mu(l_1[i_s], \lambda) + \min_{1 \le k < t}\{D(F_1[i_1, i_{s-1}], F_2[j_1, j_{k-1}]) + D(F_1[i_s], F_2[j_k, j_t])\}.$$

Case 3: the label is $(\lambda, l_2[j_t])$. Similar to Case 2. □

2.2 The algorithm

It follows from the above discussion that, for each pair of subtrees $T_1[i]$ and $T_2[j]$, we have to compute $D(F_1[i], F_2[j_s, j_t])$ for all $1 \le s \le t \le n_j$, and $D(F_1[i_s, i_t], F_2[j])$ for all $1 \le s \le t \le m_i$. That is, we need align $F_1[i]$ with each subforest of $F_2[j]$, and conversely align $F_2[j]$ with each subforest of $F_1[i]$.

For each fixed s and t, where $1 \le s \le m_i$ and $1 \le t \le n_j$, the procedure in Figure 3 computes $\{D(F_1[i_s, i_p], F_2[j_t, j_q]) | s \le p \le m_i, t \le q \le n_j\}$, assuming that all $D(F_1[i_k], F_2[j_p, j_q])$ have already been computed, where $1 \le k \le m_i$ and $1 \le p \le q \le n_j$, and all $D(F_1[i_p, i_q], F_2[j_k])$ are known, where $1 \le p \le q \le m_i$ and $1 \le k \le n_j$.

Hence we can obtain $D(F_1[i], F_2[j_s, j_t])$ for all $1 \le s \le t \le n_j$ by calling Procedure 1 n_j times, and $D(F_1[i_s, i_t], F_2[j])$ for all $1 \le s \le t \le m_i$ by calling Procedure 1 m_i times. Our algorithm to compute $D(T_1, T_2)$ is given in Figure 4.

2.3 The time complexity

For an input $F_1[i_s, i_{m_i}]$ and $F_2[j_t, j_{n_j}]$, the running time of Procedure 1 is bounded by

$$O((m_i - s) \cdot (n_j - t) \cdot (m_i - s + n_j - t)) = O(m_i \cdot n_j \cdot (m_i + n_j)).$$

```
Input: F₁[iₛ,iₘᵢ] and F₂[jₜ,jₙⱼ].
D(F₁[iₛ,iₛ₋₁],F₂[jₜ,jₜ₋₁]) := 0;
for p := s to mᵢ
    D(F₁[iₛ,iₚ],F₂[jₜ,jₜ₋₁]) := D(F₁[iₛ,iₚ₋₁],F₂[jₜ,jₜ₋₁]) + D(T₁[iₚ],θ);
for q := t to nⱼ
    D(F₁[iₛ,iₛ₋₁],F₂[jₜ,jq]) := D(F₁[iₛ,iₛ₋₁],F₂[jₜ,jq₋₁]) + D(θ,T₂[jq]);
for p := s to mᵢ
    for q := t to nⱼ
        Compute D(F₁[iₛ,iₚ],F₂[jₜ,jq]) as in Lemma 3.
Output: {D(F₁[iₛ,iₚ],F₂[jₜ,jq])|s ≤ p ≤ mᵢ, t ≤ q ≤ nⱼ}.
```

Fig. 3. Procedure 1: Computing $\{D(F_1[i_s,i_p],F_2[j_t,j_q])|s \le p \le m_i, t \le q \le n_j\}$ for fixed s and t.

```
Input: T₁ and T₂.
D(θ,θ) := 0;
for i := 1 to |T₁|
    Initialize D(T₁[i],θ) and D(F₁[i],θ) as in Lemma 1;
for j := 1 to |T₂|
    Initialize D(θ,T₂[j]) and D(θ,F₂[j]) as in Lemma 1;
for i := 1 to |T₁|
    for j := 1 to |T₂|
        for s := 1 to mᵢ
            Call Procedure 1 on F₁[iₛ,iₘᵢ] and F₂[j];
        for t := 1 to nⱼ
            Call Procedure 1 on F₁[i] and F₂[jₜ,jₙⱼ];
        Compute D(T₁[i],T₂[j]) as in Lemma 2.
Output: D(T₁[|T₁|],T₂[|T₂|]).
```

Fig. 4. Algorithm 1: Computing $D(T_1,T_2)$.

So, for each pair i and j, Algorithm 1 spends $O(m_i \cdot n_j \cdot (m_i + n_j)^2)$ time. Therefore, the time complexity of Algorithm 1 is

$$\sum_{i=1}^{|T_1|}\sum_{j=1}^{|T_2|} O(m_i \cdot n_j \cdot (m_i + n_j)^2) \le \sum_{i=1}^{|T_1|}\sum_{j=1}^{|T_2|} O(m_i \cdot n_j \cdot (\deg(T_1) + \deg(T_2))^2)$$

$$\le O((\deg(T_1) + \deg(T_2))^2 \cdot \sum_{i=1}^{|T_1|} m_i \cdot \sum_{j=1}^{|T_2|} n_j)$$

$$\le O(|T_1| \cdot |T_2| \cdot (\deg(T_1) + \deg(T_2))^2)$$

If both T_1 and T_2 have degrees bounded by some constant, the time complexity becomes $O(|T_1| \cdot |T_2|)$. In fact the algorithm actually computes $D(T_1[i], T_2[j])$, $D(F_1[i], F_2[j])$, $D(F_1[i_s, i_t], F_2[j])$ and $D(F_1[i], F_2[j_s, j_t])$. With these data, an actual optimal alignment can be easily found using a simple back-tracking technique. The complexity will remain the same.

Whether the above complexity can be improved is a rather hard question. A direct approach would be to prove that the alignment distance between two forests satisfies either quadrangle or inverse quadrangle inequality. If this is true, then one can reduce the complexity for computing $D(F_1[i], F_2[j_s, j_t])$ and $D(F_1[i_s, i_t], F_2[j])$, using a matrix search technique. Unfortunately, we can show that neither of these inequalities hold.

3 The alignment of unordered trees

Here we consider unordered labeled trees, *i.e.*, the order among the siblings is insignificant. It is known that computing the edit distance for unordered trees is MAX SNP-hard even if both trees have bounded degrees [15]. We will show that aligning unordered trees with bounded degrees can be done in polynomial time, and give a simple algorithm to align unordered binary trees. Finally, we prove that aligning unordered trees becomes NP-hard if one of the input trees can have an arbitrary degree.

3.1 Unordered trees with bounded degrees

When the degrees are bounded, we can compute the alignment distance using a modified version of Algorithm 1. Lemmas 1 and 2 still work. The only difference is in the computation of $D(F_1[i], F_2[j])$. We have to revise the recurrence relation in Lemma 3 as follows: for each (forest) $A \subseteq \{T_1[i_1], \ldots, T_1[i_{m_i}]\}$ and each (forest) $B \subseteq \{T_2[j_1], \ldots, T_2[j_{n_j}]\}$,

$$D(A, B) =$$

$$\min \begin{cases} \min_{T_1[i_p] \in A, T_2[j_q] \in B} D(A - \{T_1[i_p]\}, B - \{T_2[j_q]\}) + D(T_1[i_p], T_2[j_q]), \\ \min_{T_1[i_p] \in A, B' \subseteq B} D(A - \{T_1[i_p]\}, B - B') + D(F_1[i_p], B') + \mu(l_1[i_p], \lambda), \\ \min_{A' \subseteq A, T_2[j_q] \in B} D(A - A', B - \{T_2[j_q]\}) + D(A', F_2[j_q]) + \mu(\lambda, l_2[j_q]) \end{cases}$$

Since m_i and n_j are bounded, $D(A, B)$ can be computed in polynomial time. If T_1 and T_2 are both in fact binary trees, the algorithm can be much simplified, as shown in Figure 5. It is easy to see that the time complexity of this algorithm is $O(|T_1| \cdot |T_2|)$.

3.2 The hardness of aligning unordered trees

Theorem 4. *Computing the alignment distance between ordered trees is NP-hard if one of the trees can have an arbitrary degree.*

Proof: The reduction is from exact cover by 3-sets (X3C), which is NP-hard [1]. Instance: A set $A = \{a_1, \ldots a_m\}$ with $m = 3q$ and a collection C of 3-element subsets of A.

```
Input: T₁ and T₂
for i := 1 to |T₁|
   for j := 1 to |T₂|
      D(F₁[i], F₂[j]) := min{  μ(l₁[i₂], λ) + D(F₁[i₂], F₂[j]) + D(T₁[i₁], θ),
                               μ(l₁[i₁], λ) + D(F₁[i₁], F₂[j]) + D(T₁[i₂], θ),
                               μ(λ, l₂[j₂]) + D(F₁[i], F₂[j₂]) + D(θ, T₂[j₁]),
                               μ(λ, l₂[j₁]) + D(F₁[i], F₂[j₁]) + D(θ, T₂[j₂]),
                               D(T₁[i₁], T₂[j₁]) + D(T₁[i₂], T₂[j₂]),
                               D(T₁[i₁], T₂[j₂]) + D(T₁[i₂], T₂[j₁]) };

      D(T₁[i], T₂[j]) := min{  μ(l₁[i], l₂[j]) + D[F₁[i], F₂[j]),
                               μ(l₁[i], λ) + D(T₁[i₁], T₂[j]) + D(T₁[i₂], θ),
                               μ(l₁[i], λ) + D(T₁[i₂], T₂[j]) + D(T₁[i₁], θ),
                               μ(λ, l₂[j]) + D(T₁[i], T₂[j₁]) + D(θ, T₂[j₂]),
                               μ(λ, l₂[j]) + D(T₁[i], T₂[j₂]) + D(θ, T₂[j₁]) };
Output: D(T₁[|T₁|], T₂[|T₂|]);
```

Fig. 5. Algorithm 2: Aligning unordered binary trees.

Question: Does C contain an exact cover of A, i.e., a subcollection $C' \subseteq C$ such that every element of A occurs in exactly one member of C'?

Let $A = \{a_1, \ldots, a_m\}$, and $C = \{c_1, \ldots c_n\}$, where $c_i = \{c_{i,1}, c_{i,2}, c_{i,3}\}$ and $c_{i,j} \in A$, be an instance of X3C. We construct two trees as in Figure 6. The alphabet of labels is $A \cup \{r, s, p\}$, assuming letters r, s, p are not contained in A. Let $\mu(a, b) = 0$ if $a = b$ or 2 if $a, b \neq \lambda$ and $a \neq b$, $\mu(a, \lambda) = 1$, and $\mu(\lambda, b) = 1$. The degree of T_1 is bounded by 3. Each $T_{1,i}$ is a subtree, which corresponds to the subset c_i in C. The sequence of nodes with label s in each $T_{1,i}$ is called the *upper segment* and the three branches are called the *lower segment*. The root of T_2 has $n + m$ children.

Now, we want to show that C contains an exact cover of A if and only if the alignment distance between T_1 and T_2 is $(n-2) + 5q + 6(n-q)$. Observe that for each subtree $T_{1,i}$, either its entire upper segment or its entire lower segment is "matched" with the corresponding parts of T_2 in an optimal alignment. Matching the upper segment saves a cost of 5, whereas matching the lower segment saves a cost of 6. Thus, an optimal alignment matches as many lower segments as possible.

If C contains an exact cover of A, say, C', then we can align the two trees such that the q lower segments given by C' are all matched. This yields an alignment with cost $(n-2) + 5q + 6(n-q)$, since the q $T_{1,i}$'s, where $c_i \in C'$, contribute a cost of 5 each, the rest $n - q$ $T_{1,i}$'s contribute a cost of 6 each, and $n - 2$ p's contribute a cost of 1 each. Conversely, if the alignment distance between T_1 and T_2 is $(n-2) + 5q + 6(n-q)$, there must exist q $T_{1,i}$'s each contributing a cost of 5. These $T_{1,i}$'s induce an exact cover of A. □

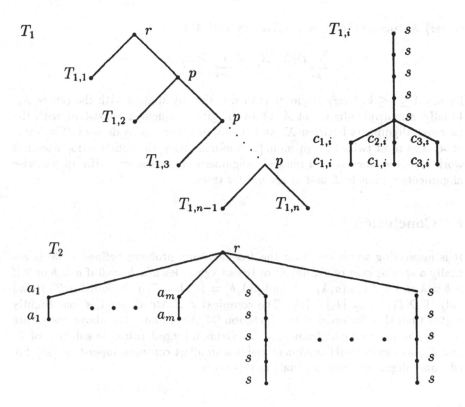

Fig. 6. The reduction.

4 Multiple alignment of ordered trees

For the purpose of finding highly conserved structural motifs, the comparison of multiple labeled ordered trees is required [4, 11]. The alignment of two trees can be easily extended to multiple trees. Many existing formulations of multiple sequence alignment are also applicable to multiple alignment of trees. For example, one can define the SP-score of a multiple alignment $\mathcal{A}$ of trees as the sum of the values of pairwise alignments induced by $\mathcal{A}$. (Here, SP stands for sum of all pairs.)

Since multiple sequence alignment with SP-score is NP-hard [14], clearly multiple alignment of ordered trees with SP-score is also NP-hard. So it would be of interest to investigate approximation algorithms for multiple alignment of ordered trees. It is known that the *center star method* approximates a multiple sequence alignment with ratio 2[2]. The method first finds a sequence (the center) X_c among the k given sequences, such that

$$\sum_{j \neq c} D(X_c, X_j) \leq \sum_{j \neq i} D(X_i, X_j),$$

for any $1 \leq i \leq k$. Every sequence is then optimally aligned with the center X_c. Finally, a multiple alignment $\mathcal{A}_c$ of all the given sequences consistent with the pairwise alignments between X_c and the other sequences is derived. The value of $\mathcal{A}_c$ is at most twice the optimum [2]. Unfortunately, the construction does not work for trees since such a multiple alignment $\mathcal{A}_c$ consistent with all pairwise alignments cannot be found in general for trees.

5 Conclusion

It is interesting to observe that the *tree inclusion* problem defined in [3] is actually a special case of alignment of trees. Again, let $\mu(a, b) = 0$ if $a = b$ or 2 if $a, b \neq \lambda$ and $a \neq b$, $\mu(a, \lambda) = 1$, and $\mu(\lambda, b) = 1$. Then T_1 is included in T_2 if and only if $D(T_1, T_2) = |T_2| - |T_1|$. The complexity of Algorithm 1 is just slightly higher than that for ordered tree inclusion [3]. Also, under the above cost/score scheme, an optimal edit from T_1 to T_2 yields a largest common sub-tree of T_1 and T_2, and an optimal alignment yields a smallest common super-tree [15]. So, edit and alignment form a "dual" in this sense.

References

1. M. R. Garey and D. S. Johnson, *Computers and Intractability: A Guide to the Theory of NP-Completeness*, W. H. Freeman, 1979.
2. D. Gusfield, Efficient methods for multiple sequence alignment with guaranteed error bounds, *Bulletin of Mathematical Biology* 55, pp. 141-154, 1993.
3. P. Kilpelainen and H. Mannila, Ordered and unordered tree inclusion, Report A-1991-4, Dept. of Comp. Science, University of Helsinki, August 1991; to appear in *SIAM J. on Computing*.
4. S.-Y. Le, J. Owens, R. Nussinov, J.-H. Chen B. Shapiro and J. V. Maizel, RNA secondary structures: comparison and determination of frequently recurring sub-structures by consensus, *Comp. Appl. Biosci.* 5, 205-210, 1989.
5. S.-Y. Le, R. Nussinov, and J.V. Maizel, Tree graphs of RNA secondary structures and their comparisons, *Computers and Biomedical Research*, 22, 461-473, 1989.
6. S.Y. Lu, A tree-tree distance and its application to cluster analysis, *IEEE Trans. Pattern Anal. Mach. Intelligence* 1, 219-224, 1979.
7. D. Sankoff and J. Kruskal (Eds), *Time Warps, String Edits, and Macromolecules: the Theory and Practice of Sequence Comparison*, Addison Wesley, Reading Mass., 1983.
8. B. Shapiro, An algorithm for comparing multiple RNA secondary structures, *Comput. Appl. Biosci.* 387-393, 1988.
9. F.Y. Shih, Object representation and recognition using mathematical morphology model, *J. System Integration*, vol. 1, pp.235-256, 1991.

10. F.Y. Shih and O.R. Mitchell, Threshold decomposition of grayscale morphology into binary morphology, *IEEE Trans. Pattern Anal. Mach. Intell.*, vol. PAMI-11, pp.31-42, 1989.

11. B. Shapiro and K. Zhang, Comparing multiple RNA secondary structures using tree comparisons, *Comput. Appl. Biosci.* vol. 6, no. 4, pp.309-318, 1990.

12. Y. Takahashi, Y. Satoh, H. Suzuki and S. Sasaki, Recognition of largest common structural fragment among a variety of chemical structures, *Analytical Science*, vol. 3, pp23-28, 1987.

13. K.C. Tai, The tree-to-tree correction problem, *J. ACM*, 26, 422-433, 1979.

14. L. Wang and T. Jiang, On the complexity of multiple sequence alignment, 1993, to appear in *Journal of Computational Biology*.

15. K. Zhang and T. Jiang, Some MAX SNP-hard results concerning unordered labeled trees, 1993, To appear in *Information Processing Letters*.

16. K. Zhang and D. Shasha, Simple fast algorithms for the editing distance between trees and related problems, *SIAM J. Comput.* 18, 1245-1262, 1989.

17. K. Zhang, R. Statman, and D. Shasha, On the editing distance between unordered labeled trees, *Information Processing Letters*, 42, 133-139, 1992.

Parametric Recomputing in Alignment Graphs

Xiaoqiu Huang[1] *, Pavel A. Pevzner[2] ** and Webb Miller [2] ***

[1] Department of Computer Science
Michigan Technological University
Houghton, MI 49931

[2] Computer Science Department
The Pennsylvania State University
University Park, PA 16802

Abstract. DNA/protein sequence alignments in computational molecular biology depend heavily on the settings of penalties for substitutions, insertions/deletions and gaps. Inappropriate choice of parameters causes irrelevant matches ("noise") to be reported, thus obscuring biologically relevant matches. In practice, biologists frequently compare sequences in a few iterations, starting from a vague idea about appropriate parameters, then refining parameters to reduce noise. This procedure often helps to delineate biologically interesting similarities and to substantially reduce laborious analysis. This paper provides a computational underpinning for such iterative noise filtration in alignment graphs. Our main results assume that a preliminary "noisy" alignment, computed with reasonable but *ad hoc* parameters, is given; the problem is to modify the parameters to reduce noise. We present fast algorithms to refine penalty parameters and describe an application of these algorithms.

1 Introduction

Sequence alignment algorithms frequently produce results that depend heavily on the choice of alignment-scoring parameters. While the choice of parameters strongly influences the alignment's quality, it is often made in an *ad hoc* manner. Moreover, despite many studies dealing with substitution scores (Dayhoff *et al.*, 1983; Altschul, 1991) and insertion/deletion/gap penalties (Fitch and Smith, 1983; Rechid *et al.*, 1989; Gotoh, 1990; Vingron and Waterman, 1994), there is considerable disagreement among molecular biologists concerning the correct choice.

Recently Gusfield *et al.* (1992) and Waterman *et al.* (1992) suggested *parametric* sequence comparison to efficiently compute optimal alignments for *all* values of parameters. Their methods determine a decomposition of a parameter space into regions such that the optimal alignment is fixed within each region. However, even

---

* The research was supported in part by the National Science Foundation under grant DIR-9106510.

** The research was supported in part by the National Science Foundation under grant CCR-9308567 and the National Institutes of Health under grant R01 HG00987.

*** The research was supported in part by the National Institutes of Health under grant R01 LM05110.

for the case of relatively short sequences, the number of regions in the decomposition may be very large: Gusfield *et al.* (1992) gave upper bounds $O(n^{\frac{2}{3}})$ and $O(n^2)$ for the number of regions for *global* and *local* alignments, respectively, for sequences of length n. These estimates indicate a potential problem facing application of parametric sequence alignment for long DNA sequences (e.g., tens of thousands of base pairs). Optimal alignment of long DNA sequences requires hours on a workstation even for fixed parameters, thus making parameter space decomposition for such sequences computationally infeasible. Moreover, biologists might well find that analyzing many thousands of sequence alignments for different parameter regions is too time-consuming to be practical. Practitioners of sequence alignment are thus faced with a conundrum: parametric alignment may be too expensive, while a "fixed-parameters" alignment may inspire little confidence. If a family of pairwise alignments among several sequences is given, then methods such as there of Vingron and Pevzner, 1993 can be applied, but frequently only two sequences are available.

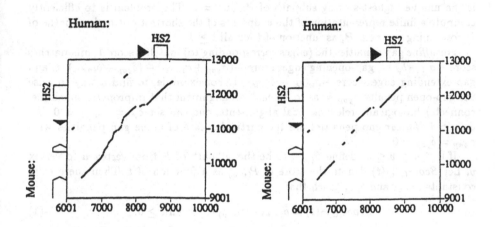

Fig. 1. Refining alignment graphs.

This study provides a way out of this dilemma. Given a preliminary "noisy" alignment that is optimal for reasonable but *ad hoc* chosen parameters, we show how to efficiently modify parameters to reduce noise. A typical example of this approach is presented in Figure 1. A preliminary alignment with *ad hoc* chosen parameters of human and mouse HS2 fragments of the β-globin locus control region clearly indicates the presence of noise which is likely to be reduced by an appropriate choice of parameters (Figure 1a). Figure 1b presents an alignment graph for modified parameters with greatly reduced noise (Hardison *et al.*, 1993) and a clear partition of the long (biologically irrelevant) alignment path into 5 subpaths.

The example above presents parametric recomputing of a *fixed* alignment path. Section 2 describes recomputing of a fixed alignment for *all* values of a parameter in linear time. Section 3 gives an algorithm that solves the more general problem of parametric recomputing of the entire alignment graph, and observes that it is too time-consuming to be practical. Section 4 extends the algorithm to handle affine gap penalties. Section 5 describes a linear algorithm for decomposing an alignment path into its highest-scoring, non-overlapping subpaths for a given value of the parameter. Sections 6 presents an efficient algorithm for decomposition of an alignment for *all* values of the parameter. Finally, Sections 7 and 8 illustrate use for these algorithms.

2 Parametric recomputing of a fixed alignment path

As many authors have noted, there is a natural correspondence between the alignment problem and the problem of computing optimal paths in a graph (see e.g. Miller and Myers, 1988); we assume familiarity with the correspondence. Fix a path P in the *alignment graph* for two sequences. Let the vertices along P be $0, 1, \ldots, n$, and let e_v denote the edge from $v - 1$ to v for $1 \leq v \leq n$. An alignment path's score is the sum of edge scores, where edge e_v has score $\alpha_v - \beta_v \times t$, α_v and $\beta_v \geq 0$ are constants depending on e_v, and t is a parameter. Without loss of generality, P is the unique highest-scoring subpath of P for $t = 0$. The problem is to efficiently compute a finite representation of the endpoints of the shortest optimal subpaths of P containing vertex $v \epsilon P$, as function of t for all $t \geq 0$.

For *affine* gap penalties the (edges corresponding to) matches score 1, mismatches score $-\sigma_{\neq} - \delta_{\neq} \cdot t$, gap-opening edges score $-(\sigma_{gap} + \sigma_{indel}) - (\delta_{gap} + \delta_{indel}) \cdot t$, and gap-extending edges score $-\sigma_{indel} - \delta_{indel} \cdot t$. For example, to find a way to raise the gap-open penalty σ_{gap} so as to "break" an alignment that improperly spans (i.e. connects) biologically relevant local alignments, one can set $\delta_{\neq} = \delta_{indel} = 0$ and $\delta_{gap} = 1$. *Linear* gap penalties are the particular case of affine gap penalties with $\sigma_{gap} = \delta_{gap} = 0$.

If $0 \leq u \leq v \leq n$, define $P_{(u,v)}$ to be the subpath of P from vertex u to vertex v. Let $Score_{(u,v)}(t)$ denote the score of $P_{(u,v)}$ as a function of t. Then there exist constants $a_{(u,v)}$ and $b_{(u,v)}$ such that

$$Score_{(u,v)}(t) = a_{(u,v)} - b_{(u,v)} \cdot t \text{ for all } t \geq 0. \tag{1}$$

In particular, $a_{(v,v)} = b_{(v,v)} = 0$, and if $u < v$ then $a_{(u,v)} = \sum_{w=u+1}^{v} \alpha_w$ and $b_{(u,v)} = \sum_{w=u+1}^{v} \beta_w$. For fixed $v \epsilon [0,n]$ we define $Score_v(t) = \max_{w \epsilon [0,v]} Score_{(w,v)}(t)$. For $t \geq 0$, vertex u is said to *maximize the score to v at t* if u is the largest index not exceeding v and satisfying $Score_{(u,v)}(t) = Score_v(t)$.

Fix $v \epsilon [0,n]$. For every $u \leq v$ Equation (1) corresponds to a line (Figure 2). The *upper envelope* of these lines presents the values of optimal alignments ending in v for all $t \geq 0$. A naive approach to compute the upper envelope of n lines requires $O(n^2)$ time. Note that the problem of computing the upper envelope is equivalent to the *half-planes intersection* problem for which *optimal* $O(n \log n)$ algorithms are known (Preparata and Shamos, 1985). Note that in our case the lines $Score_{(u,v)}(t) = a_{(u,v)} - b_{(u,v)} \times t$ are *ordered* by decreasing slopes since $\beta_w \geq 0$ implies $b_{(u,v)} \geq b_{(u',v)}$ for $u \leq u'$. This observation leads to a linear time algorithm for computing the upper envelopes of lines with ordered slopes.

The *list of maximizers at v* is the list of quadruples (u_k, t_k, a_k, b_k), such that u_k maximizes the score to v at some t, t_k is smallest of the t's where u_k maximizes the score, and $Score_{(u_k,v)}(t) = a_k - b_k t$; the quadruples occur in order of increasing t_k. (To be more rigorous, we could have defined t_k using the notion of "greatest lower bound" instead of "smallest", then shown that u_k actually maximizes the score to v at t_k.) Since the slopes b_k of the segments in the upper envelope are decreasing and $b_{(u,v)} = \sum_{w=u+1}^{v} \beta_w$, the entries u_k are strictly increasing in the list of maximizers at v.

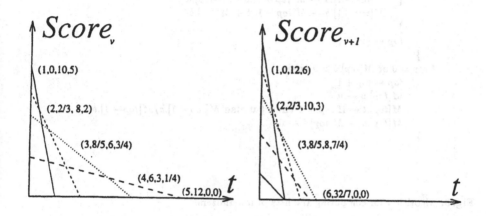

Fig. 2. Updating upper envelopes

Given a list $\mathcal{L}_{v-1}$ of maximizers at $v-1$ we can compute the list $\mathcal{L}_v$ of maximizers at v. A maximizer (u_k, t_k, a_k, b_k) from $\mathcal{L}_{v-1}$ either corresponds to a maximizer $(u_k, t_k, a_k + \alpha_v, b_k + \beta_v)$ in $\mathcal{L}_v$ or is deleted from the list if it is dominated by other lines defined by Equation 1, i.e. if $(a_k + \alpha_v) - (b_k + \beta_v)t_k \leq 0$. For example, maximizers $(4,6,3,1/4)$ and $(5,12,0,0)$ are deleted from $\mathcal{L}_v$ in Figure 2 as they are dominated by other lines. This observation leads to an algorithm for computing the list of maximizers for each vertex along P. To achieve $O(1)$ amortized time for updating $\mathcal{L}_{v-1}$ to $\mathcal{L}_v$, we use an incremental representation of a and b. That is, stack locations $M[top].a$ and $M[top].b$ hold coefficients of $Score_{(M[top].u,v)}(t)$, while for $1 \leq k < top$, $M[k].a$ and $M[k].b$ are the coefficients of $Score_{(M[k].u, M[k+1].u)}(t)$.

Theorem 1 *The algorithm of Figure 3 correctly computes the list of maximizers for each vertex along the path and runs in linear time.*

3 Parametric recomputing of alignment graph

In Section 2, we considered the parametric recomputing of optimal *subpaths* of a given alignment path. This section considers parametric computing of optimal alignment paths without restricting the paths considered to subpaths of a given path.

```
top ← 0
for v ← 0 to n do
    {  if top ≥ 1 then
        {   M[top].a ← M[top].a + αᵥ
            M[top].b ← M[top].b + βᵥ
        }
    while top ≥ 1 and M[top].a − M[top].b × M[top].t ≤ 0.0 do
        /* pop a dominated line */
        {  if top > 1 then
            {   M[top − 1].a ← M[top − 1].a + M[top].a
                M[top − 1].b ← M[top − 1].b + M[top].b
            }
            top ← top − 1
        }
    if top = 0 or M[top].b > 0 then
        {   top ← top + 1
            M[top].u ← v
            M[top].t ← if top = 1 then 0.0 else M[top − 1].a/M[top − 1].b
            M[top].a ← M[top].b ← 0
        }

    }
```

Fig. 3. Algorithm for computing the lists of maximizers.

The alignment graph for sequences of lengths n and m contains $(n + 1)(m + 1)$ vertices (i, j) with $0 \leq i \leq n$, $0 \leq j \leq m$. Fix a vertex (i, j) in the alignment graph and consider the set of paths $\mathcal{P}$ ending at (i, j). For every path $P \epsilon \mathcal{P}$ there exist constants a_P and b_P such that the score of the path is $a_P - b_P t$ for all $t \geq 0$. We are interested in the upper envelope of all lines $\{a_P - b_P \cdot t : P \epsilon \mathcal{P}\}$. The maximizer list of this envelope is denoted $M(i, j)$ and is represented by 5-tuples (I, J, t, a, b), where the path starts at (I, J), has score $a - bt$, and is highest-scoring for some parameter interval starting at t; the 5-tuples are ordered by decreasing b values. For linear gap penalties, $M(i, j)$ is computed from $M(i − 1, j)$, $M(i, j − 1)$ and $M(i − 1, j − 1)$ as shown in Figure 4. The extension to affine penalties is described in the next section.

$M \oplus (\Delta_a, \Delta_b)$ is obtained by adding Δ_a and Δ_b to the a and b entries, respectively, of each 5-tuple of the list M. The $Merge$ operation merges lists, producing a composite list that is ordered by decreasing b values.

The algorithm of Figure 4 requires $O(|M(i, j−1)| + |M(i−1, j)| + |M(i−1, j−1)|)$ time to compute $M(i, j)$. Since there are at most $O(n^2)$ regions in the decomposition of parameter space for local alignment of sequences of length n (Gusfield et al., 1992), every upper envelope contains at most $O(n^2)$ segments and, therefore, the length of every maximizer list $M(i, j)$ is bounded by $O(n^2)$ (note that every segment in upper envelope is represented by exactly one 5-tuple (I, J, t, a, b) in the algorithm of Figure 4). Thus a dynamic programming algorithm for computing $M(i, j)$ for all i, j runs in $O(n^4)$ time.

$$M \leftarrow Merge \ (\quad \begin{array}{ll} M(i-1,j) \oplus (\sigma_{indel}, \delta_{indel}) & \text{if } i > 0, \\ M(i,j-1) \oplus (\sigma_{indel}, \delta_{indel}) & \text{if } j > 0, \\ M(i-1,j-1) \oplus (1,0) & \text{if } i,j > 0 \text{ and } a_i = b_j, \\ M(i-1,j-1) \oplus (\sigma_{\neq}, \delta_{\neq}) & \text{if } i,j > 0 \text{ and } a_i \neq b_j, \\ < (i,j,0,0,0) > \end{array}$$
$$)$$

```
top ← 0
for  k ← 1 to |M| do
    { while top > 0 and  M[top].a − M[top].b × M[top].t ≤ M[k].a − M[k].b × M[top].t do
        top ← top − 1 /* pop a dominated line */
      if  top = 0 or M[top].b > M[k].b then
        {  top ← top + 1
           M[top].I ← M[k].I
           M[top].J ← M[k].J
           M[top].a ← M[k].a
           M[top].b ← M[k].b
           M[top].t ← if  top = 1 then  0.0 else
               (M[top].a − M[top − 1].a)/(M[top].b − M[top − 1].b
        }
    }
|M| ← top
M(i,j) ← M
```

Fig. 4. Computing the maximizer list at (i,j) (linear gap penalties).

4 Parametric recomputing of alignment graph (affine gap penalties)

Let $f_1(t)$ and $f_2(t)$ be two piecewise linear concave functions with n_1 and n_2 segments given by the lists of maximizers M_1 and M_2 correspondingly. The piecewise linear concave function $\max\{f_1(t), f_2(t)\}$ can be computed in $O(n_1 + n_2)$ time using *intersection of convex polygons* algorithms (Preparata and Shamos,1985). We denote $Intersection(M_1, M_2)$ the list of maximizers of $\max\{f_1(t), f_2(t)\}$. For a list $M = \{(I, J, t, a, b)\}$ of linear functions $a - bt$ corresponding to scores of all paths ending at (i,j) we denote $Envelope(M)$ the list of maximizers of this set defining the upper envelope of M.

Partition the entire set of paths $\mathcal{P}$ ending at (i,j) into 3 subsets $\mathcal{P}_\rightarrow$, $\mathcal{P}_\uparrow$ and $\mathcal{P}_\nearrow$ consisting of paths ending in horizontal (insertions), vertical (deletions) and diagonal (matches and mismatches) edges correspondingly. We assume that the "zero-length" path starting and ending at (i,j) belongs to $\mathcal{P}_\nearrow$. Denote maximizer lists corresponding to the upper envelopes of $\mathcal{P}$, $\mathcal{P}_\rightarrow$, $\mathcal{P}_\uparrow$ and $\mathcal{P}_\nearrow$ as M, $M_\rightarrow$, $M_\uparrow$ and $M_\nearrow$. Clearly

$$M = Intersection(M_\rightarrow, M_\uparrow, M_\nearrow) \tag{2}$$

For affine gap penalties the algorithm in Figure 5 computes the maximizer list $M(i,j)$ for the scores of *all* paths ending at (i,j) for $i > 0$ and $j > 0$. We assume that $M_\rightarrow(i,j) = M_\uparrow(i,j) = \emptyset$ and $M(i,j) = M_\nearrow(i,j) = \{< (i,j,0,0,0) >\}$ if $i = 0$ or $j = 0$.

$$
\begin{aligned}
M_\rightarrow(i,j) \leftarrow Merge \quad (\quad & M_\rightarrow(i-1,j) \oplus (\sigma_{indel}, \delta_{indel}) \\
& M_\uparrow(i-1,j) \oplus (\sigma_{gap} + \sigma_{indel}, \delta_{gap} + \delta_{indel}) \\
& M_\nearrow(i-1,j) \oplus (\sigma_{gap} + \sigma_{indel}, \delta_{gap} + \delta_{indel}) \\
&)
\end{aligned}
$$

$$
\begin{aligned}
M_\uparrow(i,j) \leftarrow Merge \quad (\quad & M_\rightarrow(i,j-1) \oplus (\sigma_{gap} + \sigma_{indel}, \delta_{gap} + \delta_{indel}) \\
& M_\uparrow(i,j-1) \oplus (\sigma_{indel}, \delta_{indel}) \\
& M_\nearrow(i,j-1) \oplus (\sigma_{gap} + \sigma_{indel}, \delta_{gap} + \delta_{indel}) \\
&)
\end{aligned}
$$

$$
\begin{aligned}
M_\nearrow(i,j) \leftarrow Merge \quad (\quad & M_\rightarrow(i-1,j-1) \oplus (1,0) && \text{if } a_i = b_j, \\
& M_\uparrow(i-1,j-1) \oplus (1,0) && \text{if } a_i = b_j, \\
& M_\nearrow(i-1,j-1) \oplus (1,0) && \text{if } a_i = b_j, \\
& M_\rightarrow(i-1,j-1) \oplus (\sigma_{\neq}, \delta_{\neq}) && \text{if } a_i \neq b_j, \\
& M_\uparrow(i-1,j-1) \oplus (\sigma_{\neq}, \delta_{\neq}) && \text{if } a_i \neq b_j, \\
& M_\nearrow(i-1,j-1) \oplus (\sigma_{\neq}, \delta_{\neq}) && \text{if } a_i \neq b_j, \\
& < i,j,0,0,0 > \\
&)
\end{aligned}
$$

$$
\begin{aligned}
M_\rightarrow(i,j) &= Envelope(M_\rightarrow(i,j)) \\
M_\uparrow(i,j) &= Envelope(M_\uparrow(i,j)) \\
M_\nearrow(i,j) &= Envelope(M_\nearrow(i,j)) \\
M(i,j) &= Intersection(M_\rightarrow(i,j), M_\uparrow(i,j), M_\nearrow(i,j))
\end{aligned}
$$

Fig. 5. Computing the maximizer list at (i,j) (affine gap penalties).

$Envelope(M)$ and $Intersection(M_1, M_2)$ can be computed in $O(|M|)$ and $O(|M_1| + |M_2|)$ time, respectively, and the number of elements in every maximizer list is bounded by $O(n^2)$ (Gusfield et al., 1992). Therefore the algorithm of Figure 5 requires at most $O(n^2)$ time. It implies that for affine gap penalties a dynamic programming algorithm for computing $M(i,j)$ runs in $O(n^4)$ time.

5 Complete decomposition of a path

Another useful capability is to efficiently decompose a path (i.e., an alignment) into its highest-scoring, non-overlapping subpaths, given a fixed $t \geq 0$. Informally, we seek first a highest-scoring subpath, then repeatedly seek a highest-scoring subpath not overlapping with the ones already found. Note the superficial similarity with the Waterman-Eggert approach to defining the k best local alignments.

However, in this case we can give a non-procedural definition of what is being computed. A subpath $P_{(u,v)}$ is *full* for t, if for each subpath $P_{(u',v')}$ contained in $P_{(u,v)}$, $u \leq u' \leq v' \leq v$, we have $Score_{(u',v')}(t) \leq Score_{(u,v)}(t)$. A full subpath $P_{(u,v)}$ is *optimal* for t (with respect to P) if no subpath of P that properly contains $P_{(u,v)}$ is full for t. Note that the definition of an optimal subpath given here is slightly different from the one used in Huang (1994), where optimality is taken over overlapping subpaths in addition to contained subpaths.

The algorithm of Figure 6 decomposes a given path into optimal subpaths. The kth of these subpaths extends from vertex $Lpos[k]$ to vertex $Rpos[k]$; $Lscore[k]$ and $Rscore[k]$ are the cumulative path scores at those vertices. Let $Lower[k]$ be $\max\{j < k : Lscore[j] \leq Lscore[k]\}$ if the set $\{j < k : Lscore[j] \leq Lscore[k]\}$ is non-empty and 0 otherwise.

```
top ← 0
Score(0) ← 0
for v ← 1 to n do
  {  edge_score ← αᵥ − βᵥ × t
     Score(v) ← Score(v − 1) + edge_score
     if edge_score ≥ 0 then
        {  if top > 0 and Rpos[top] = v − 1 then
              /* add edge to top subpath */
              {  Rpos[top] ← v
                 Rscore[top] ← Score(v)
              }
        else
              /* create a one-edge subpath */
              {  top ← top + 1
                 Lpos[top] ← v − 1
                 Lscore[top] ← Score(v − 1)
                 Rpos[top] ← v
                 Rscore[top] ← Score(v)
                 Lower[top] ← top − 1
                 while Lower[top] > 0 and Lscore[Lower[top]] > Lscore[top] do
                    Lower[top] ← Lower[Lower[top]]
              }
        /* merge subpaths */
        while top > 1 and Lower[top] > 0 and Rscore[Lower[top]] ≤ Rscore[top] do
           {  Rpos[Lower[top]] ← Rpos[top]
              Rscore[Lower[top]] ← Rscore[top]
              top ← Lower[top]
           }
        }
  }
```

Fig. 6. Linear algorithm for decomposition of path (t fixed).

Theorem 2 *The algorithm of Figure 6 correctly decomposes an alignment path into optimal subpaths in linear time.*

6 Decomposition tree of a path

In this section we decompose a path into the optimal subpaths for *all* $t \geq 0$. It is easy to see that the optimal subpaths of P for a larger value of t are contained in the optimal subpaths of P for a smaller value of t. This suggests that an ordered tree be used to represent the optimal subpaths of P for all $t \geq 0$. The *decomposition tree* of a subpath $P_{(v-1,v)}$ is a single node s associated with the interval $[v-1,v]$. If $\alpha_v \geq 0$, then the *value* of the node s is α_v/β_v if $\beta_v > 0$ and ∞ otherwise, meaning that the subpath $P_{(v-1,v)}$ is optimal with respect to itself for all t between 0 and the value of s. If $\alpha_v < 0$, then the value of the node s is -1, meaning that the subpath $P_{(v-1,v)}$ is not optimal for any $t \geq 0$. In general, the decomposition tree of a subpath $P_{(u,v)}$ is an ordered tree rooted at a node s associated with the interval $[u,v]$. If the subpath $P_{(u,v)}$ is optimal for $t = 0$ with respect to itself, then the value of the node s is the largest number $r \geq 0$ such that $P_{(u,v)}$ is optimal for $t = r$ with respect to itself. Otherwise, the value of s is -1. Let $s.value$ denote the value of s. If there are no optimal subpaths of $P_{(u,v)}$ for any t, $t > s.value$ and $t \geq 0$, with respect to $P_{(u,v)}$, then the node s has no children. Otherwise, let q, $q > s.value$ and $q \geq 0$, be a number such that there are optimal subpaths of $P_{(u,v)}$ for $t = q$ with respect to $P_{(u,v)}$ and that each of the subpaths is still optimal for any t, $s.value < t \leq q$ and $t \geq 0$. Let those subpaths be denoted by $P_{(u_1,v_1)}, P_{(u_2,v_2)}, \ldots, P_{(u_m,v_m)}$, with $u \leq u_1 < v_1 < u_2 < v_2 < \ldots < u_m < v_m \leq v$. In the decomposition tree of $P_{(u,v)}$, the children of the root s from left to right are $s_1, s_2, \ldots, s_m$, where for each i, $1 \leq i \leq m$, the subtree rooted at s_i is the decomposition tree of $P_{(u_i,v_i)}$. Since for each i, $1 \leq i \leq m$, $P_{(u_i,v_i)}$ is full for $t = q$, the value of s_i is greater than or equal to q and hence is greater than the value of s. This means that in any decomposition tree, the value of any node is strictly larger than that of its father. It can be easily shown by induction that the number of nodes in the decomposition tree of $P_{(u,v)}$ is at most $v - u$.

We present a divide-and-conquer algorithm for computing the decomposition tree of $P_{(0,n)}$. Let $w = \lfloor n/2 \rfloor$. The decomposition tree of $P_{(0,n)}$ is constructed by merging the decomposition trees of $P_{(0,w)}$ and $P_{(w,n)}$. The merge is based on the following observation. If $P_{(u,v)}$ is optimal for t with respect to $P_{(0,w)}$, $P_{(u',v')}$ is optimal for t with respect to $P_{(w,n)}$, $score_{(u,x)}(t) \geq 0$ for each $x \epsilon [u,w]$, $score_{(x,v')}(t) \geq 0$ for each $x \epsilon [w,v']$, $score_{(v,v')}(t) \geq 0$, and $score_{(u,u')}(t) \geq 0$, then $P_{(u,v')}$ is full for t. In addition, if $P_{(u,v)}$ and $P_{(u',v')}$ are the leftmost and rightmost, respectively, to satisfy the conditions, then $P_{(u,v')}$ is optimal for t with respect to $P_{(0,n)}$. Note that there may be optimal subpaths between v and u' with respect to $P_{(0,w)}$ or $P_{(w,n)}$.

We consider how to determine efficiently if $score_{(u,x)}(t) \geq 0$ for each $x \epsilon [u,w]$ and if $score_{(x,v')}(t) \geq 0$ for each $x \epsilon [w,v']$ for any t. For fixed $u \leq w$, define $leftlow_u(t) = \min_{x \epsilon [u,w]} Score_{(u,x)}(t)$. Then we have $leftlow_u(t) \leq Score_{(u,u)}(t) = 0$ for each $t \geq 0$. Observe that $leftlow_u(t)$ is non-increasing in t. The function $leftlow_u(t)$ is represented by the lower envelope of the lines $Score_{(u,x)}(t)$, $x \epsilon [u,w]$. For each $u \leq w$, the lower envelope can be computed in linear time using an algorithm similar to that of Section 2. If $leftlow_u(0) < 0$, let $B[u] < 0$; otherwise, let $B[u]$ be the largest value $q \geq 0$ such that $leftlow_u(q) = 0$. It is easy to see that for any $t \geq 0$, $B[u] \geq t$ if and only if $score_{(u,x)}(t) \geq 0$ for each $x \epsilon [u,w]$. Given the finite representation of $leftlow_u(t)$, $B[u]$ can be computed in $O(1)$ time. Thus the vector B can be obtained in linear time. Similarly, we define $C[v']$ for each $v' \geq w$ such that for any $t \geq 0$, $C[v'] \geq t$ if and only if $score_{(x,v')}(t) \geq 0$ for each $x \epsilon [w, v']$.

The recursive procedure for computing the decomposition tree of $P_{(u,v)}$ is shown in Figure 7. The procedure produces the tree of $P_{(u,v)}$ rooted at node s and an auxiliary list $Slist$ of all nodes in the tree sorted by their values. The sorted list is used to construct a larger decomposition tree efficiently. If $u < v - 1$, let $w = \lfloor (u+v)/2 \rfloor$. The trees of $P_{(u,w)}$ and $P_{(w,v)}$ along with their sorted node lists are built recursively. The tree of $P_{(u,v)}$ is constructed in a top-down manner by combining the trees of $P_{(u,w)}$ and $P_{(w,v)}$. For a node x, its interval is $[x.left, x.right]$. Node sw is the most recently created node of the tree of $P_{(u,v)}$ whose interval contains w. Initially, sw is set to the root s. Two working lists $Wllist$ and $Wrlist$ are used to keep those nodes from the trees of $P_{(u,w)}$ and $P_{(w,v)}$, respectively, that are being considered for merge. Initially, $Wllist$ contains only the root of the tree of $P_{(u,w)}$ if its value is nonnegative; otherwise, $Wllist$ contains the children of the root ordered from left to right. The list $Wrlist$ is initialized similarly for the tree of $P_{(w,v)}$. The procedure $ComputeB(u, w)$ calculates $B[z]$ for $u \leq z \leq w$ and $ComputeB(w, v)$ calculates $C[z]$ for $w \leq z \leq v$. The pseudo-codes for both procedures are similar to that of Figure 3 and hence omitted.

The algorithm frequently performs the following two operations: set q to the minimum value of the nodes in $Wllist$ and $Wrlist$, and replace each node of the value q in $Wllist$ and $Wrlist$ by its children. The operations are efficiently supported by using the auxiliary node lists $Sllist$ and $Srlist$ for the trees of $P_{(u,w)}$ and $P_{(w,v)}$. For a node z in the tree of $P_{(u,w)}$ (or $P_{(w,v)}$) $z.status$ is 'in' if z is in $Wllist$ (or $Wrlist$) and 'out' otherwise. The lists $Sllist$ and $Srlist$ contain the pointers to the nodes in the trees of $P_{(u,w)}$ and $P_{(w,v)}$, respectively. The working lists $Wllist$ and $Wrlist$ are implemented as doubly linked lists of pointers to the nodes in the trees of $P_{(u,w)}$ and $P_{(w,v)}$, respectively It should be pointed out that the status of the node sw has a different meaning: $sw.status$ is 'in' if sw is not yet in $Slist$ and 'out' otherwise.

Let $T(n)$ be the time required by the algorithm to compute the tree of $P_{(0,n)}$. Assume that n is a power of 2. Then the two recursive calls together take $2T(n/2)$ time. The calls $ComputeB(0, n/2)$ and $ComputeC(n/2, n)$ take $O(n)$ time. The **while** statement in the procedure $BuildTree$ requires at most n iterations since there are at most $n/2$ nodes in each of the trees of $P_{(0,n/2)}$ and $P_{(n/2,n)}$ and each node enters $Wllist$ or $Wrlist$ at most once. The macros $MinValue$ and $Replace$ have an amortized cost of $O(1)$ per node, and the rest of the code in the **while** loop costs $O(1)$ per iteration. So the total time on the non-recursive part is $O(n)$. We conclude that $T(n)$ is $O(n \log n)$. By the discussion given above, we have

Theorem 3 *The algorithm of Figure 7 correctly computes the decomposition tree of $P_{(0,n)}$ and runs in time $O(n \log n)$.*

7 Example

For pairwise alignments between the β-like globin gene clusters of mammals we frequently penalize 1 for a mismatch and $6.0 + 0.2k$ for a gap of length k, i.e.,

procedure *BuildTree(u,v)*
    Create node s
    $(s.left, s.right) \leftarrow (u, v)$
    $s.value \leftarrow -1.0$
    Initialize *Slist* to *null*
    if $u = v - 1$ **then**
        { $s.status \leftarrow$ 'out'
            if $\alpha_v \geq 0.0$ **then**
                $s.value \leftarrow$ **if** $\beta_v > 0$ **then** α_v/β_v **else** ∞
            Append s into *Slist*
        }
    else
        { $w \leftarrow \lfloor (u + v)/2 \rfloor$
           $(sl, Sllist) \leftarrow BuildTree(u, w)$
           $(sr, Srlist) \leftarrow BuildTree(w, v)$
           $Compute B(u, w)$
           $Compute C(w, v)$
           Initialize *Wllist* to *sl* and *Wrlist* to *sr*
           $sw \leftarrow s$
           $sl.status \leftarrow sr.status \leftarrow sw.status \leftarrow$ 'in'
           $Replace(-1.0)$
           while both *Wllist* and *Wrlist* not empty **do**
                {   Let x be the first node in *Wllist* and y be the last node in *Wrlist*
                    $MinValue(q)$
                    if $B[x.left] < q$ **then**
                        $Update(x,$'left'$)$
                    else
                        if $C[y.right] < q$ **then**
                            $Update(y,$'right'$)$
                        else
                            if $score_{(x.right,y.right)}(q) < 0.0$ **then**
                              $Update(x,$'left'$)$
                            else
                                if $score_{(x.left,y.left)}(q) < 0.0$ **then**
                                  $Update(y,$'right'$)$
                                else
                                    { $Join()$
                                        $Replace(q)$
                                }
                }
           Merge *Sllist* and *Srlist* and append the result to *Slist*
        }
    Return $(s, Slist)$

/* Set p to the minimum value of nodes in *Wllist* and *Wrlist*. */
macro *MinValue(p)*
    repeat
        Let x_1 be a first node in *Sllist* or *Srlist* with the minimum value
        if $x_1.status =$ 'out' **then**
            Remove x_1 from *Sllist* or *Srlist* and append x_1 to *Slist*
    until $x_1.status =$ 'in'
    $p \leftarrow x_1.value$

```
/* Remove z from Wllist or Wrlist. */
macro Update(z,d)
    z.status ← 'out'
    if d = 'left' then
        Make z a left child of sw and remove it from Wllist
    else
        Make z a right child of sw and remove it from Wrlist
    if sw.status = 'in' then
        {   sw.status ← 'out'
            Append sw into Slist
        }

/* Combine optimal subpaths of P(u,w) and P(w,v) into that of P(u,v). */
macro Join()
    if sw.left < x.left or sw.right > y.right then
        {   Create node s₁
            (s₁.left, s₁.right) ← (x.left, y.right)
            Make s₁ the middle child of sw
            if sw.status = 'in' then
                {   sw.status ← 'out'
                    Append sw into Slist
                }
            sw ← s₁
            sw.status ← 'in'
        }
    sw.value ← q

/* Replace each node in Wllist and Wrlist with the value r by its children */
macro Replace(r)
    while Sllist or Srlist not empty do
        {   Let x₁ be a first node in Sllist or Srlist with the minimum value
            if x₁.value ≤ r then
                {   if x₁.status = 'in' then
                        if x₁ has children then
                            {   Replace x₁ by its children in Wllist or Wrlist
                                Set the status of each of x₁'s children to 'in'
                            }
                        else
                            Remove x₁ from Wllist or Wrlist
                    Remove x₁ from Sllist or Srlist
                    if x₁.status = 'out' then
                        Append x₁ to Slist
                }
            else
                break
        }
```

Fig. 7. Algorithm for constructing the decomposition tree of $P_{(u,v)}$.

$\sigma_{\neq} = 1$, $\sigma_{gap} = 6.0$ and $\sigma_{indel} = 0.2$ (Hardison and Miller, 1993). When these parameters are used to compare human and galago sequences, the highest scoring local alignment extends between positions 15754 and 37673 of the human sequence. This encompasses the ε-globin and γ-globin genes, as well as a region of length nearly 10,000 that lies between those genes in the human sequence and that is clearly not related to any of the galago sequence. (The region in question corresponds to multiple insertions of L1 interspersed repeat sequences.) The reason for the inappropriately-long alignment is that humans and galagos are more closely related than are other pairs of mammals that we consider (such as humans and rabbits), and alignment penalties should be increased accordingly.

Consider the problem of finding how to raise the mismatch penalty so as to break this improper alignment into sensible pieces. Since we are considering only changes to the mismatch penalty, it is appropriate to take $\delta_{\neq} = 1$, $\delta_{indel} = \delta_{gap} = 0$. We applied the algorithm of Section 2 and an inverted version of the algorithm (i.e., that computes backwards along the path), stopping at human position 20,000, which is within the ε gene. The computation required about a second on a Sun Sparc2 workstation. The two maximizer lists revealed the following results, which show that raising the mismatch penalty to 1.08 will yield an optimal subalignment that spans human positions 15754-22123 (Figure 8). This subalignment is optimal (among all *subalignments* containing human position 20,000) for mismatch penalties between 1.08 and 1.65, and correctly identifies the ε-globin gene and its conserved flanking regions.

8   Discussion

The presented algorithms for parametric recomputing represent a compromise between time-consuming parametric alignment and fast "fixed-parameters" alignment (see Panjukov (1993) for a different approach to parametric recomputing of alignments). These methods were developed as part of a project to produce a software toolkit for aligning long DNA sequences (Schwartz *et al.*, 1991; Boguski *et al.*, 1992), where the alignments are used to improve our understanding of regulation of gene transcription (Hardison and Miller, 1993; Miller *et al.*, 1994). It is frequently the case that the first run of an alignment algorithm (with fixed and, perhaps, "wrong" parameters!) provides an intuition for a biologist on how the "correct" alignment might look like. The described parametric recomputing helps a biologist to check the intuition by changing alignment parameters. Ideally, the full-size parametric alignment (Gusfield *et al.*, 1992, Waterman *et al.*, 1992) should be used to assist a biologist in recomputing parameters. However, in many cases (Miller *et al.*, 1994) the compared fragments are so long that the full-size parametric alignment is unfeasible thus forcing a biologist to use the suggested parametric recomputing.

In particular, the methods of Sections 2 and 5 are proposed for use in contexts like the following. We currently generate a single multiple alignment that covers a sequence of length 73,000 from the human β-like globin gene cluster; matching (i.e., *orthologous*) sequences from five other mammals are aligned to their counterpart in the human sequence (Hardison *et al.*, 1993). To produce this alignment, we begin by computing pairwise alignments using the program *sim* (Huang and Miller, 1991),

$\sigma_{\neq} + \delta_{\neq}$	alignment starts
1.00	(15754,748)
1.82	(18615,3485)
1.94	(19321,4160)
2.58	(19364,4199)
3.94	(19558,4394)
4.33	(19622,4458)
4.69	(19767,4601)
5.33	(19786,4620)
8.67	(19815,4649)
9.00	(19835,4669)
9.89	(19933,4767)
32.00	(19999,4833)
	alignment ends
1.00	(37673,15180)
1.08	(22123,6989)
1.65	(22053,6921)
1.84	(21096,5894)
2.20	(21048,5846)
2.51	(20006,4840)

Fig. 8. Tables of mismatch penalties versus alignment end-points.

which is a linear-space version of the method of Waterman and Eggert (1987). The resulting pairwise alignments are used (1) to decide which portions of sequences are indeed similar to part of the human sequence and (2) to give approximate positions for matches between sequences. The sequences are then aligned by a multiple alignment procedure using those constraints (paper in preparation). Any overly-long pairwise local alignment can be decomposed into more sensible alignments, for example when an alignment spans a known *Alu* repeat in the human sequence. If this is done using the approach described in this paper, then different local (pairwise) alignments are optimal for different scoring parameters, even for the same sequence pair. Our multiple alignment procedure accommodates these variations using the sum-of-pairs scores, where the summands can depend on sequence position.

References

1. Boguski, M., R. Hardison, S. Schwartz and W. Miller (1992) Analysis of conserved domains and sequence motifs in cellular regulatory proteins and locus control regions using new software tools for multiple alignment and visualization. *The New Biologist* 4, 247-260.
2. Dayhoff, M., W. Barker and L. Hunt (1983) Establishing homologies in protein sequences. *Methods in Enzymology* 91, 524-545.
3. Fitch, W., and T. Smith (1983) Optimal sequence alignments. *Proc. Natl. Acad. Sci. USA* 80, 1382-1386.

4. Gotoh, O. (1990) Optimal sequence alignment allowing for long gaps. *Bull. Math. Biol.* **52**, 359-373.
5. Gusfield, D., K. Balasubramanian and D. Naor (1992) Parametric optimization of sequence alignment. *Proceedings of the Third Annual ACM-SIAM Symposium on Discrete Algorithms, January 1992*, 432-439.
6. Hardison, R., K.-M. Chao, M. Adamkiewicz, D. Price, J. Jackson, T. Zeigler, N. Stojanovic and W. Miller (1993) Positive and negative regulatory elements of the rabbit embryonic ε-globin gene revealed by an improved multiple alignment program and functional analysis. *DNA Sequence*, **4**, 163-176.
7. Hardison, R., and W. Miller (1993) Use of long sequence alignments to study the evolution and regulation of mammalian globin gene clusters. *Molecular Biology and Evolution* **10**, 73-102.
8. Huang, X., and W. Miller (1991) A time-efficient, linear-space local similarity algorithm. *Advances in Applied Mathematics* **12**, 337-357.
9. Huang, X. (1994) An algorithm for identifying regions of a DNA sequence that satisfy a content requirement. *Comput. Applic. Biosci.* (to appear).
10. Miller, W., and E. W. Myers (1988) Sequence comparison with concave weighting functions, *Bull. Math. Biol.* **50**, 97-120.
11. Miller, W., S. Schwartz and R. Hardison (1994) A point of contact between computer science and molecular biology. *IEEE Computational Science and Engineering* (to appear).
12. Panjukov V.V. (1993) Finding steady alignments: similarity and distance. *Comp. Appl. in Biol. Sci*, **9**, 285-290.
13. Preparata F., and M. Shamos (1985) *Computational geometry. An introduction.* Springer-Verlag, New York.
14. Rechid, R., M. Vingron and P. Argos (1989) A new interactive protein sequence alignment program and comparison of its results with widely used algorithms. *Comput. Appl. Biosci.* **5**, 107-113.
15. Schwartz, S., W. Miller, C.-M. Yang and R. Hardison (1991) Software tools for analyzing pairwise sequence alignments. *Nucleic Acids Research* **19**, 4663-4667.
16. Vingron, M., and P. A. Pevzner (1993) Multiple sequence alignment and n-dimensional image reconstruction. A. Apostolico, M. Crochermore, Z. Galil, U. Manber (eds.) Combinatorial Pattern Matching 1993, Padova, Italy *Lecture Notes in Computer Science* **684**, 243-253.
17. Vingron, M., and M. S. Waterman (1994) Parametric sequence alignment and penalty choice: Case studies. *J. Mol. Biol.* **235**, 1-12.
18. Waterman, M. S., and M. Eggert (1987) A new algorithm for best subsequence alignments with application to tRNA-rRNA comparisons. *J. Mol. Biol.* **197**, 723-725.
19. Waterman, M. S., M. Eggert and E. Lander (1992) Parametric sequence comparisons. *Proc. Natl. Acad. Sci. USA* **89**, 6090-6093.

A Lossy Data Compression Based on String Matching:
Preliminary Analysis and Suboptimal Algorithms

Tomasz Łuczak[1] and Wojciech Szpankowski[2]

[1] Mathematical Institute, Polish Academy of Science, 60-769 Poznań, Poland
[2] Dept. of Computer Science, Purdue University, W. Lafayette, IN 47907, USA

Abstract. A practical suboptimal algorithm (source coding) for lossy (non-faithful) data compression is discussed. This scheme is based on an approximate string matching, and it naturally extends lossless (faithful) Lempel-Ziv data compression scheme. The construction of the algorithm is based on a careful probabilistic analysis of an approximate string matching problem that is of its own interest. This extends Wyner-Ziv model to lossy environment. In this conference version, we consider only Bernoulli model (i.e., memoryless channel) but our results hold under much weaker probabilistic assumptions.

1 Introduction

Repeated patterns and related phenomena in words (sequences, strings) play a central role in many facets of telecommunications and theoretical computer science, notably in coding theory and data compression, in the theory of formal languages, and in the design and analysis of algorithms. For example: in faithful data compressions, such a repeated subsequence can be used to reduce the size of the original sequence (e.g., universal data compression schemes [15, 24, 22]); in exact string matching algorithms the longest suffix that matches a substring of the pattern string is used for "fast" shift of the pattern over a text string (cf. Knuth-Morris-Pratt and Boyer-Moore [1]; see also [8]).

However, in practice *approximate* repeated patterns are even more important. Non-faithful (or lossy) data compression and molecular sequence comparison are most notable examples. In this paper, we shall use approximate pattern matching to design suboptimal lossy (non-faithful) data compression. Hereafter, we shall think in terms of data compression, but most of our analysis and algorithms can be directly used to molecular sequences comparison (e.g., finding approximate palidroms).

We first briefly review some aspects of the distortion theory to put our results in the proper perspective. The reader is referred to [9, 10] for more details.

Consider a stationary and ergodic sequence $\{X_k\}_{k=-\infty}^{\infty}$ taking values in a finite alphabet $\mathcal{A}$. For simplicity of presentation, we consider only binary alphabet $\mathcal{A} = \{0, 1\}$. We write X_m^n to denote $X_m X_{m+1} \ldots X_n$.

In data compression, one investigates the following problem: Imagine a source of information generating a block $x_1^n = (x_1, \ldots, x_n)$ which is a realization of the process X_1^n. To send it efficiently one codes it into another sequence $y_1 \ldots y_\ell$ of length ℓ. Then, the compression factor is defined as $c(x_1^n) = \ell/n \le 1$ and its expected value is $C = Ec(X_1^n)$. What are the achievable values of C for lossless and lossy data compressions?

It is well known [9, 10, 13, 14, 24] that the average compression factor in a lossless data compression can asymptotically reach entropy rate, h. For the lossy transmission, one needs to introduce a measure of fidelity to find the achievable region of C. We restrict our discussion to Hamming distance defined as $d_n(x_1^n, \tilde{x}_1^n) = n^{-1} \sum_{i=1}^n d_1(x_i, \tilde{x}_i)$ where $d_1(x, \tilde{x}) = 0$ for $x = \tilde{x}$ and 1 otherwise $(x, \tilde{x} \in \mathcal{A})$. Let now fix $D > 0$. Roughly speaking, a data compression (or a code) is called *non-faithful* or *lossy* or *D-faithful* if any set of sequences x_1^n lying within distance D of a representative sequence $\tilde{x}_1^n$ is coded as $\tilde{x}_1^n$.

The optimal compression factor depends on the so called rate-distortion function $R(D)$. This is defined as follows (we give the definition of the operational rate-distortion function): Let $B_D(w_n)$

---

* This research was partially done while the authors were visiting INRIA in Rocquencourt, France. The authors wish to thank INRIA (project ALGO) for a generous support. Additional support for the first author was provided by KBN grant 2 1087 91 01, while for the second author by NSF Grants NCR-9206315 and CCR-9201078 and INT-8912631, and in part by NATO Collaborative Grant 0057/89.

be the set of all sequences of length n whose distance from the center w_n is smaller or equal to D, that is, $B_D(w_n) = \{x_1^n : d_n(x_1^n, w_n) \leq D\}$. We call the set $B_D(w_n)$ a D-ball. Consider now the set $\mathcal{A}^n$ of all sequences of length n, and let $\mathcal{S}_n$ be a subset of $\mathcal{A}^n$. We define $N(D, \mathcal{S}_n)$ as the *minimum* number of D-balls needed to cover $\mathcal{S}_n$. Then[1]

$$R_n(D, \varepsilon) = \min_{\mathcal{S}_n:\, P(\mathcal{S}_n) \geq 1-\varepsilon} \frac{\log N(D, \mathcal{S}_n)}{n}.$$

The operation rate-distortion is (cf. [13, 17])

$$R(D) = \lim_{\varepsilon \to 0} \lim_{n \to \infty} R_n(D, \varepsilon). \tag{1}$$

Kieffer [13, 14] and Ornstein and Shields [17] proved that the compression factor in a D-faithful data compression is asymptotically equal to $R(D)$, and this cannot be improved. (Observe that $R(0) = h$.) Note, however, that to construct an optimal data compression one needs to "guess" the optimal (i.e., minimum) cover of the set $\mathcal{A}^n$ by D-balls. (This is actually identical of guessing a probability measure on $\mathcal{A}^n$ that minimizes the mutual information [9, 13, 19] which is equally difficult task!).

In this paper, we propose a practical suboptimal lossy data compression scheme that extends the Lempel-Ziv scheme and that achieves rate $r(D)$ which is asymptotically optimal for $D \to 0$, that is, $\lim_{D \to 0} r(D) = h$. Our scheme reduces to the following approximate pattern matching problem. Let the "database" sequence x_1^n be given. Find the longest L_n such that there exists $i \leq n$ in the database satisfying $d(x_i^{i-1+L_n}, x_{n+1}^{n+L_n}) \leq D$. We shall propose an algorithm that finds L_n on average in $O(n \cdot \text{poly}(\log n))$ steps (but in $O(n^2 \log n)$ steps in the worst case). More importantly, we also propose two compression schemes that are based on this algorithm and our probabilistic analysis. Actually, the real engine behind this study (and its algorithmic issues) is a probabilistic analysis of an approximate pattern matching problem.

Our probabilistic results are confined to the Bernoulli model (however, in the forthcoming journal version of the paper we extend them to mixing models; see Remark after Theorem 2 below). Thus, we assume that: *symbols from $\mathcal{A}$ are generated independently and "0" occurs with probability p while "1" with probability $q = 1 - p$*. We prove that $L_n/\log n \to 1/r(D)$ in probability (pr.) where $r(D)$ represents the rate distortion, and in general $r(D) \geq R(D)$, except the symmetric case ($p = q = 0.5$) in which $r(D) = R(D)$. But, we shall show that $\lim_{D \to 0} r(D) = \lim_{D \to 0} R(D) = h$. Surprisingly enough, $L_n/\log n$ does not converge almost surely (a.s.) but rather oscillates between two random variables $s_n/\log n$ and $H_n/\log n$ that converge almost surely to two *different* constants. This kind of behavior was already observed in the faithful case (cf. [20, 21]).

Our results extends those of Wyner and Ziv [22] and Szpankowski [20, 21] to lossy transmission. We observe, however, that in the lossless case the natural data structure around which practical schemes could be built is a suffix tree (cf. [20, 21]) or digital search tree. The situation with lossy data compression is much more complicated since the decoder at any time has as a database a sample of the *distorted* process. We shall propose two solutions to remedy this problem.

Our paper is close in spirit to the one of Steinberg and Gutman [19] who also considered a practical data compression scheme based on a string matching. But the authors of [19] studied the so called waiting time while we concentrate on approximate prefix analysis. Furthermore, the authors of [19] obtained only an upper bound while we establish here precise asymptotic results. Finally, we should mention that there are results (cf. [13, 17]) indicating the existence of the optimal (thus, achieving the rate $R(D)$) data compression. However, these scheme are exponentially expensive in implementations (cf. [19]). Very recently, Zhang and Wei [23] proposed an asymptotically optimal lossy data compression that is based on the so called "gold washing" or "information-theoretical sieve" method.

2 Main Results

After formulating the pattern matching problem, we present some analytical (probabilistic) results. These results are of prime importance for the algorithmic issues which are discussed next.

---

[1] All logarithms in this paper are with base 2 unless otherwise explicitly stated.

2.1 Analytical Results

Let $\{X_k\}_{k=1}^{\infty}$ be a stationary ergodic sequence generated over a binary alphabet $\mathcal{A} = \{0, 1\}$. Wyner and Ziv [22] (see also [16, 20]) proposed the following mutation of the Lempel-Ziv data compression scheme: Assume first n symbols, X_1^n, are known to the transmitter and the receiver. Call it the *database* sequence. Find the longest prefix of X_{n+1}^{∞} that occurs at least once in the database. Say, this occurrence is at position $i_0 \leq n$ and it is of length $L_n - 1$. It was proved by Wyner and Ziv [22] that $L_n / \log n \to 1/h$ in probability (pr.), where h is the entropy of the alphabet. However, Szpankowski [20, 21] showed that $L_n / \log n$ does *not* converge almost surely (a.s.) to any constant but rather oscillates between two different constants.

Based on these results, Wyner and Ziv [22] proposed the following data compression scheme: The encoder sends the position i_0 in the database, the length $L_n - 1$ and one symbol, namely X_{n+L_n}. Using this information the decoder reconstructs the original message, and both the encoder and the decoder enlarge the database to the right, that is, the new database becomes $X_1^{n+L_n}$ or $X_{L_n}^{n+L_n}$ (the so called sliding window scheme). Based on the probabilistic results discussed above, one easily concludes that the compression ratio of such an algorithm is equal to the entropy, and it is asymptotically optimal. This scheme is called a faithful data compression scheme.

In this paper, we discuss a scheme that directly extends the above algorithm to a lossy data transmission with a fidelity criterion. As in the introduction, we define the Hamming distance $d(x_1^n, \tilde{x}_1^n)$ as the ratio of the number of mismatches between x_1^n and $\tilde{x}_1^n$ to the length n, and we assume that the database X_1^n is given (see Section 2.2 for a detailed discussion of this point). We construct the longest prefix of X_{n+1}^{∞} that is within distance $D > 0$ of a substring in the database. More precisely:

Let L_n be the length of the largest prefix of X_{n+1}^{∞} such that there exists $i \leq n$ so that
$$d(X_i^{i-1+L_n}, X_{n+1}^{n+L_n}) \leq D.$$

We call L_n the *depth* to mimic the name adopted in the faithful case (cf. [20, 21]).

As in the faithful case, the quality of compression depends on the probabilistic behavior of L_n. It turns out that its behavior depends on two other quantities, namely s_n and H_n defined in sequel.

The *height* H_n is the length of the longest substring in the database X_1^n for which there exists another substring in the database within distance D. More precisely: the height is equal to the largest K for which there exist $1 \leq i < j \leq n$ such that $d(X_i^{i-1+K}, X_j^{j-1+K}) \leq D$. In the proof, we shall need another definition of H_n which is presented below. Let $1(A)$ be the indicator function of the event A. Then, the following is true (this is a correct version of the definition presented in [4])

$$\{H_n \geq k\} = \bigcup_{l \geq k} \bigcup_{1 \leq i < j \leq n} \left\{ \sum_{t=1}^{l} 1(X_i^{i-1+t} \neq X_j^{j-1+t}) \leq Dl \right\}. \tag{2}$$

The *shortest path* s_n is defined as follows: Let $\mathcal{W}_k$ be the set of words of length k, and $w_k \in \mathcal{W}_k$. The shortest path s_n is the longest k such that for every $w_k \in \mathcal{W}_k$ there exists $1 \leq i \leq n$ such that $d(X_i^{i-1+k}, w_k) \leq D$.

Now, we are in a position to present our main results. As mentioned before, in this preliminary version we discuss only Bernoulli model in which "0" occurs with probability p and "1" with probability $q = 1 - p$. We also define $p_{min} = \min\{p, q\}$ and $P = p^2 + q^2$. All proofs are delayed till the next section.

Theorem 1. *Let* $h(D, x) = (1 - D) \log((1 - D)/x) + D \log(D/(1 - x))$. *Then*

$$\lim_{n \to \infty} \frac{s_n}{\log n} = \frac{1}{h(D, p_{min})} \qquad (a.s.), \tag{3}$$

and

$$\lim_{n \to \infty} \frac{H_n}{\log n} = \frac{2}{h(D, P)} \qquad (a.s.), \tag{4}$$

for $0 \leq D < p_{min}$. ∎

Remark 1. Observe that for $D \geq p_{min}$ the shortest path s_n and the height H_n with high probability grow faster than logarithmic function. However, this case is not too interesting from the algorithmic view point since the Hamming distance between the database sequence and a string consisting entirely either of zeros (when $p_{min} = q$) or of ones (when $p_{min} = p$) is smaller than D with high probability. Thus, we neglect this case in our analysis. □

The next theorem tells us about the probabilistic behavior of L_n which is really responsible for the asymptotic behavior of a lossy data compression scheme discussed below.

Theorem 2. *Let*

$$x_0 = \begin{cases} \frac{D}{2} + \frac{\sqrt{p^2q^2+D^2(p-q)^2-2Dpq(p-q)^2}-pq}{2(p-q)} & \text{for } p \neq q \\ \frac{D}{2} & \text{for } p = q = 0.5 . \end{cases} \quad (5)$$

Define $r(D) = -\log F$ *where*

$$F = \frac{p^{2p-2x_0+D}q^{2q+2x_0-D}}{x_0^{x_0}(p-x_0)^{p-x_0}(D-x_0)^{D-x_0}(q-D+x_0)^{q-D+x_0}} . \quad (6)$$

Then, for any $\varepsilon > 0$ *and large* n

$$\Pr\left\{\left|\frac{L_n}{\log n} - \frac{1}{r(D)}\right| \leq \varepsilon\right\} \geq 1 - O\left(\frac{\log n}{n^\varepsilon}\right) , \quad (7)$$

that is, $L_n/\log n \to 1/r(D)$ *(pr.). But,* $L_n/\log n$ *does not converge almost surely. More precisely,*

$$\liminf_{n\to\infty} \frac{L_n}{\log n} = \frac{1}{h(D, p_{min})} \quad \text{(a.s.)} \qquad \limsup_{n\to\infty} \frac{L_n}{\log n} = \frac{2}{h(D, P)} \quad (8)$$

provided $0 \leq D \leq p_{min}$. ∎

Remark 2. Actually, in the journal version of this paper we will prove much stronger results that we announce in this remark. Consider a stationary, ergodic *mixing model* (cf. [7, 18, 20]) in which the sequence $\{X_k\}_{k=-\infty}^\infty$ is stationary and ergodic. In addition, it is mixing in strong sense, that is, (informally speaking) for two events A and B defined respectively of σ-algebra of $\{X_k\}_{-\infty}^m$ and $\{X_k\}_{m+b}^\infty$ for some integer b, the following holds

$$(1 - \alpha(b))\Pr\{A\}\Pr\{B\} \leq \Pr\{A \cap B\} \leq (1 + \alpha(b))\Pr\{A\}\Pr\{B\}$$

for some some $\alpha(b)$ such that $\lim_{b\to\infty} \alpha(b) = 0$. Let now $P(B_D(w_n))$ and $P(B_D(X_1^n))$ denote the probabilities of all strings in the balls $B_D(w_n)$ and $B_D(X_1^n)$, respectively. Observe that $P(B_D(X_1^n))$ is in fact a random variable. Then, in the mixing model, we define three quantities as follows:

$$h_{min}(D) = \lim_{n\to\infty} \frac{-\log\left(\min_{w_n}\{P(B_D(w_n))\}\right)}{n} , \quad (9)$$

$$h_1(D) = \lim_{n\to\infty} \frac{-\log\left(EP(B_D(X_1^n))\right)}{n} , \quad (10)$$

$$r(D) = \lim_{n\to\infty} \frac{-E\left(\log P(B_D(X_1^n))\right)}{n} . \quad (11)$$

It can be proved that the above limits exist in the mixing model. Observe also that in the Bernoulli model $h_{min}(D)$ becomes $h(D, p_{min})$, and $h_1(D)$ is equivalent to $h(D, P)$. Also, $r(D)$ defined above can be computed as in Theorem 2 for the Bernoulli model.

Having these definitions in mind, we can now formulate our general results. First of all, we shall prove in the forthcoming paper that

$$\lim_{n\to\infty} \frac{s_n}{\log n} = \frac{1}{h_{min}(D)} \quad \text{(a.s.)} ,$$

provided $\alpha(b)$ decays to zero faster than any polynomial, that is, for all $m \geq 0$ we have $b^m\alpha(b) \to 0$. Secondly, the height H_n becomes

$$\lim_{n\to\infty} \frac{H_n}{\log n} = \frac{2}{h_1(D)} \quad \text{(a.s.)} ,$$

provided $\alpha^2(b)$ is summable, that is $\sum_{b=0}^{\infty} \alpha^2(b) < \infty$. Finally, Theorem 2 generalizes to the following

$$\lim_{n \to \infty} \frac{L_n}{\log n} = \frac{1}{r(D)} \qquad \text{(pr.)} ,$$

if $\alpha(d) \to 0$. But, as in Theorem 2, L_n does not converge almost surely, and only the following can be claimed

$$\liminf_{n \to \infty} \frac{L_n}{\log n} = \frac{1}{h_{\min}(D)} \quad \text{(a.s.)} \qquad \limsup_{n \to \infty} \frac{L_n}{\log n} = \frac{2}{h_1(D)}$$

under the same condition on $\alpha(b)$ as for s_n. $\square$

A lossy data compression scheme based on Theorem 2 is presented below. Observe that such a scheme is more intricate than in the lossless case due to the fact that the decoder and encoder have different database (i.e., the decoder has as a database a sample of the distorted process). Before we discuss algorithmic issues concerning such schemes, we first estimate the compression factor.

It is rather clear that any compression scheme based on Theorem 2 should have compression factor C equal to $r(D)$. Indeed, we observe that, as in the faithful case, any non-faithful data compression scheme based on the approximate string matching needs a pointer to the database and the length of the approximate matching. The former information costs $\log n$ while the latter can be decoded in $O(\log \log n)$ bits. In other words, instead of sending $(1/r(D)) \cdot \log n$ bits of L_n, one transmits $\log n + O(\log \log n)$ bits, thus the compression factor is $r(D)$.

In view of the above, one may ask how close is the rate (compression factor) $r(D)$ of our scheme to the optimal compression factor equal to $R(D)$ as defined in (1). An explicit formula for $R(D)$ seems to be unknown except for the Bernoulli case. In this case [9], $R(D) = h - h(D)$ where $h = -p \log p - q \log q$ is the entropy of the memoryless channel, and $h(D) = -D \log D - (1 - D) \log(1 - D)$. Note that $R(0) = h$. From Theorem 2 formulæ (5)-(6) we conclude that the scheme is:

– asymptotically optimal in the limiting case, namely

$$\lim_{D \to 0} R(D) = \lim_{D \to 0} r(D) = h , \tag{12}$$

– asymptotically optimal in the symmetric Bernoulli case ($p = q = 0.5$) since

$$r(D) = R(D) = \log 2 - h(D) . \tag{13}$$

In general, $r(D) \geq R(D)$, however, a numerical study shows that the discrepancy between $R(D)$ and $r(D)$ is not too big as one may conclude from Figure 1 which presents the gains in compression factors, namely, $h/r(D)$ and $h/R(D)$ versus D.

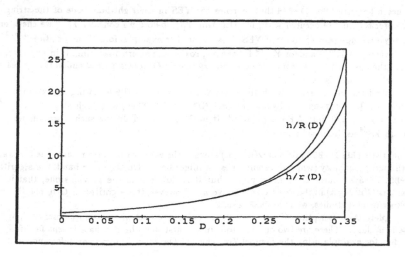

Fig. 1. Comparison of gains in the compression factors for $p = 0.4$

2.2 Algorithmic Results

As mentioned above, the non-faithful data compression is much more intricate than the faithful one due to two reasons: In the faithful case, the prefix of length L_n can be found in $O(n)$ time-complexity by a simple application of the suffix tree structure (cf. [20]). Secondly, the encoder and the decoder have different view on the database. These two problems must be solved in order to obtain an efficient lossy data compression based on Theorem 2, and we discuss them in sequel.

We start with an approximate pattern matching algorithm that finds the longest prefix of X_{n+1}^{∞} that is within distance D of a substring in the database. We shall write below lowercase letter x_m^n to denote a realization of the process X_m^n.

The following algorithm is an adaptation of the idea already applied in Atallah *et al.* [3] to another problem.

Algorithm **PREFIX**

begin
    For $i = n$ to 1 **do**
        Apply Fast Fourier Transform (FFT) to compute matches between x_{n+1}^{n+i} and $\{x_j^{j-1+i}\}_{j=1}^n$,
        Select $j \leq n$ that gives the longest substring with $(1 - D)\%$ of matches,
    doend
end

Clearly this algorithm works in $O(n^2 \log n)$ time-complexity since the FFT needs $O(n \log n)$ to compute matches between a string and *all* substrings of another string.

Although, $O(n^2 \log n)$ algorithm sounds like a good solution, it is too expensive in most applications when **PREFIX** is expected to be run very often. One needs an algorithm that *most* of the time is linear or poly-linear. Below, we discuss two possible solutions.

The problem with the previous **PREFIX** algorithm is the do-loop which requires n iterations. One possible solution (suggested by M. Atallah) is to apply binary search. The idea of the new algorithm **PREFIX-BS** is as follows. Let $Y_1^n = X_{n+1}^{2n}$. Using FFT we check if Y_1^n has $(1 - D)\%$ of matches with any substrings of X_1^n. If the answer is YES we stop, otherwise we continue the binary search. That is, we divide the substring Y_1^n into two halves, and check whether $Y_1^{n/2}$ approximately occurs (i.e., with less than $D\%$ mismatches) in X_1^n. Again if answer is YES, we are fine and start investigating $Y_1^{3n/4}$. The only problem arises, however, when the algorithm returns NO. Say, it happens when checking $Y_1^{n/2}$. This, unfortunately does not mean – as in the classical binary search – that we can proceed to $Y_1^{n/4}$ since still there is a possibility that $Y_1^{3n/4}$ almost occur in X_1^n. There are two possibilities:

(A) We use a heuristic **PREFIX-BSH** that searches for YES in the right-hand side of the string Y. More precisely, if NO occured when investigating $Y_1^{n/2}$ before we consider $Y_1^{n/4}$ we check only few, say two, up-searches to see if YES does occur. For example, for NO at position $Y_1^{n/2}$ we only investigate $Y_1^{3n/4}$ and/or $Y_1^{5n/8}$ for the approximate pattern matching. If in any case, we receive the answer YES, we continue exact binary search. Otherwise, we abandon the up-search, and the next check is at $Y_1^{n/4}$.

(B) We append the binary search with the exact search to obtain the following algorithm that is further called **PREFIX-BSE**. As before, consider NO at $Y_1^{n/2}$. Then, we search all prefixes $Y_1^{n/2+i}$ with $i = 1, \ldots, n/2$ until YES is obtained. If no YES occured during such a search, we then move to $Y_1^{n/4}$ as discussed above.

Our heuristic **PREFIX-BSH** works in $O(n \log n)$ steps in the worst case, but it returns the true value only with high probability (whp) and sometimes we might be off. On the other hand, the algorithm **PREFIX-BSE** always returns the longest prefix, but it is slower than the previous one, that is, its complexity is $O(n^2 \log n)$ in the worst case. On average, however, this algorithm is $O(n \cdot poly(\log n))$. In our experimental studies, we used **PREFIX-BSE**.

To complete the description of our lossy data compression scheme we must describe how the database is updated. There are two options, too. In the first one, the database is sent faithfully by the encoder, for example using the Lempel-Ziv scheme. The lossy compression refers now only to the new transmissions and the references are made to the common copy of the database. We also systematically measure the compression ratio, and once it falls below some specified level, a new faithful transmission of database is required. This procedure might be on-line.

The above scheme seems to be appropriate for situations when the database is kept unchanged for some time. For example, when sending pictures from a satellite, usually several pictures have the same background, hence the same database, so clearly our scheme is suitable for such transmission.

In the case when the database is varying quickly, another algorithm is needed. We suggest the following one. Instead of sending lossly a faithful database, we rather send faithfully (e.g., by Lempel-Ziv scheme) a non-faithful (distorted) database that is maintained simultaneously by the encoder and decoder.

We only briefly present the main idea of this scheme leaving details to a journal version (cf. [19]). When a new prefix of length L_n of X_{n+1}^∞ is constructed, it is *not* added directly to the database but rather we add the center $w_1^{L_n}$ of a ball $B_D(w_1^{L_n})$ to which the prefix falls. For example, this can be accomplished by finding the prefix of length L_n by approximate pattern matching, say PREFIX-BSE, in the distorted database $\widetilde{X}_1^n$ that stores only the *centers* of balls $B_D(\cdot)$. Then, the encoder transmits faithfully the distorted version of the database $\widetilde{X}_1^n$ (i.e., the centers of D-balls). More precisely, the encoder sends only the pointer to the distorted database (maintained the same by the encoder and decoder) and the length L_n. Since the pointer costs $\log n$ and by Theorem 2 we have $L_n \sim 1/r(D)\log n$, so one can conclude that the compression factor is still asymptotically equal to $r(D)$.

3 Probabilistic Analysis

In this section, we present a sketch of proofs for Theorems 1 and 2. To simplify our analysis, we observe that the following formulation of the problem turns out to be asymptotically equivalent to our original model (see [20, 21] where a similar approach is used).

Let us generate unbounded sequences $X(1), X(2), \ldots, X(m+1)$ according to the original distribution, *independently* from each other. Let $\hat{L}_m$ denote the length of the longest prefix of $X(m+1)$ that lies within the distance D from the prefix of $X(i)$ for some $i = 1, 2, \ldots, m$. Let also $\hat{H}_m$ be the largest k such that

$$d(X_1^k(i), X_1^k(j)) < D \quad \text{for some} \quad i, j, 1 \le i < j \le m .$$

Finally, let $\hat{s}_m$ denote the length of the shortest string that has no approximate match among prefixes of $X(1), X(2), \ldots, X(m)$.

One can show (cf. [20, 21]) that the behaviour of random variables $\hat{L}_m$, $\hat{H}_m$ and $\hat{s}_m$ defined for the above *independent model* asymptotically resembles that of L_m, H_m and s_m in the original model provided $m = O(n/\log n)$. Thus, throughout the following section we shall work within the framework of the independent model, and we do not distinguish between these two cases writing L_m, H_m and s_m instead of $\hat{L}_m$, $\hat{H}_m$ and $\hat{s}_m$.

3.1 The Shortest Length

We first prove Theorem 1 for s_n, that is, (3). Let us introduce some additional notation. To recall, we define $\mathcal{W}_k$ as the set of words of length k. For a $w_k \in \mathcal{W}_k$ we write $P(w_k)$ for the probability of w_k. Let $w_{\min} \in \mathcal{W}_k$ be such that $P(w_{\min}) = \min_{w \in \mathcal{W}_k}\{P(w)\}$. We also write $P(B_D(w_{\min}))$ as the probability of a D-ball centered at $w_{\min}$. It is easy to verify that $P(B_D(w_{\min})) = \min_{w_k \in \mathcal{W}_k} \Pr\{B_D(w_k)\}$.

As defined before, the shortest path s_m is the longest k such that for every $w_k \in \mathcal{W}_k$ there exists $1 \le i \le m$ such that $d(X_1^k(i), w_k) \le D$. Clearly, the following is true

$$\Pr\{s_m > k\} \le m \min_{\mathcal{W}_k} \Pr\{d(X_1^k(i), w_k) \le D\} = mP(B_D(w_{\min})) . \tag{14}$$

To estimate the above probability, one needs to assess $P(B_D(w_{\min}))$. Let $p_{\min} = \min\{p, q\}$. We note that $w_{\min}$ is a string that consists of all zeros or all ones depending whether $p < q$ or $p > q$, hence

$$P(B_D(w_{\min})) = \sum_{j=0}^{kD} \binom{k}{j} p_{\min}^{k-j} (1 - p_{\min})^j \,.$$

By Stirling's formula we have

$$\binom{k}{kD} \sim \left(\frac{1}{((1-D)^{1-D} D^D)} \right)^k \,.$$

Thus, for large k and $D \le p_{\min}$

$$\left(\left(\frac{p_{\min}}{1-D} \right)^{1-D} \left(\frac{1 - p_{\min}}{D} \right)^D \right)^k \le P(B_D(w_{\min})) \le k \left(\left(\frac{p_{\min}}{1-D} \right)^{1-D} \left(\frac{1 - p_{\min}}{D} \right)^D \right)^k \,.$$

In view of the above, we obtain $P(B_D(w_{\min})) \sim 2^{-kh(D,p_{\min})}$ where $h(D, p_{\min}) = (1-D)\log((1-D)/p_{\min}) + D\log(D/(1-p_{\min}))$. Thus, for $k = \lfloor (1+\varepsilon)h^{-1}(D, p_{\min}) \log m \rfloor$ we conclude that $\Pr\{s_m > (1+\varepsilon)h^{-1}(D, p_{\min}) \log m\} \le 1/m^\varepsilon$, which proves the upper bound for the convergence in probability of s_m.

To get the lower bound for s_m, we proceed as follows. Note that

$$\Pr\{s_m < k\} \le \sum_{w_k}(1 - P(B_D(w_k)))^m \le 2^k(1 - P(B_D(w_{\min})))^m \,.$$

Using the above estimate for $P(B_D(w_{\min}))$ and setting $k = \lfloor (1-\varepsilon)h^{-1}(D, p_{\min}) \log m \rfloor$ we finally obtain

$$\Pr\{s_m < k\} \le \exp(-m^{\varepsilon/2})$$

which is the desired lower bound.

From the above, we conclude that $s_m/\log m \to 1/h(D, p_{\min})$ (pr.) but the rate of convergence (upper bound) does not yet warrant direct application of the Borel-Cantelli Lemma. Nevertheless, one can use Kingman's idea as in [18, 20, 21] to extend this result to the almost sure convergence. Indeed, one selects a subsequence like $m_r = s2^r$ along which $s_{m_r}/\log m$ converge almost surely (a.s.), and then by noting that s_m is a nondecreasing sequence with respect to m one can extend the last assertion to all m. This completes the proof for s_m, and actually for s_n since $m = O(n/\log n)$, hence all the results above easily extend to this case, too.

3.2 The Height

The height was already treated by Arratia and Waterman [4] (cf. Theorem 1 in [4]) for the independent model, and the string model can be analyzed along the same lines.

For completeness, we only present the derivation of the upper bound, which also corrects a minor problem of [4]. From the definition (2) we have

$$\{H_m \ge k\} = \bigcup_{l \ge k} \bigcup_{1 \le i < j \le m} \left\{ \sum_{t=1}^{l} 1\{X_t(i) = X_t(j)\} \ge al \right\}$$

$$= \bigcup_{l \ge k} \bigcup_{1 \le i < j \le m} \{d(X(i), X(j)) \le D\} \,.$$

(In [4] the first union symbol was missing.)

Now, we consider $M = m(m-1)/2$ new sequences $Y(1), \ldots Y(M)$ such that $Y_k(t) = 1$ ($t = 1, \ldots, M$, $k = 1, \ldots$) if and only if for $1 \le i, j \le m$ resulting in t there is a match between $X_k(i)$ and $X_k(j)$, i.e., $X_k(i) = X_k(j)$; otherwise $Y_k(t) = 0$. Note that $\Pr\{Y_k(t) = 1\} = P = p^2 + q^2$.

The rest is easy, and we obtain

$$\Pr\{H_m \ge k\} \le m^2 \sum_{l \ge k} P(B_D(Y(1), w_l)) \,,$$

for some $w_l \in \mathcal{W}_l$. From our previous estimate of the probability of a D-ball, we observe that $P(B_D(Y(1), w_l)) \sim 2^{-lh(D,P)}$. Thus, for $k = \lfloor (1 + \varepsilon)h^{-1}(D, P) \log m \rfloor$ and a constant B

$$\Pr\{H_m \geq k\} \leq Bn^2 2^{-kh(D,P)} = B/m^{2\varepsilon}$$

which is the desired upper bound. The lower bound can be derived by using the "second moment method" in a similar fashion as in [4] (cf. also [20, 21]).

So, far only convergence in probability was derived. But using again the Kingman trick, and noting that H_m is nondecreasing, we prove Theorem 1.

3.3 The Depth

Now, we prove Theorem 2, and we begin with the convergence in probability, that is, we establish (7). To accomplish our task, we need to show that a prefix of an independently generated string $X(m+1)$ of length L_m is within distance D of $X(i)$ for some $1 \leq i \leq m$, that is, $d(X_1^{L_m}(m+1), X_1^{L_m}(i)) \leq D$. We prove that $L_m / \log m \to 1/r(D)$ (pr.) where $r(D)$ is defined in Theorem 2.

Let w_k be a *given* and *typical* word of length k. More precisely, $w_k \in \mathcal{W}_k$ and by Shannon-McMillan-Breiman Theorem (cf. [7, 9, 10]) $P(w_k) \sim 2^{-kh}$ where h is the entropy of the alphabet. In the above $P(w_k)$ has the meaning of probability of w_k occurrence, that is, $P(w_k) = p^{|0|_w} |1|_w$ where $|0|_w$ ($|1|_w$) denotes the number of zeros (ones) in w_k. For the Bernoulli model, we can say that with high probability the number of "0" and "1" in w_k is approximately equal to $kp \mp j$ and $kq \mp j$ where $j = o(k)$, respectively. Below, to simplify further discussion we assume that these numbers are $\lfloor kp \rfloor$ and $\lfloor kq \rfloor$ respectively (and actually we ignore the floor function). Naturally, $-\log P(w_k) \sim kh$.

We should stress that the word w_k is deterministic, but since it is also typical, the prefix of $X(m + 1)$ of length k is close in probability to w_k. More specifically, for any $\varepsilon > 0$

$$\lim_{k \to \infty} \Pr\{|k^{-1} \log P(X_1^k(m + 1)) - k^{-1} \log P(w_k)| \geq \varepsilon\} = 0 . \tag{15}$$

The above implies that instead of working with random string $X(m + 1)$ we can work with deterministic word w_k provided the bounds on L_m hold uniformly for all w_k.

Let now Z_k be a random variable denoting the number of strings $X_1^k(1), \ldots, X_1^k(m)$ that lie within distance D from w_k, that is $Z_k = |\{1 \leq i \leq m; \; d(X_1^k(i), w_k) \leq D\}|$. Due to our deterministic choice, Z_k has the binomial distribution with parameter m and $P_k \equiv P(B_D(w_k))$, i.e.,

$$\Pr\{Z_k = \ell\} = \binom{m}{l} P_k^\ell (1 - P_k)^{k-\ell} .$$

The rest is a simple application of the *first moment method* and the *second moment method* (cf. [2]). Indeed,

$$\Pr\{Z_k = 0\} = \Pr\{D_n < k\} \leq \frac{var\, Z_k}{(EZ_k)^2} = \frac{1 - P_k}{mP_k} \tag{16}$$

$$\Pr\{Z_k > 0\} = \Pr\{D_n \geq k\} \leq EZ_k = mP_k . \tag{17}$$

To complete the proof, we need to estimate the probability P_k which is discussed next.

Clearly, the following is true (for $P(w_k) = p^{\lfloor kp \rfloor} q^{\lfloor kq \rfloor}$)

$$P_k = P(B_D(w_k)) = \sum_{0 \leq l+r \leq kD} \binom{kp}{l} \binom{kq}{r} p^{kp+l-r} q^{kq-l+r}$$

where we assumed above for simplicity that kp and kq are integers. Let now $x = l/k$, and define

$$P_x = \binom{kp}{xk} \binom{kq}{(D-x)k} p^{k(p-x)+(D-x)k} q^{kx+k(q-D+x)} . \tag{18}$$

From the above, we immediately observe that

$$C \max_x \{P_x\} \leq P(B(w_k)) \leq Ck^2 \max_x \{P_x\}$$

where C is a constant. Thus, by the above $\log P(B_D(w_k)) \sim \log(\max_x\{P_x\})$, and it suffices to compute $\max_x \{P_x\}$.

Observe that by Stirling's formula

$$\binom{kp}{xk} \sim \left(\frac{p^p}{x^x (p-x)^{p-x}} \right)^k .$$

Thus, $P_x \sim (F(x))^k$ where

$$F(x) = \frac{p^{2p-2x+D} q^{2q+2x-D}}{x^x (p-x)^{p-x} (D-x)^{D-x} (q-D+x)^{q-D+x}} .$$

We need the following restriction on x: $\min\{0, p-q\} \leq x \leq \max\{p, D\}$.

Finally, to maximize $F(x)$ with respect to x, we are looking for x_0 such that $F'(x_0) = 0$. It turns out that this x_0 must solve the following quadratic equation

$$x^2(p-q) + x(pq + D(q-p)) - pq^2 = 0 .$$

The solution x_0 of the above is given by (5) in Theorem 2.

In summary, we have just proved that $P(B_D(w_k) \sim 2^{-kr(D)}$. Thus, by (15) and (16) with $k = \lfloor (1-\varepsilon) \frac{\log m}{r(D)} \rfloor$ we obtain the lower bound, while by (15) and (17) with $k = \lfloor (1+\varepsilon) \frac{\log m}{r(D)} \rfloor$ we derive the upper bound, which complete the proof of the convergence in probability of L_m.

To establish the second part of Theorem 2, namely (8), we proceed along the lines of [18, 20, 21]. More specifically, we note that $s_n \leq L_n \leq H_n$, and infinitely often (i.o.) $L_n = s_n$ as well as $L_n = H_n$. This, and Theorem 1, suffice to derive (8).

References

1. A.V. Aho, Algorithms for Finding Patterns in Strings, in *Handbook of Theoretical Computer Science. Volume A: Algorithms and Complexity* (ed. J. van Leeuwen), 255-300, The MIT Press, Cambridge (1990).
2. N. Alon and J. Spencer, *The Probabilistic Method*, John Wiley&Sons, New York (1992).
3. M. Atallah, P. Jacquet and W. Szpankowski, Pattern matching with mismatches: A probabilistic analysis and a randomized algorithm, *Proc. Combinatorial Pattern Matching*, Tucson, Lecture Notes in Computer Science, 644, (eds. A. Apostolico, M. Crochemore, Z. Galil, U. Manber), pp. 27-40, Springer-Verlag 1992.
4. R. Arratia and M. Waterman, The Erdös-Rényi Strong Law for Pattern Matching with Given Proportion of Mismatches, *Annals of Probability*, 17, 1152-1169 (1989).
5. R. Arratia, L. Gordon, and M. Waterman, The Erdös-Rényi Law in Distribution for Coin Tossing and Sequence Matching, *Annals of Statistics*, 18, 539-570 (1990)
6. T. Berger, *Rate Distortion Theory: A Mathematical Basis for Data Compression*, Englewood Cliffs, NJ: Prentice-Hall, 1971.
7. P. Billingsley, *Convergence of Probability Measure*, John Wiley & Sons, New York, 1968.
8. W. Chang, and E. Lawler, Approximate String Matching in Sublinear Expected Time, *Proc. of 1990 FOCS*, 116-124 (1990).
9. T.M. Cover and J.A. Thomas, *Elements of Information Theory*, John Wiley&Sons, New York (1991).
10. I. Csiszár and J. Körner, *Information Theory: Coding Theorems for Discrete Memoryless Systems*, Academic Press, New York (1981).
11. J. Feldman, r-Entropy, Equipartition, and Ornstein's Isomorphism Theory in R^n, *Israel J. Math.*, 36, 321-345 (1980).
12. P. Jacquet and W. Szpankowski, Autocorrelation on Words and Its Applications. Analysis of Suffix Tree by String-Ruler Approach, *J. Combinatorial Theory. Ser. A*, (1994); to appear.
13. J.C. Kieffer, Strong Converses in Source Coding Relative to a Fidelity Criterion, *IEEE Trans. Information Theory*, 37, 257-262 (1991).
14. J. C. Kieffer, Sample Converses in Source Coding Theory, *IEEE Trans. Information Theory*, 37, 263-268 (1991).
15. A. Lempel and J. Ziv, On the Complexity of Finite Sequences, *IEEE Information Theory* 22, 1, 75-81 (1976).
16. D. Ornstein and B. Weiss, Entropy and Data Compression Schemes, *IEEE Information Theory*, 39, 78-83 (1993).
17. D. Ornstein and P. Shields, Universal Almost Sure Data Compression, *Annals of Probability*, 18, 441-452 (1990).
18. B. Pittel, Asymptotic Growth of a Class of random Trees, *Annals of Probability*, 13, 414 - 427 (1985).

19. Y. Steinberg and M. Gutman, An Algorithm for Source Coding Subject to a Fidelity Criterion, Based on String Matching, *IEEE Trans. Information Theory*, 39, 877-886 (1993).
20. W. Szpankowski, Asymptotic Properties of Data Compression and Suffix Trees, *IEEE Trans. Information Theory*, 39, 1647-1659 (1993).
21. W. Szpankowski, A Generalized Suffix Tree and Its (Un)Expected Asymptotic Behaviors, *SIAM J. Computing*, 22, 1176-1198 (1993).
22. A. Wyner and J. Ziv, Some Asymptotic Properties of the Entropy of a Stationary Ergodic Data Source with Applications to Data Compression, *IEEE Trans. Information Theory*, 35, 1250-1258 (1989).
23. Z. Zhang and V. Wei, An On-Line Universal Lossy Data Compression Algorithm via Continuous Codebook Refinement, submitted to a journal.
24. J. Ziv and A. Lempel, A Universal Algorithm for Sequential Data Compression, *IEEE Trans. Information Theory*, 23, 3, 337-343 (1977).

A TEXT COMPRESSION SCHEME THAT ALLOWS FAST SEARCHING DIRECTLY IN THE COMPRESSED FILE

Udi Manber[1]

Department of Computer Science
University of Arizona
Tucson, AZ 85721

ABSTRACT. A new text compression scheme is presented in this paper. The main purpose of this scheme is to speed up string matching by searching the compressed file directly. The scheme requires no modification of the string-matching algorithm, which is used as a black box; any string-matching procedure can be used. Instead, the *pattern* is modified; only the outcome of the matching of the modified pattern against the compressed file is decompressed. Since the compressed file is smaller than the original file, the search is faster both in terms of I/O time and processing time than a search in the original file. For typical text files, we achieve about 30% reduction of space and slightly less of search time. A 30% space saving is not competitive with good text compression schemes, and thus should not be used where space is the predominant concern. The intended applications of this scheme are files that are searched often, such as catalogs, bibliographic files, and address books. Such files are typically not compressed, but with this scheme they can remain compressed indefinitely, saving space while allowing faster search at the same time. A particular application to an information retrieval system that we developed is also discussed.

1. Introduction

Text compression is typically used to save storage or communication costs. We suggest an additional purpose in this paper. By reducing the size of a text file in a special way, we reduce the time it takes to search through it. Our scheme improves the speed of string matching and we also save space in the process. The savings are not spectacular, but they are in a sense for free. Files that are usually not compressed because they are often read, can now be compressed and at the same time the speed of searching is improved. The improvement is independent of current technology, because it comes from the fact that the compressed files are smaller than the original and therefore less work is done. The same improvement will hold for faster CPU or I/O. Another important advantage of our scheme is that it is independent of the actual string-matching program. There is no need to modify the search program in any way; we only add a preprocessing step and a postprocessing step. As will be discussed in section 3.1, in our implementation the name of the search program is part of the input to the search.

Searching in compressed files was studied, in a theoretical framework, by several authors. Eilam-Tsoreff and Vishkin [EV88] looked at run-length compression, Amir, Landau and Vishkin [ALV92] and Amir and Benson [AB92a, AB92b] looked at two-dimensional matching, and very recently Amir, Benson and Farach [ABF94] looked at

[1] Supported in part by NSF grants CCR-9002351 and CCR-9301129, and by the Advanced Research Projects Agency under contract number DABT63-93-C-0052. Part of this work was done while the author was visiting the University of Washington.

LZW compression. While these authors have attacked and solved some fundamental questions, the practical side has not been sufficiently addressed yet. It is not clear, for example, whether in practice the compressed search in [ABF94] will indeed be faster than a regular decompression followed by a fast search. As far as we know [Fa93], no implementation of these algorithms exist. In this paper we concentrate on practical issues in this area. In particular, we compare our algorithms to Boyer-Moore pattern matching, which is typically 5-7 times faster in practice (for natural language texts) than the Knuth-Morris-Pratt approach (which is taken in [ABF94]). Speeding up the fast Boyer-Moore searching is a real challenge. For this reason we also rejected Huffman's and other bit-level compression, because we want to keep the search on a byte level for efficiency.

There have been attempts to compress full-text information-retrieval systems ([KBD89] [WBN92]). Such systems contain an inverted index, and the main problem is how to implement random access into a compressed file. Although our scheme allows random access and can be used to compress indexed text, this is not our main goal. We will assume throughout the paper that the search is sequential. Sequential search occurs in many other applications, and in particular, it plays a key role in an information-retrieval system that we designed [MW94]. We discuss this in section 4.

We first present the idea behind the compression scheme, and then describe in detail the algorithms involved and their implementation. Although an NP-complete problem needs to be solved as part of the compression algorithm, we present a solution to it that turns out to be fast and effective in practice. We then discuss how the search is done and present experimental results for different search routines. We conclude with a discussion of possible applications.

2. The Compression Algorithm

The basis of the compression is a very simple pattern-substitution method that has been used by Jewell [Je76] and most likely reinvented by many others. The goal is to substitute common pairs of characters with special symbols that are still encoded in one byte. A byte allows 256 possible encodings, but typical text uses much fewer. ASCII encoding, for example, uses only 128 characters, leaving us 128 additional encodings to be used for pairs. The compression consists of substituting each of the common pairs with the special byte allocated for it, and decompression is achieved by reversing this procedure. Both are very fast. This kind of compression is not as good as adaptive compression techniques (such as the Lempel-Ziv based algorithms [ZL77] or context-modeling algorithms [BCW90]). However, with adaptive compression one cannot perform string matching without keeping track of the compression mechanism at the same time, because the compression depends on previous text. Our goal is to perform the string matching directly on the compressed text so that it takes *less time* than a matching to the original text. If we perform even a small amount of additional processing to do decompression on the fly, we defeat the purpose. Not only do we revert to the original string matching procedure with the original text, but we spend extra time for the decompression. Speeding up string matching is a difficult goal because string matching can be done very efficiently; with Boyer-Moore filtering [BM77], the matching procedure skips many characters and only a fraction of the text is looked at. As we will show, a pattern-substitution method can be modified to allow a direct search in the compressed file. Essentially, the search for a given pattern will work by modifying the pattern, but not the search routine, and matching the modified pattern to the compressed file. There is, however, one major problem with this idea.

Suppose that the pattern we are looking for is the word *wood*. Suppose further that both *wo* and *od* belong to the list of common pairs. We'll use $\boxed{ij}$ to denote the encoding of the pair ij, and □ to denote a blank. The pattern *wood* is thus translated into $\boxed{wo}\,\boxed{od}$, but that is not necessarily how the word *wood* appears in the compressed file. For example, the pairs □w, oo, and d□ could be common pairs as well, in which case it is possible that the word *wood* was compressed into $\boxed{□w}\,\boxed{oo}\,\boxed{d□}$, which are 3 totally different characters. But, under the same assumptions, it could be that the word preceding *wood* in the text ended with d, and the encoding is $\boxed{d□}\,\boxed{wo}\,\boxed{od}$. And these are not the only two possibilities. The word *wood* could start a new line, it could be a part of a larger word (say, *deadwood* in which case the encoding could be $\boxed{dw}\,\boxed{oo}\,\boxed{d□}$), or it could be followed by a period — $\boxed{□w}\,\boxed{oo}\,\boxed{d.}$. Each of these combinations leads to a different encoding of the word *wood* in the compressed text.

The reason for the many different combinations of the encoded word *wood* is that, even if we know all the common pairs, we cannot determine ahead of time whether the first character w will belong to the end of a pair, the beginning of a pair, or to no pair; it depends on the text. Suppose for a minute that w is not a second character in any common pair. In that case, the encoding for *wood* must start with either w or $\boxed{wo}$ depending on whether or not wo is a common pair. Since the common pairs are known, we could determine when we see the pattern *wood* what to look for: the modified pattern will either be $\boxed{wo}\,\boxed{od}$, wo $\boxed{od}$, w $\boxed{oo}$ d, and so on, depending on the common pairs. (Notice however that if $\boxed{od}$ is not part of the encoding, then we may still not be able to determine the end of it. For example, $\boxed{wo}$ o $\boxed{d□}$ and $\boxed{wo}$ o $\boxed{d.}$ are two possibilities that depend on the actual text.) But, of course, all that depends on w not being a second character in *any* pair. We want to have many common pairs, so many characters will be second characters of those pairs.

The main idea of this paper is to devise a way to restrict the number of possible encodings for any string, and still achieve good compression. One possible approach would be to start the compression always from the beginning of a word. In other words, compress the words separately, and leave all other symbols intact. This will partially solve our problem, but it has two major weaknesses. First, it restricts the search to whole words. If the pattern is *compression* and the text has *decompression*, it may not be found (because of the problem mentioned above). Second, blanks, commas, and periods are very common characters. Removing them from consideration for compression limits the compression rate significantly. We present a different approach.

We choose the common pairs in a special way, not just according to their frequency. The main constraint that we add is that no two chosen pairs can overlap. An equivalent way to pose this constraint is that no character can be a first character in a pair and a second character in a (possibly another) pair. This constraint does not completely make the encoding unique but it reduces the ambiguities to a minimum that we can handle and, as we will show, it allows a speedy matching. But before we see the matching part, let's see how to enforce this constraint in the best possible way. To do that we abstract the problem.

2.1. Finding the Best Non-Overlapping Pairs

Let's assume that we first scan the text and find, for each possible pair of characters, the number of times it appears in the text. We denote by $f(ab)$ the frequency of the pair ab. We now construct a directed graph $G = (V, E)$, such that the vertices of G correspond to the different characters in the text, one vertex for each unique character, and the edges correspond to character pairs. The weight of each edge ab is $f(ab)$. We want to find the "best" edges (corresponding to pairs) such that no edges "overlap." More precisely, we want to partition the vertices into two sets V_1 and V_2, such that the sum of the weights of the edges that go from V_1 to V_2 (these are the edges we choose) is maximized. We also need to restrict the number of edges we select (there is a limit on the number of unused byte combinations, which we assume is 127), but let's leave this aside for now. The sum of weights of the edges from V_1 to V_2 gives us the exact compression savings. Each of these edges corresponds to a pair that will be encoded with one byte, thus it contributes one byte of savings.

The formulation above is a very clean abstraction of our problem, but unfortunately, it is also a very difficult problem. It turns out that the problem is NP-complete even for unit weights and undirected graphs [GJ79]. But, on the other hand, the graphs we are dealing with are not too large. The important edges correspond to common pairs and not too many characters appear in common pairs. In fact, the number of unique characters overall is quite limited. So there is a chance to be able to solve such problems in practice, even though the problem for large graphs is intractable. We experimented with several approaches to solving this problem. The following simple heuristic seems to be the most effective. It is fast, and in all the cases we verified, it gave the optimal solution. More on the verification later. It is worth to note that standard methods of algorithm analysis would not be applicable here, because the size of the problem (e.g., the number of unique characters) is small and fixed. We believe that the only practical way to evaluate different approaches for such small-scale problems is to try them on typical inputs, which is what we did.

The main step of the algorithm is a local optimization heuristic that can be called a 1-OPT procedure. Given an initial partition of the vertices into V_1 and V_2, we examine each vertex to see whether switching this vertex to the other set would improve the total. We continue with such switches until no more are possible (since each switch improves the total, this procedure converges). We also tried a 2-OPT heuristic, in which we attempt to switch 2 vertices at the same time. This is different from switching each one separately, because the edges connecting these two vertices count differently. The main question is how to select the initial partition. We tried several greedy methods, but found that the simplest method is the most effective: use a random partition. We repeat the algorithm several times with different random partitions. The solutions were almost uniformly good, even without using 2-OPT. The best solution was typically found in 5-20 random trials, which led us to set the number of trials to 100 to be on the safe side. In fact, the *average* total compression over 100 single random trials was 98.6% of the best total for one text, 96.1% for the second text, and 96.9% for the third text. The standard deviation ranged from 2.3% to 4.7%. The *worst* total we found in 300 random trials for 3 different texts was 80% of the optimal. Running 1-OPT 100 times can be done reasonably fast, as we will show shortly.

The last problem is to limit the number of selected edges to the available unused byte combinations. We set this number to be 127, leaving 128 characters for the regular ASCII set plus one special delimiter (with value 128). We need this delimiter for the (hopefully) rare cases of characters in the text with values of > 128 (which may occur with non-ASCII text). In these cases, the compressed file contains the delimiter followed by the non-ASCII character. We could have achieved a slightly better compression rate by figuring out the exact set of characters used by the text, and using all other unused byte combinations, but for simplicity, we decided not to do that. Also, for simplicity, we did not use the newline symbol in any of the pairs, leaving it always uncompressed.

We could not see any intelligent way to incorporate the limit on the number of pairs during the pairs selection. Instead, we simply find the best partition, ignoring the limit, and then take the best (highest weight) 127 edges from V_1 to V_2. We also tried to limit the number of edges initially; for example, start with, say, the 400 highest-weight edges of the graph and find the best partition for them, and then take the 127 best of those. Or, alternatively, start with, say 30 vertices of highest degree. The differences in the total compression were usually small (2-3%), although the running time was, of course, improved because the graphs were smaller. However, there were added complications because if the graph is reduced too much, the total number of edges becomes less than the optimal 127. So we decided to stay with the original scheme. The algorithm, using only 1-OPT switches, is given in pseudo-code in Figure 1.

We leave it to the reader to fill in the details of the implementation. The graphs we handle are small (hundreds of edges at the most), and for English texts they will always be small (the size of the graph depends on the number of unique characters in the language, not the size of the text). Therefore, as long as x (the number of randomizations) is relatively small, the algorithm is quite fast. In particular, the running time of this algorithm is a small fraction of the running time of the entire compression algorithm (unless the files are very small). Sample of compression savings, running times, and the percentage of the running times used to find the best pairs are given in Figure 2 for 5 different texts: Roget's Thesaurus, the 1992 CIA world facts, and King James' Bible from Project Gutenberg; a collection of articles from the Wall Street Journals, and the archives of the Telecom Digest, a discussion group about telecommunication. The running times are in seconds on a DEC 5000/240.

To verify that the solutions obtained from our algorithm are not too far from optimal, we also implemented a deterministic algorithm that guarantees the best solution. This algorithm is exponential, but by using binary reflected Gray codes [BER76] it was possible to run experiments for up to 30 vertices, which is not too far from real data. In

Algorithm Best_Non_Overlapping_Pairs(G: weighted graph)

repeat x times { x is a constant; we used 100 for ASCII texts }
   randomly assign each vertex to either V_1 or V_2 with equal probability;
   for each vertex $v \in V$ do
      put v on the queue;
   loop until the queue is empty
      pop v from the queue;
      if switching v to the opposite set improves the sum of weights then
         switch v;
         if switching v caused other vertices, not already on the
         queue, to prefer to switch then put them on the queue;
   store the best solution to date;
  output the highest weight 127 edges from the best partition;

Figure 1: The algorithm to select the partition.

Text	size original	compression savings	compression time	uncompression time	% of time for pairs
Roget Thesaurus	1.38	31%	10.3	0.9	14.6%
CIA world facts	2.42	28%	11.4	1.7	13.2%
King James' Bible	4.85	30%	13.3	3.5	9.0%
Wall Street J.	15.84	33%	23.9	11.2	6.7%
Telecom archives	69.21	28%	104.0	52.1	1.9%

Figure 2: Statistics for the compression and decompression algorithms.
Files sizes are given in Mbytes; running times are given in seconds.

all the tests we ran the solution found by the random algorithm was indeed the optimal. We also compared the best 127 edges obtained from the random algorithm on the whole graph (usually about 40-60 vertices) vs. the best 127 edges from the optimal solution for a "good" induced subgraph of 30 vertices, and those solutions were within 1-2% apart.

Another interesting experiment was done with random text. The compression rates of this algorithm for random text of uniform distribution are quite predictable. Suppose that the text consists of S characters each with equal probability, and that it is large enough. A partition of the vertices into equal-size sets would yield $(S/2)^2$ edges from the first set to the second set. (if the text is large all possible pairs will most likely appear). All edges occur with approximately the same probability, so one would expect that the selected edges will have about the same weight as any other set of $(S/2)^2$ edges, which is one fourth of the total number of edges. The best partition will be better than the average, but again, if the text is large enough, the variance will be small. Therefore, if about one fourth.^{2} of the edges can be chosen, we expect the savings to be close to 25%.^{2} Since we can use only 127 edges, the saving would be reduced by a factor of $127/(S/2)^2$. For example, for $S=30$ we get approximately 14% savings for random text (which we verified by experiments) compared to about 30% for natural text. Just another indication that text is not random.

2.2. Some Additional Implementation Issues

A couple of other points are worth mentioning. The compression algorithm is a two-stage algorithm; the first stage reads the file and computes the best pairs, and the second stage reads the file again and performs the compression. To improve this process, we use only a sample of the file in the first stage, usually the first 1 million characters. We keep that part in memory for the next stage, so it is not too much of a overhead. It is possible to use "standard" pairs that are computed on a large sample of text. This will save time for the compression, but for most cases, it is probably not cost effective, as it leads to a less effective compression. For example, we compressed the version of the Bible we had using the best pairs computed from the Wall Street Journal text and obtained a compression that required 3.9% more space. We also did it the other way around and found the Wall Street Journal to be much more sensitive: it required 6.0% more space. We support an option that reads the pairs from another compressed file rather than computing them for the given file, because it allows one to concatenate to a compressed file. If file1 is a compressed file and we want to add to it file2, we need only to compress file2 using file1's pairs and concatenate; there is no need to decompress, to concatenate, and to compress again. We can do that in the middle of the file as well. It is thus possible to adapt a text editor to read and edit a compressed file directly.

Our compressed files save about 30% space. If searching is not needed, there are much better compression algorithms, such as UNIX *compress*, based on the Lempel-Ziv-Welch algorithm [We84], which can achieve 50-60% reduction. We tested whether compressing our compressed files further with UNIX compress lead to the same compression rates as compressing the original file, and to our surprise we found that we got *better* compression rates. Typically using both programs together (provided that our scheme is used first) improves the compression by an extra 6%. For example, the Wall Street Journal text we had was compressed to 6.96MB with compress, and to 6.55MB using our scheme in addition to compress.

---

^{2} At some point we used the UNIX old *rand* procedure to generate random text and obtained a perfect compression, for our algorithm, of 50%. The reason was that the old rand procedure generates "random" numbers that go from odd to even to odd with no exception. Our heuristic was good enough (it uses much better random numbers) to catch the best partition, which was, of course, all the odd characters in one set and the even in the other. All edges were then used. This is another example of the risk of using bad random number generators, or as in our case, copying from old code.

3. Searching Compressed Files

3.1. The Search Algorithm

The search program consists of five parts. In the first part the common pairs used in the compression are read (they are stored at the beginning of the compressed file) and two translation tables are constructed. One table maps all new characters (identified by having a value of ≥ 129) to their original pairs of characters. The second table is the inverse; it maps pairs of original characters into new characters. In the second part, the pattern is translated (compressed) using the inverse table. This translation is unique as we showed in the previous section, except possibly for the first and last character of the pattern. If either the first character belongs to V_2 in the partition or the last character belongs to V_1, then they are removed from the pattern for now. The third part is to use the string-matching program — as a black box — to search for the compressed pattern in the compressed file and output all matches. This is done by creating a new process[3] and using UNIX's pipe facilities to divert the output to the fourth part, in which the output of the match is decompressed using the translation table. If either the first or last character of the pattern were removed in the second part, then the output must be further filtered. We do that by creating another process and running the same string-matching program on the (decompressed) output of the fourth part. This can be done more efficiently during the search because we know the exact location of the match, but since in most cases the output size is small, it makes little difference. (Adding it to the search would require an access to the source code of the search procedure, which we do not assume.) An example of a search is given in Figure 3.

The five parts may seem like a lot of work, but in fact only the third part is substantial. The first part depends on the common pairs and there are at most 127 of them. The second part depends on the pattern, which is typically very small. The fourth and fifth parts depend on the size of the output, which again is typically very small. The third part therefore dominates the running time of the program.

One of the main features of our scheme is that it can work with almost any string-matching program. The only requirement is that the program can handle any character encoded in a byte (i.e., that it is 8-bit clean). We encountered no problem using any (UNIX) string-matching program. We believe that this is a very important design decision. The modularity allows using our compression without having to modify existing software. This makes the whole scheme much more reliable, general, and convenient. In addition, we mark a compressed file with a distinct signature at its beginning. If our search program does not see this signature it reverts to the original search procedure. So, one can use our program as a default on all files and get the speedup whenever the file is compressed. (A move to a two-byte representation of text will most probably create even more unused byte combinations so our scheme will still work effectively; finding the best non-overlapping pairs may be more of a problem.)

---

[3] It is possible, of course to use the string-matching procedure directly as a procedure in the search rather than to create a new process. We chose to create another process to make the whole program more general by allowing any string-matching program.

---

command: `cgrep pattern input_file`

First part: read the list of common words;
for example, ⌷pa⌷ ⌷te⌷ and ⌷n␣⌷ may be common pairs.

Second part: translate *pattern* into ⌷pa⌷ t ⌷te⌷ r;
(Notice that we removed the last character n, because it is a first character in a common pair.)

Third part: search for ⌷pa⌷ t ⌷te⌷ r in input_file using your favorite string-matching program.

Fourth part: decompress what you find above; for example,
⌷pa⌷ t ⌷te⌷ r ⌷n␣⌷ ap ⌷pe⌷ a ⌷rs⌷ ⌷␣h⌷ e ⌷re⌷ — pattern appears here

Fifth part: run the string matching again on the output to filter a possible match to only *patter* (recall that we removed the last n).

Figure 3: An example of the search.

3.2. Experiments

We used texts of varying sizes (these texts were also used in Figure 1), and several search programs. We present the results for three large texts and two representative search programs. The first program is fgrep, which does not use Boyer-Moore filtering, and the second is our own agrep [WM92], which does use it and is much faster as a result. We selected 100 random words from the dictionary (the effectiveness of a Boyer-Moore search depends somewhat on the pattern), and ran 100 searches for each combination of search and text. We give the average running times in Figure 4. All experiments were run on a DECstation 5000/240 running Ultrix and times were obtained from the UNIX time routines (which are not very precise), and given in seconds.

4. Applications

Our scheme can be used for many applications that require searching. For example, we have a rather large bibliographic database which we often search. Compressing it saves time and space. (Of course, one can build an inverted index and be able to search much quicker, but that requires much more space.) The same holds for other large information files. We highlight two other possible applications of this scheme.

Text	compression savings	original file fgrep	original file agrep	compressed file fgrep	compressed file agrep	improvements fgrep	improvements agrep
Bible	30%	7.5	1.4	5.4	1.0	28%	29%
WSJ	33%	24.9	5.0	17.4	3.8	30%	24%
Telecom	28%	109.2	21.2	80.0	16.4	27%	23%

Figure 4: Running times for agrep and fgrep for various texts.

In [MW94] we present a design of a multi-level information retrieval system called *glimpse*. We cannot describe the system in detail here, but the main idea behind it is to partition the information space (e.g., a personal file system) into blocks, and build an index that is similar to an inverted index but much much smaller (typically 2-4% of the total size, which is an order of magnitude smaller than the best full-text indexes). The index contains all distinct words, but for each word there are pointers to only the blocks in which it appears (rather than to its exact location as in an inverted index). Searching for a word consists of finding all its blocks and then searching those blocks sequentially. Sequential search is fast enough for medium-size files, therefore this search strategy, if implemented carefully, is reasonably fast. Its main advantage is the tiny space occupied by the index (an inverted index typically occupies 30%-150% of the total). Having a small index allows for many improvements in the search. Using the compression described here, one can keep many files compressed and still be able to perform the sequential search. That will again save time and space. In particular, the index itself can be compressed.

Another application is in filtering information. For example, the information may consist of some newsfeed and all articles containing certain patterns are supposed to be extracted. With our scheme the newsfeed can come in compressed, resulting in faster filtering. (Although we discussed searching single patterns here, our scheme will work for searching multiple patterns with either the Aho-Corasick algorithm used in fgrep [AC75] or with agrep's algorithm [WM92].) If the setting is such that compression is already used, then, as we said in section 2.2, the original compression can come on top of our compression and the search can be performed after the original compression is removed.

5. Future Work

A natural improvement attempt to our compression scheme is to consider more than just two characters at a time. This is typical of fixed dictionary schemes (e.g., [BCW90]). The problem is to find some restrictions on the common strings that will prevent overlaps and will be easy to maintain.

The same approach may also be used for approximate matching, although the problem is more difficult. An error in the original text may be transformed into two errors in the compressed text (e.g., by changing a common pair to a pair that is not common), and two errors in the original text may be transformed into one error in the compressed text (e.g., by substituting one common pair for another). Therefore, we cannot simply match the compressed pattern to the compressed file. However, there are efficient techniques for reducing an approximate matching problem to a different exact matching problem. A simple example [WM92] is to divide the pattern into $k+1$ parts such that any k errors will leave at least one part intact. An exact search is performed for all parts, and the output, which is hopefully much smaller, is then filtered appropriately. The scheme presented in this paper can be used for the exact matching part.

Recently [GM94] we implemented a different scheme that achieves better compression (45-50%) and faster search, but requires a much larger dictionary. It essentially translates many words into short representations and uses a fixed dictionary for the translation. A search first translates the pattern (assuming it is a complete word) into its encoding and then searches for the encoding directly. Patterns that are not in the dictionary are stored verbatim and the search for those turns out to be even faster (in the best case we achieved 4 fold speedup for unsuccessful search). The main problem with this approach is portability: unless everyone uses the same dictionary, the dictionary has to accompany the text when it is sent to another location.

Acknowledgements

Thanks to Dan Hirschberg, Richard Ladner, Martin Tompa, and Sun Wu for helpful discussions, and to Jan Sanislo for help in improving the code.

References

[AB92a]   Amir, A, and G. Benson, 'Two-dimensional periodicity and its application," *Proc. of the 3rd Symp. on Discrete Algorithms*, Orlando Florida (January 1992), pp. 440–452.

[AB92b]   Amir, A, and G. Benson, "Efficient two dimensional compressed matching," *Proc. of the Data Compression Conference*, Snowbird Utah (March 1992), pp. 279–288.

[ABF94]   Amir, A, G. Benson, and M. Farach, "Let sleeping files lie: pattern matching in Z-compressed files," *Proc. of the 5rd Symp. on Discrete Algorithms*, (January 1994), to appear.

[AC75]   Aho, A. V., and M. J. Corasick, "Efficient string matching: an aid to bibliographic search", *Communications of the ACM*, **18** (June 1975), pp. 333–340.

[BER76]   Bitner J. R., G. Erlich, and E. M. Reingold, "Efficient generation of the binary reflected Gray code and its applications," *Communications of the ACM*, **19** (September 1976), pp. 517–521.

[BCW90]   Bell, T. G., J. G. Cleary, and I. H. Witten, *Text Compression*, Prentice-Hall, Englewood Cliffs, NJ (1990).

[BM77]   Boyer R. S., and J. S. Moore, "A fast string searching algorithm," *Communications of the ACM*, **20** (October 1977), pp. 762–772.

[EV88]   Eilam-Tsoreff T., and U. Vishkin, "Matching patterns in a string subject to multilinear transformations," *Proc. of the Int. Workshop on Sequences, Combinatorics, Compression, Security, and Transmission*, Salerno, Italy (June 1988).

[Fa93]   Farach M., private communication (October 1993).

[GJ79]   Garey M. R., and D. S. Johnson, *Computers and Intractability, A Guide to the Theory of NP-completeness*, W. H. Freeman, San Francisco, CA, 1979.

[GM94]   B. Gopal, and U. Manber, "A Fixed-Dictionary Approach to Fast Searching in Compressed Files," submitted for publication.

[Je76]   Jewell G. C., "Text compaction for information retrieval systems," *IEEE SMC Newsletter*, **5** (February 1976).

[KBD89]   Klein, S.T., A. Bookstein, and S. Deerwester, "Storing text retrieval systems on CD-ROM: compression and encryption considerations," *ACM Trans. on Information Systems*, **7** (July 1989), pp. 230–245.

[MW94]   Manber U. and S. Wu, "GLIMPSE: A Tool to Search Through Entire File Systems," *Usenix Winter 1994 Technical Conference*, San Francisco (January 1994), pp. 23–32.

[WBN92]   Witten, I. H., T. C. Bell, and C. G. Nevill, "Models for compression in full-text retrieval systems," *Proc. of the Data Compression Conference*, Snowbird, Utah (April 1991), pp. 23–32.

[We84]   Welch, T. A., "A technique for high-performance data compression," *IEEE Computer*, **17** (June 1984), pp. 8–19.

[WM92]   Wu S., and U. Manber, "Fast Text Searching Allowing Errors," *Communications of the ACM* **35** (October 1992), pp. 83–91.

[ZL77]   Ziv, J. and A. Lempel, "A universal algorithm for sequential data compression," *IEEE Trans. on Information Theory*, **IT-23** (May 1977). pp. 337–343.

An Alphabet–Independent Optimal Parallel Search for Three Dimensional Pattern

Marek Karpinski*   Wojciech Rytter**

Abstract. We give an alphabet-independent optimal parallel algorithm for the searching phase of three dimensional pattern-matching. All occurrences of a three dimensional pattern P of shape $m \times m \times m$ in a text T of shape $n \times n \times n$ are to be found. Our algorithm works in $\log m$ time with $\mathcal{O}(N/\log(m))$ processors of a *CREW PRAM*, where $N = n^3$. The searching phase in three dimensions explores classification of two-dimensional periodicities of the cubic pattern. Some new projection techniques are developed to deal with three dimensions. The periodicites of the patern with respect to its faces are investigated. The nonperiodicities imply some sparseness properties, while periodicities imply other special useful properties (*i.e.* monotonicity) of the set of occurrences. Both types of properties are useful in deriving an efficient algorithm.

The search phase is preceeded by the preprocessing phase (computation of the witness table). Our main results concern the searching phase, however we present shortly a new approach to the second phase also. Usefullness of the dictionaries of basic factors (*DBF*'s), see [7], in the computation of the three dimensional witness table is presented. The *DBF* approach gains simplicity at the expense of a small increase in time. It gives a (nonoptimal) $\mathcal{O}(\log(m))$ time algorithm using m processors of a *CRCW PRAM*. The alphabet-independent optimal preprocessing is very complex even in the case of two dimensions, see [9]. For large alphabets the *DBF*'s give assymptotically the same complexity as the (alphabet-dependent) suffix trees approach (but avoids suffix trees and is simpler). However the basic advantage of the *DBF* approach is its simplicity of dealing with three (or more) dimensions.

The algorithm can be easily adjusted to the case of unequally sided patterns.

* Department of Computer Science, University of Bonn, 53117 Bonn, and the International Computer Science Institute, Berkeley, California. Research supported in part by the DFG Grant KA 673/4-1, by the ESPRIT BR Grants 7097 and ECUS030, and by the Volkswagen-Stiftung.
** Institute of Computer Science, Warsaw University, 02–097 Warsaw. Research supported in part by the Grant KBN 2–1190–91–01.

1 Introduction

The problem of *three dimensional matching* (3d-matching, in short) is to find all occurrences of a three dimensional pattern array P in a text array T. By an occurrence we mean the position of the specified corner of P in T in a full exact-match of P against T. For simplicity of exposition we assume that all sides are equal, sides of P are of length m and sides of T are of length n. Assume $m < n$. The total size of T is $N = n^3$ and the total size of P is $M = m^3$. The 3D-matching is a natural generalization of the classical string matching and two-dimensional pattern-matching problems, and aside of applications, of independent algorithmic interest.

The pattern-matching usually consists of two quite independent parts: pre-processing and searching phase. The main role of the preprocessing is the computation of the so called *witness table* (defined later). Let Σ be the underlying alphabet. In two dimensions there are two approaches to compute this table efficiently: use the suffix trees (see [2]), which is a factor $\log|\Sigma|$ slower than linear time, and the linear time alphabet independent algorithms of [9] and [6]. The alphabet independent algorithms are extremely complicated. They would be even more complicated in three dimensions. On the other hand if Σ is large then we can replace $\log|\Sigma|$ by $\log m$. We show a simple approach through the *dictionary of basic factors* (*DBF*, in short). This is a useful data structure introduced in [12]. It has received the name *DBF* and its usefulness in string algorithms was shown in [7]. The advantage of the DBF is that it can be very easily extended to the three dimensional situation. For large alphabets the complexity of the DBF approach is not inferior to that of the sufix trees. In the three dimensional case the DBF works in much simpler way as the suffix trees approach.

In the paper we concentrate mostly on the first phase of the pattern-matching: the searching phase. Amir, Benson and Farah were the first to give alphabet-independent linear time searching phase, see [2]. They have also given in [3] an alphabet-independent searching in $logM$ time with $\mathcal{O}(M/\log(M))$ processors of a *CREW PRAM*. We refer to the latter algorithm as the algorithm ABF. The algorithm ABF needs only the *witness table* from the preprocessing phase. An $O(1)$ time optimal algorithm was given recently in [6], however it needs additional data structure from the preprocessing phase: so called *deterministic sample*. The basic precomputed data structure needed in our algorithm is (similarly as in the algorithm ABF) the *witness table* WIT. The entries of WIT correspond to vectors (potential periods). The components of each vector are integers, the size of the vector $\alpha = (\alpha_1, \alpha_2, \alpha_3)$ is $|\alpha| = \max(|\alpha_1|, |\alpha_2|, |\alpha_3|)$.

Usually only (potential periods) vectors of size at most $c \times m$ are considered, assume here that $c = 1/8$. We call such vectors *short*. The vector α is a *period* of P iff $P(x) = P(x + \alpha)$ for each position x in P, whenever both sides of the equation are defined (correspond to positions in the pattern). If α is not a period then $WIT(\alpha) = x$ is a witness (to this fact) if :

$$P(x) \neq P(x + \alpha).$$

If α is a period then by convention let $WIT(\alpha) = 0$.

We say that P is *1D-nonperiodic* iff it has no short period parallel to one of edges of the pattern cube. Let H be a face of P, it is an $n \times n$ square parallel to two of the three axes of the coordinates. By a face of a given cube we mean a set of its points with one of the coordinates fixed. The faces can be boundary faces or internal faces of the cube. If we consider all (global) periods of P parallel to H then we can classify them in the same way as periodicities in two dimensions. So the face H can be:

nonperiodic, lattice periodic, radiant periodic or line periodic.

We say also that P has a given (one of four) periodicity type w.r.t. face H. We emphasize that we consider *global* periods, so the period w.r.t. H is parallel to H but works globally in the cube P. We refer to [2] for definitions of periodicity types. Our three dimensional matching uses in essential way the classification of (two-dimensional) periodicities of the pattern cube P with respect to its faces.

2 An Alphabet–Independent Optimal Parallel Algorithm for the Searching Phase

Throughout this section assume that the *witness table* WIT has been already precomputed. Using WIT we can easily decide if P is *1D-periodic*. Also we know (one of four) types of the periodicity for each face of P.

Lemma 1. *The 3D-matching can be reduced in* $\log M$ *time using* $\mathcal{O}(N/\log(M))$ *processors (independently of the alphabet) to the case of* 1D-nonperiodic *patterns.*

Proof. We can decompose the cube P into smaller subcubes if P is *1D-periodic*. These smaller subcubes will be *1D-nonperiodic*. The same argument as reducing periodic to nonperiodic case in one dimensional matching can be applied, see [8]. We omit the details.

Recall that by short vectors we mean vectors whose size is at most $c \cdot m$. Let us partition the whole text array T into *cubic windows*, each of the same shape $c \cdot m \times c \cdot m \times c \cdot m$. It is enough to show how to find all occurrences in a fixed window in $\mathcal{O}(M)$ time. We have $O(N/M)$ windows. Let us fix one window W to the end of the section. The occurrence in W does not mean that the whole P is in W, it just means that the specified corner of an instance of P is in W. Assume this specified corner is fixed, let it be for example the lower left corner of the top face.

We say that two positions $x, y \in W$ are consistent (and write *consistent(x, y)*), iff overlaps of copies of P placed in positions x, y agree each with other (though they could disagree with the actual parts of T). The searching phase has two basic subphases:

Subphase (I):
compute a consistent set $CAND$ of candidate positions in the window W: if $x, y \in CAND$ then *consistent(x, y)*. No position outside $CAND$ can be an occurrence of P.

Subphase (II):
Verification of $CAND$: check which positions $x \in CAND$ are real occurrences (P placed at x matches the corresponding part of the text cube T).

Subphase (II) is rather simple compared with (I), so we concentrate later mostly on the first subphase.

Lemma 2. *Subphase (II) can be implemented to work in $\mathcal{O}(\log M)$ time with $O(N/\log(M))$ processors of a CREW PRAM.*

Proof. The basic point here is the reduction to the search of a *unary pattern* P' in a *binary text* T'. *Unary* means that P' is a cube consisting of the same symbol "1" repeated. The computation of such patterns essentially reduces to the calculation of runs of consecutive 1's, or to the computation of the first "0" (which is easy in parallel). It has to be done in each window independently.

The reduction to the unary case works in three dimensions essentially in the same way as in two dimensions, see [2]. Each position in the text "finds" any element of $CAND$ which "covers" this position. The important point is that *any* covering position represents all such positions due to consistency. We place '1' if the symbol on a given position agrees with the pattern placed at the covering element of $CAND$. We omit details.

The nonperiodicity is explored using the operation of a *duel*. If two positions x, y are related through the vector $\alpha = x - y$ and α is not a period then the operation $DUEL(x, y)$ "kills" one of positions in constant time. The witness table is used. If α is a period then we know that x and y are consistent. We refer the reader to [2] for the details about the *duelling*. The rough idea how to construct $CAND$ is: start with $C = W$, then use more and more duels to reduce the size of C, if no duel kills any element of C then C is the required set $CAND$. However we cannot make too many duels. We are allowed to make in total $\mathcal{O}(M) = O(m^3)$ duels in a fixed window.

Observe that if we know a set C such that $CAND \subseteq C$ and C is small ($|C| \leq m^{3/2}$) then we can perform duels between each pair in C simultaneously and we are done. We perform at most M duels in total.

Due to lemma 1 we can assume that P is *1D-nonperiodic*. Let us make duels between positions on each line in W parallel to some edge of the cube W. There are $\mathcal{O}(m)$ positions on one line. They can be eliminated except at most one position per line by processing each line independently. A given line needs $\mathcal{O}(m/\log(m))$ processors to process it in $\log(m)$ time. There are $\mathcal{O}(m^2)$ lines, altogether the computation is optimal.

Therefore we can assume now that in an initial set C of candidates, each line (parallel to an edge of W) contains at most one position of C.

Remark. Unfortunately C can be too large in this moment. We construct such large set C of candidates using the *latin square* strategy. Each row and each column of such square is a permutation of integers $1 \ldots n$. *Pile* m squares one on the other. The first one has candidates on positions containing 1 in the latin square, the second on positions containing 2, etc. Hence the set C has quadratic number of points in a cube and no two points are on the same line (parallel to one of three orthogonal directions).

Lemma 3. (simple case)
Assume that at least one of the faces of P is nonperiodic. Then we can find all occurrences of P in $\log M$ time with $\mathcal{O}(N/\log(M))$ processors (independently of the alphabet).

Proof. Assume the face H is nonperiodic. Partition the window W into disjoint squares of shape $c \cdot m \times c \cdot m$. Each square is an external or internal face parallel to H. On each of them we can apply the two-dimensional algorithm ABF. Only one candidate remains on any face, due to nonperiodicity w.r.t. H. We have now together (on all internal faces) at most m candidates in the set C of survivors. For each pair $x, y \in C$ we perform $DUEL(x, y)$. The whole computation has $\mathcal{O}(M)$ work since C is small. C is a set of pairwise consistent positions. This completes subphase (I). The second subphase can be done optimally due to Lemma 2.2.

Assume to the end of the section that P is periodic w.r.t. each of its faces (otherwise we could apply the lemma above). Assume also P is nonperiodic w.r.t. each line parallel to an edge of P. We can preprocess each set of points on each face independently using the two-dimensional algorithm ABF. Then we have reduced the situation to the one satisying the following:

1. we have an initial set of candidates $C \subseteq W$. Positions outside C in W are known to be nonoccurrences
2. for each two points x, y if $x, y \in C \cap H$ for some face H, then $consistent(x, y)$
3. there are no two elements of C on the same line parallel to an edge
4. P is globally periodic w.r.t. each of its faces.

Let H be an (external) face of the window W. Assume w.l.o.g. that
$$H = \{x = (x_1, x_2, x_3) : 0 \leq x_1, x_2 < cm \text{ and } x_3 = 0\}.$$
Let us project the set C onto the face H. Assume that H is parallel to the first two axes. The point (x_1, x_2, x_3) is projected onto the point $x = (x_1, x_2)$ of H. The third component is associated with x as its *weight*. We have $weight(x) = x_3$. Let
$$\Gamma = project_H(C)$$
be the collection of projected points on H together with their weights. We write (x, k), for a point with weight k. We write also $weight(x) = k$. Due to the fact that no two points in C are on the same line each point in C is projected onto a different point in H. However Γ can contain many points on the same line, though none two of them can have the same weights.

In a certain sense we reduced the problem to a two-dimensional one. We have a collection Γ of points on the two-dimensional square H. Also we have a witness table for them. It refers to three dimensions but all we need is the operation $DUEL$ which works in constant time for any two points. Hence the *duelling* can be treated as two-dimensional since it involves points on a two-dimensional array. We have to eliminate some points from Γ and be left with the subset of pairwise consistent element, which means that for any two points x, y $DUEL(x, y)$

will eliminate none of x, y. One could try to apply in this situation the two-dimensional algorithm ABF. Unfortunately it doesn't work in a straightforward way. The algorithm ABF is based on some partial transitivity properties of the consistency relation. These properties are here more complicated due to weights which correspond to the third dimension (and which cannot be neglected). At this moment we can assume that all faces are periodic, otherwise Lemma 2.4 can be applied. The searching depends on the type of periodicity. The *lattice-periodicity* means that a 2D-pattern has a short periodic vector in quadrant (I) and a short periodicity in quadrant (II). However it could happen that both periods are equal, but then the pattern is line periodic according to one of the axes. In the lemma below such possibility is excluded by assuming that the whole pattern is not *1D-periodic*.

Lemma 4. *Assume that a* 1D-nonperiodic *cubic pattern* P *is lattice-periodic w.r.t. one of its faces* H. *Then we can find all occurrences of* P *in time* $\log M$ *with* $\mathcal{O}(N/\log(M))$ *processors (independently of the alphabet).*

Proof. Let Γ_k be the set of points in Γ of weight k. We know that all elements of Γ_k are pairwise consistent for a fixed k. If P is *1D-nonperiodic* and lattice-periodic w.r.t. H then the following properties can be proved:

Claim A
Assume $x \in \Gamma_k$ and $y \in \Gamma_l$ for $k \neq l$. Then

1. If in $DUEL(x, y)$ x is "killed" then all positions in Γ_k can be also "killed". They certainly do not start any occurrence of the pattern.
2. If in $DUEL(x, y)$ both elements survive then all positions in $\Gamma_k \cup \Gamma_l$ are pairwise consistent.

We omit the proof of the claim. It is reduced to the following observation: if a two-dimensional pattern P' is *lattice-periodic* but not *line-periodic* then for each position in the left-upper window of shape $c \cdot m \times c \cdot m$ there is a witness in the central subarray of shape $(1 - cm) \times 1 - cm)$, if there is any witness at all. This is the only place where we need the constant c to be rather small, $c = 1/8$ is sufficiently small.

Due to the claim the computation of the consistent set $CAND$ can be solved by choosing a representative from each set Γ_k and then by making duels between all possible pairs of representatives. Each killed representative in some group Γ_k consequently kills all memebres of Γ_k. Let Γ' be the set of remaining elements. Each element in fact corresponds to a three dimensional point (we are undoing the projection). This gives the required set $CAND$ as the three-dimensional version of Γ'. Then $CAND$ is the input to the subphase (II). The whole searching can be done optimally due to Lemma 2.2. This completesthe proof.

Lemma 5. *Assume that the pattern* P *is quadrant-periodic or line-periodic w.r.t. each face of* P. *Then we can find all occurrences of* P *in time* $\log M$ *with* $\mathcal{O}(N/\log(M))$ *processors independently of the alphabet.*

Proof. Define Γ to be *row-monotonic* if the weights of points in Γ are increasing in each row or are decreasing in each row of H. Analogously define *column-monotonicity* of Γ.

If the two-dimensional pattern is *line-* or *radiant-periodc* then it is known, see [1], that any set of consitent candidates in the 2D-text is monotonic in an unweighted-sense. this means that one of the coordinates is a monotonic function of the second one. Such property holds for all faces orthogonal to H. The successive points are at agrowing or decreasing distance from H. This implies the following property.

Claim 1. If P is *line-periodic* or *quadrant-periodic* w.r.t. each of its faces then Γ is *row-monotonic* and *column-monotonic*.

Assume w.l.o.g. that the weights of points in Γ are increasing in rows left-to-right and decreasing in columns top-down. Consider a point x in Γ. We refer the reader to Figure 1. We explain how to make duells between x and all points in Γ to the right of x. There are $\mathcal{O}(m^2)$ such points, hence making all possible duells needs quadratic work for a single point x. Altogether it gives $\mathcal{O}(m^4)$ work as there are possible $\mathcal{O}(m^2)$ points x. However we make duells in an implicit way. The processing for X is done in three seprate areas denoted by A, B and C in Figure 1. In area A weights are smaller than $weight(x)$ and in two other areas the weights are larger. We explain only how we process part A, other parts are processed similarly. Each column in A is processed independently. Let us fix some column L, see Figure 1. Let y, z be two points on L, where z is further from x. It is easy to observe the following:
$$P(x) \cap P(z) \subseteq P(y) \cap P(z).$$
where $P(x), P(y)$ and $P(z)$ denote the copies of the cubic pattern P which start at the three dimensional points in T corresponding to x, y and z, respectively (they do not have to match T).

This observation is due to the fact that $weight(x) > weight(y) > weight\,(z)$ and due to the way how x, y, z are situated on H. It implies the following properties:
(not $consistent(x,z)$) $\implies$ (not $consistent(x,y)$) ;
$consistent(x,y) \implies consistent(x,z)$.

Denote by $top\_incons(x, L)$ the topmost *point* $y \in L \cap \Gamma$ which is inconsistent with x and which is in a row not below x. The above two properties imply that for each $z \in L \cap \Gamma$:
(z is *above* $top\_incons(x,L)$) $\implies consistent(x,z)$;
(z is *below* $top\_incons(x,L) \implies$ (not $consistent(x,z)$).

In a certain sense $top\_incons(x, L)$ is the "stronger fighter" for x in $L \cap A$. We perform:
if $top\_incons(x,L)$ "kills" x in a single duell then x is removed;
otherwise we know (without more duells) that x "kills" all $z \in L \cap \Gamma$ which are below $top\_incons(x, L)$ and it does not kill any z which is above.

All that can be done easily by a parallel algorithm for all x's together. Hence it is essentially enough to compute the values $top\_incons(x, L)$ for each column L to the right of x. We start with an algorithm which is optimal within logarithmic factor, afterwards we explain how to remove this factor.

Almost optimal algorithm: assume x is fixed. Assign one processor to each column L in area A. This processor finds $top\_incons(x, L)$ in $\log m$ time by a binary search method. For a single x we need m processors. Altogether it works in $\log m$ time with $\mathcal{O}(m^3)$ processors.

Optimal algorithm: we have waisted the work by a logarithmic factor due to independency in computing values of $top\_incons(x, L)$. Let us now compute these values together for all $x \in K$, where k is some column preceeding L, see Figure 2. If we can do it for all $x \in K$ in $\log m$ time with $\mathcal{O}(m/\log(m))$ processors then we will have optimality. We use a standard trick of partitioning columns into $\mathcal{O}(m/\log m)$ small segments and processing each of them by one processor in logarithmic time. We refer the reader to Figure 2. The column L is partitioned into logarithmic segments, denote the partitioning points by y_k's.. Each partitioning point y_k computes its value $x_k = bottom\_incons(y, K)$, where $bottom\_incons$ is the table analogous to top_incons but working for column K against L and in a bottom-down manner. Then for $x \in L$ we know the following:

if x is between x_{k-1} and x_k in K then $top\_incons(x)$ is between y_{k-1} and y_k in L.

Unfortunately the segments implied in column K can be larger than logarithmic. We overcome it by refining L with horizontal lines containing points y_k's, see Figure 2. In this way K is divided into $\mathcal{O}(m/\log(m))$ segments, each of logarithmic size and the values top_incons for points in a given segment are contained in a known segment of logarithmic size. Then we asign one processor to each segmenet in K which in top-down way computes required values for all members of the segment sequentially in logarithmic time. We use $\mathcal{O}(m/\log(m))$ processors and time is logarithmic. There is the quadratic number of pairs K, L. Altogether $\mathcal{O}(m^3/\log(m))$ processors are enough.

The computation for the area C is carried out in the same way. For the array B we group points x in rows, instead of columns. We omit technical details. The computation of the tables top_incons and their counterparts for areas B and C were the bottleneck. Other parts can be easily done by an optimal algorithm. This completes the proof.

The series of lemmas above implies immediately our main result:

Theorem 6. *Assume that the witness table is precomputed. Then the 3D-matching problem can be solved by an optimal parallel algorithm working in $\log(M)$ time on a CREW PRAM, the complexity does not depend on the size of the alphabet.*

3 Preprocessing the Pattern: The DBF Approach.

Let S be a set of strings. Each subword of a word in S is specified by two integers: a position p, where it starts, and the length l. (All words of S can be concatenated, so a single position can determine where and in which word a given subword starts.) *Basic factors* are subwords whose length is a power of two.

$DBF(S)$ is a data structure which assigns to each basic factor corresponding to a pair (p, l) a unique name $ID(p, l)$. The names are integers in the range $1 \ldots |S|$ and two words of the same length are equal (as strings) if and only if their names are the same. The following fact was shown in [7].

Lemma 7. $DBF(S)$ can be computed in $\log |S|$ time with $\mathcal{O}(|S|)$ processors of a CRCW PRAM.

The power of the DBF relies on two facts:

1. *DBF* is small, it stores explicitly information only about $\mathcal{O}(|S| \log (|S|))$ objects.
2. Implicitly *DBF* gives information about $\mathcal{O}(|S|^2)$ objects. Each subword can be split into at most two (maybe overlapped) basic factors and get a constant sized name (composed of at most two smaller ones). Equality of two subwords can be checked with $\mathcal{O}(1)$ work.

We demonstrate first usefullnes of the DBF on the 1D-matching and 2D-matching.

1D-matching: Assume we want to compute the value of $WIT[i]$ for each position i in a given string P for which the DBF is computed. We can do it with one processor per each position i in logarithmic time by a kind of binary search. For a given position i names of basic factors whose lengths are decreasing powers of two are compared successively. Each position has one processor (assigned to this position) which finds a *witness* (if there is any) in $\log m$ time.

2D-matching: Assume we are to compute the witness table for a 2D-pattern P. Consider a fixed k-th column of P. We linearize the problem. Compute $DBF(S)$ for the set S of all rows of P. Place at each position in the k-th row the name of the horizontal word of length $m - k$ starting at this position. Observe that $m - k$ can be a nonpower of two (but then it can be decomposed into such powers and have a composed name). Do the same with the first column. In this way we have two strings. We compute witnesses in the second string w.r.t. the first one by the 1D-method. Consider a fixed position x in the k-th column of P. After linearization it becomes some position x' in 1D-string. If the witness for x' is in some position j, then we know that the horizontal strings of length k' starting in the first column and the k-th column in row j are unequl. The mismatch is found by the binary search method mentioned-above. A *witness* for the position x is found. This approach extends to three dimensions automatically.

Theorem 8. *The three dimensional witness table can be computed in* $\log M$ *time with* $\mathcal{O}(M)$ *processors of a CRCW PRAM.*

Proof. Consider the (whole) faces
$$P_k = \{x = (x_1, x_2, x_3) : 0 \le x_1, x_2 < m \text{ and } x_3 = k\}$$
for $0 \le k < m$. (Previously we considered only faces of the window W, the windows are not relevant here.) We show how the computation of witnesses for

points in P_k can be reduced to a two-dimensional case for a given k. It works in the same way as the reduction of *2D-case* to *1D-case*.

Let us fix k. Assume that the third coordinate corresponds to the *horizontal direction*. Compute the DBF for all horizontal strings in the cube P. Place at each position in P_0 and P_k the name of the string of size $m-k$ which starts at this position and *goes* in the horizontal direction. We receive the two-dimensional arrays $\widetilde{P_0}$ and $\widetilde{P_k}$. Compute the witnesses of all positions in $\widetilde{P_k}$ against the pattern $\widetilde{P_0}$ using the two-dimensional method described above. If the witness for position (x_1, x_2) in $\widetilde{P_k}$ is found at (y_1, y_2) then we know that the witness for (x_1, x_2, k) is at a horizontal string starting at (y_1, y_2, k). We apply the one-dimensional method to two strings of size $m - k$ going into the *horizontal direction*. The binary search described before can be applied to find a witness of one horizontal string against the other. In this way we reduce the computation of the three dimensional witness table to the independent computation of m two-dimensional witness tables. This completes the proof.

$\square$

References

1. A. Amir, G. Benson. *Two dimensional periodicity in rectangular arrays*. SODA'92, 440-452

2. A. Amir, G. Benson, M. Farach. *Alphabet independent two dimensional matching*. STOC'92, 59-68.

3. A. Amir, G. Benson, M. Farach. *Parallel two dimensional matching in logarithmic time*. SPAA'93, 79-85.

4. T .J. Baker. *A technique for extending rapid exact-match string matching to arrays of more than one dimension*. SIAM J. Comp. 7 (1978) 533-541.

5. R. S. Bird. *Two dimensional pattern matching*. Inf. Proc. letters 6, (1977) 168-170.

6. R. Cole, M. Crochemore, Z. Galil, L. Gasieniec, R. Hariharan, S. Muthukrishnan, K. Park, W. Rytter. *Optimally fast parallel algorithms for preprocessing and pattern matching in one and two dimensions*. FOCS'93.

7. M. Crochemore, W. Rytter. *Usefullness of the Karp-Miller-Rosenberg algorithm in parallel computations on strings and arrays*. Theoretical Computer Science 88 (1991) 59-62.

8. Z. Galil. *Optimal parallel algorithms for string matching*. Information and Control 67 (1985) 144-157.

9. Z. Galil, K. Park. *Truly alphabet independent two dimensional matching*. FOCS'92, (1992) 247-256.

10. Z. Kedem, G. Landau, K. Palem. *Optimal parallel prefix-suffix matching algorithm and application*. SPAA'89 (1989) 388-398.

11. R. Karp, R. Miller, A. Rosenberg. *Rapid identification of repeated patterns in strings, trees and arrays*. STOC'72 (1972) 125-136.

12. R. Karp, M. O. Rabin. *Efficient randomized pattern matching algorithms*. IBM Journal of Res. and Dev. 31 (1987) 249-260.

13. U. Vishkin. *Optimal pattern matching in strings*. Information and Control 67 (1985) 91-113.

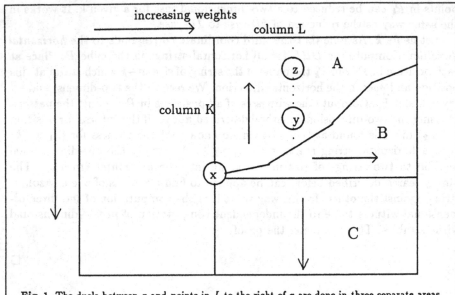

Fig. 1. The duels between x and points in L to the right of x are done in three separate areas.

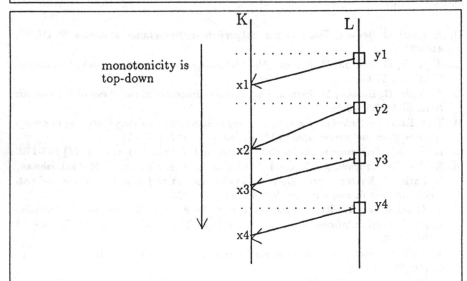

Fig. 2. The column L is partitioned into small segments. The inverse pointers partition K. This is refined by the dotted lines.

Unit Route Upper Bound for String-Matching on Hypercube

L. Lestrée[1]

L.I.T.P, Université Paris 7, 2 Place Jussieu, Paris Cedex 05.

Abstract. We give here an algorithm of string matching on an hypercube with constant memory in time $\lceil \log n \rceil + \frac{5}{2} \lceil \log m \rceil + 5$ counted in number of unit routes and with a constant number of operations by communication. This algorithm is very close to the lower bound of the problem for this architecture. It uses $2nm$ processors and it is based on combinatorial properties on the hypercube network such as the constant time shift of line for a length power of two and constructions of arrays.

1 Introduction

The string-matching problem is one of the most studied problems in the sequential complexity theory [1, 5, 11]. It is defined as follows. Given two strings T and M, of lengths n and m respectively, we want to find all occurrences of M in T.

In parallel computation, several solutions have been found on the PRAM-model [7, 14, 8, 10]; the time lower bound is $O(\log \log n)$ [3] and it is achieved by an existing algorithm [2]. Nevertheless, results are not so advanced in the Network models. For the hypercube network, the main results are obtained by the simulation of PRAM parallel algorithms; the classic simulation cost is $O(\log^2 n)$ [13] but it can be speeded up to $O(\log n \log \log^2 n)$ but with huge constants [6]. The best complexity is given by a random algorithm with $O(\log n)$ time for $O(n/\log n)$ processors proposed by Chung [4].

This article proposes a faster solution for the string-matching problem in a hypercube network model with constant memory. This algorithm performs in a deterministic way with $\lceil \log n \rceil + \frac{5}{2} \lceil \log m \rceil + 5$ communications. This result is very close to the lower bound of the problem which is $\lceil \log n \rceil + k \lceil \log m \rceil + C$ with $0 \leq k \leq \frac{5}{2}$; indeed $\lceil \log n \rceil$ is the lower bound to check for an occurrence of the pattern at a given position. Intuitively, k seems to be at least 2, that is, the time needed to transmit information and compute the existence of occurrences.

The number of processors used is twice the product of the length of the text by the length of the pattern. This algorithm is in fact an equivalent to the naive algorithm in a EREW-PRAM[9].

The memory of each processor is limited to a small number of integers.

The model chosen here is an SIMD hypercube with free communication. The same technique can be performed with a time complexity of $O(\log m \log n)$ in a model with restricted communication (at each step, each communication is performed on the same dimension of the hypercube).

An adaptation of this algorithm is given for nm processors in time $\lceil \log n \rceil + 4 \lceil \log m \rceil + 4$ communications.

1.1 Notation

Let T and M be two words of length n and m respectively. We want to find all occurrences of M in T. M is said to appear in T iff $M = T[i \cdots i + m - 1]$ with $i \in \{0, \ldots, n - m\}$.

1.2 Hypercube Model

The model chosen there is an SIMD Hypercube Network with constant memory and free communication. We have the following points :

- We have identical processors having their own local memory. This memory is only composed with few integers.
- Processors are synchronous and execute the same instruction (or may be inactive).
- Processors are connected by an Hypercube network (Figure 1). In a *unit route*, one unit of data may be transmitted between two processors directly connected by the hypercube.
- During a communication, each processor chooses the dimension where it will send information.

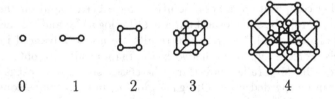

$$0 \qquad 1 \qquad 2 \qquad 3 \qquad 4$$

Fig. 1. hypercube examples.

In this article, P_i represents the processor whose hypercube coordinate is the binary representation of i. We will also use a multi-dimensional notation : $P_{d_1,\ldots,d_r}$ with $0 \le d_i < 2^{l_i}$. This notation represents the processor of hypercube coordinate :

$$\sum_{i=1}^{r} d_i \times 2^{\sum_{j=i+1}^{r} l_j} .$$

Programs will be written in an extended pseudo-Pascal.

The indexation will denote local variable; T_i is the local variable T of the processor of number i.

Parallel control is done by the instruction : **For all** P_i **in** //.

Instructions of this block will be executed simultaneously by every processor. In such a block, the instruction IF, testing on local variables, will inactivate processors which do not satisfy the condition, before executing the THEN statement. Variables appearing in this statement are local variables.

The communication between two neighbours on the hypercube is performed by the operator $\xleftarrow{e}$.

$S_j \xleftarrow{e} T_i$ means that every active processor will send the value of its local variable T to the local variable S of the j-numbered processor.

We will use the bit operators: $\otimes, \wedge, \vee$, respectively the exclusive OR, the AND and the inclusive OR. We will also use the function cg, the code of Gray function defined as follows:

$$cg(i) = \frac{i \otimes 2i}{2} .$$

2 Product Algorithm

This method consists in performing the "product" of the text by the pattern. We generate all the couples (T_i, M_j) of possible comparisons. Each one of these couples will be associated to a processor. Finally, we will compute the logic AND of the couples corresponding to an occurrence of the pattern.

The main difficulty of this algorithm is the generation of couples based on the calculus of the m first left shifts of T. On a PRAM model, generation is obvious, each processor determines by calculation its couple and reads it. In the case of hypercube networks, the cost of a general communication makes inefficient this method.

We assume without loss of generality that n and m are powers of 2. Otherwise we extend lengths of M and T to the nearest power of 2, M by a character matching everything called $\star$, and T by a character matching nothing called ϵ.

The initial situation is the text and the pattern distributed over the first processors of our hypercube:

Let $\mathcal{H}$ be an Hypercube of dimension $\log nm$. The processor P_i have in local memory $T[j]$ if $i = cg(j)$ with $i \in \{0, \ldots, n-1\}$ and it has $M[j]$ if $i = cg(j)$ with $i \in \{0, \ldots, m-1\}$.

By using the Gray code, the processors containing respectively $T[k]$ and $T[k+1]$ are physically linked.

We want to achieve the final state where each processor P_i with $i \in \{0, \ldots, n-1\}$ has its local variable O_i set as follows:

$$O_i = \begin{cases} 1 & \text{If there is an occurrence of } M \text{ at position } j \text{ in } T \text{ with } i = cg(j) \\ 0 & \text{Otherwise} \end{cases} .$$

3 Overview

The hypercube is structured in two arrays of m lines of n processors called A_0 and A_1. This decomposition is obtained by splitting the hypercube coordinate of the processors in three parts of respective length 1, $\log m$ and $\log n$.

In a first part, each processor $P_{0,i,j}$ of the first array will store a character of T and each processor $P_{1,i,j}$ will store a character of M. Processors initially containing T constitute the first line of the first array.

After several computations operated in parallel on both arrays, the couples will be stored in the first array for further calculations.

This description leads to a four step algorithm:

I) Distribution of M and T.

We proceed in two steps:
   a. M is distributed over the initial line. We obtain a line containing $M^{n/m}$.
   b. In parallel, T is distributed over A_0 and $M^{n/m}$ over T_1.

II) Generation of the Couples.

The couples are generated thanks to the calculation of left shifts of T. Every left shift of T of length smaller than m is calculated in parallel. The shift of length i is computed by the processors of line i of A_0 and A_1.

III) Verification of M Occurrences.

We group the couples in the first array. Then each processor $P_{i,j}$ of A_0 compares the local value of $T_{i,j}$ and $M_{i,j}$. Finally, we compute the logical AND of these results for groups of processors corresponding to an occurrence of M (processors with the same value of i and the same quotient of j divided by m).

IV) Ascension of Results.

The results are appended during an ascension to the initial line.

4 Steps of the Algorithm

4.1 Distribution of M and T

First of all, M is distributed over the initial line. The sub-hypercube of dimension $\log m$ corresponding to M (we call it M-hypercube) sends the pattern through the $\log m + 1$ dimension. Precisely, each processor P_j sends its character of M then M_j to its symetric on the $\log m + 1$ dimension. This operation is iterated with the new subhypercube filled until the complete distribution of M (figure 2).

The pattern will be reversed on some parts. The first line will contain: $M\overline{M}M\overline{M}\ldots M\overline{M}$ where $\overline{M}$ denotes the word composed by the letters of M in reverse order. The block [1] puts in right order the part reversed in the array, i.e. M-hypercubes of odd position (counted from 0).

At this point, the initial line stores T and $M^{n/m}$. In parallel, we distribute T over the first array and $M^{n/m}$ over the second one. Like previously, we operate dimension by dimension (figure 3).

Lemma 1. *The number of unit routes of step I is $\log n + 1$.*

Proof. We need $\log n - \log m + 1$ communications to complete the distribution of M on the initial line and $\log m$ communications for the distribution over the two arrays. □

```
For all Pⱼ in //
do
  For i = 0 to log n − log m − 1
    If M ≠ 0
    Then
        |Mₘ×2ⁱ⁺ⱼ ⟵ᵉ M

    If (cg⁻¹ (⌈ i mod n / m ⌉) mod 2 = 1)   [1]
    Then
        |M_{j⊗ m/2} ⟵ᵉ M
```

Fig. 2. step Ia: distribution of M

```
For all P_{i,j,k} in //
with i < 2, j < m and k < n
do
    If (i = 0) C_{1,j,k} ⟵ᵉ M
    Else C = T

    For l = 0 to log m
      If (C ≠ 0)
      Then
          |C_{i,j+2ˡ,k} ⟵ᵉ C
```

Fig. 3. step Ib: distribution of M and T

4.2 Generation of Couples

At this step, each line of A_0 contains T and each line of A_1 contains $M^{n/m}$. Let consider the couples $(T_{0,i,j}, M_{0,i,k})$. By shifting the text to the left by every length smaller than m without moving the pattern, we will generate all the couples of possible comparisons. To perform all the text left shifts, we use a combinatorial property of the hypercube : a power of two length shift, which we call basic shift, can be computed in constant time (in fact, 2 communications with n processors).

Power of 2 Length Shift We consider circular shifts. Thus we can possibly find "circular occurrences" after the position $n - m - 1$ in T. They are cancelled by systematically giving to $O[n - m + 1 \ldots n - 1]$ the value 0 after all operations.

Thanks to combinatorial properties of the hypercube, a basic shift can be achieved in constant time; in fact, it is a copy between sub-hypercubes.

Theorem 2. *A basic shift can be computed with two unit routes.*

Proof. Let l be the relative length of the shift, with l positive for a right shift and negative for a left one. Each processor $P_{i,j}$ of sub-hypercube k of dimension $\log_2 |l|$ is connected with the processor $P_{i,next(j)}$ of sub-hypercube $k+1$ with:

$$next(j) = (j \bmod |l|) + |l| \times \left[cg \left(\left(cg^{-1} \left(\left\lfloor \frac{j}{|l|} \right\rfloor \right) + \frac{|l|}{l} \right) \bmod \frac{n}{|l|} \right) \right]$$

After this communication, we notice that the order of the characters is reversed. The correct order is obtained by splitting in 2 each sub-hypercube and by exchanging the character between the two. □

In conclusion, the function $shift(v)$, where v is the local variable to shift, is the following:

For all $P_{i,j}$ in //
If $l \neq 0$
Then
$$v_{i,next(j)} \xleftarrow{e} v$$
$$v_{i,j \otimes \frac{|l|}{2}} \xleftarrow{e} v$$

Fig. 4. *shift(v)*

This function can be used in parallel with several shift lengths for different lines.

General Shift A general shift will be done by a succession of basic shifts. Requiring at most $\log m$ steps for a shift of length at most m. A basic shift can be accomplished in 2 unit routes, consequently the parallel generation of the $m-1$ shift will be effectuated with $2\log m$ unit routes.

Theorem 3. *A shift can be decomposed in no more than $\lceil \log(m)/2 \rceil + 1$ basic shifts.*

Proof. We notice that a circular left shift of k positions in a line of n processors with $k < m$ is equivalent to a left shift of m positions followed by a right shift of $m-k$ positions. let $N(i)$ be the function counting the number of 1 in the binary decomposition of i. We have $N(k) + N(m-k) \leq \log(m) + 1$. So if $N(k)$ is greater than $\lceil \log(m)/2 \rceil$ then $N(m-k)$ is inferior or equal to $\lceil \log(m)/2 \rceil$. □

Solution with $2nm$ Processors The previous shift algorithm is composed of two parts:

– A communication transfering the element in the right subhypercube.

- A communication ordering the elements of each subhypercube (in fact it reverses the order).

For each shift, we will use only one communication: nm processors operating the first communication on the text and nm processors exchanging the order of the pattern.

Precisely, for a shift of length l:

- Each processor $P_{0,i,j}$ of A_0 will send his character of T to $P_{0,i,next(j)}$.
- Each processor $P_{1,i,j}$ of A_1 will send his character of M to $P_{1,i,j\otimes\frac{l}{2}}$.

Let us prove the validity of such a method.

Theorem 4. *Performing separately this two communications in parallel for each shift of a serie of basic shifts of lengths $s_1, \ldots, s_l$ applied in increasing order will generate every couple $(T_{(i-s) \bmod n}, M_{i \bmod m})$ with $s = \sum_{i=1}^{l} s_i$.*

Proof. It's a recurrent proof on the number of shift applied in increasing order.

- For one shift, it's obvious that reversing the text or the pattern by the second communication gives the same couples.
- We suppose the validity for n basic shifts applied in increasing order of length.
- We apply a $(n+1) - th$ shift of length 2^p, with 2^p greater than all other shift lengths.
  Let $H_{i,j,k}$ (with $i < 2, j < m$ and $k < \frac{n}{2^p}$) be the class of subhypercube containing the processors $P_{i,j,\alpha}$ with $k2^p < ch^{-1}(\alpha) < (k+1)2^{p+1}$.
  Each $H_{i,j,k}$ contains the characters of M of position $k2^p \ldots (k+1)2^p$ (modulo m). Indeed each previous inversion on the pattern have been effectuated on blocks of size at most 2^{p-1}.
  Secondly, every $H_{i,j,k}$ has the same order of index for the pattern. As every block has received the same transformations, we have that if $P_{i,j,cg(k)}$ stores $M[v]$ then $P_{i,j,cg(k+2^p)}$ stores $M[(v+2^p) \bmod m]$.
  So we just need to operate a left shift of the text of size 2^p to have the correct couples. But operating a left shift of the text can be done equivalently on the couples in one unique communication simultaneously on the text and the pattern (case of a unique shift). $\square$

This analysis leads to a new shift algorithm (figure 5).

To be able to find the lowest power of two to apply, we use the function $last(i)$ that gives the position of the last digit one in the binary decomposition of i ($last$ can be computed in constant time with the classical integer operations).

The figure 6 gives the algoritm of the couple generation with $2nm$ processors.

Lemma 5. *The number of unit routes for the step II is $\log(m)/2 + 3$.*

Proof. Two unit routes are needed for the possible left shift of length m plus $\log(m)/2 + 1$ for the decomposition loop. $\square$

```
For all P_{i,j,k} in //
with i < 2, j < m and k < n
 |If l ≠ 0
 |Then
 | |If i = 0 Then envoi = next(j)
 | |Else envoi = j ⊗ |l|/2
 | |v_{i,envoi} ←e— v
```

Fig. 5. parshift(v)

```
For all P_{i,j,k} in //
with i < 2, j < m and k < n
 |If N(j) > ⌈log m/2⌉
 |Then
 | |l_{i,j,k} = (¬i) × -m , Shift(C)
 | |valshift_{i,j,k} = m - j
 | |dir = 1
 |Else
 | |valshift_{i,j,k} = j
 | |dir = -1
 |
 |For k = 0 to ⌈log(m)/2⌉
 |do
 | |If (valshift ≠ 0)
 | |Then
 | | |p = 2^{last(valshift)}
 | | |l = dir × p , Parshift(C)
 | | |valshift = valshift - p
```

Fig. 6. step II : shifts generation

4.3 Occurrences Computation

We group the data in the first array. So each processor of A_0 has in local memory one character of T and one of M to compare. An occurrence of M is recognized iff the comparison succeeds for all the processors corresponding to a M sub-hypercube i.e. for all the processors $P_{i,j}$ having the same value of i and the same value of $\lfloor j/m \rfloor$. We proceed then to the logical AND of all these results. The final result is stored by the processor of coordinate $cg(i)$ in a M-hypercube on the $i-th$ line (the processor $P_{i,cg(i)+m \times \lfloor j/m \rfloor}$ in global coordinate). This is done by operating the AND dimension by dimension until reaching the processor P_i. It's a reduction with P_i as root (figure 7).

This operation needs $\log m + 1$ communications.

```
For all P_{i,j,k} in //
with i < 2, j < m and k < n
 |If (i = 1) Then C'_{0,j,k} ←ᵉ C
For all P_{i,j} in //
with i < m and j < n
 |O = (C == C')
 |
 |For k = 0 to log m − 1
 | |If (j ∧ 2^k ≠ cg(i) ∧ 2^k)
 | |Then
 | | |O'_{i,j ⊗ 2^k} ←ᵉ O
 | | |O = 0
 | |Else
 | | |O = O AND O'
```

Fig. 7. step III : occurrences computation

4.4 Ascension of Results

The results are cumulated dimension by dimension until the first line. It's a reduction applied to an hypercube of lines (figure 8).

```
For all P_{i,j} in // with i < m and
j < n
 |For k = 0 to log m − 1
 | |If (i AND 2^k)
 | |Then
 | | |O'_{i,j ⊗ 2^k} ←ᵉ O
 | | |O = 0
 | |Else
 | | |O = O OR O'
```

Fig. 8. step IV : ascension of results.

This reduction is performed in $\log n$ communications.

5 Conclusion

This algorithm operates in $\lceil \log(n) \rceil + \frac{5}{2} \lceil \log(m) \rceil + 5$ communications with $2nm$ processors.

With nm processors the four steps become :

- Separate distribution of M and T in time $\log n$ for M and $\log m$ for T.
- Generation of couples with 2 unit route by shift, then $\log m + 3$ unit routes is needed.
- Verification of occurrences with $\log m + 1$ unit routes.
- Ascension of results with $\log m$ communications.

Giving a total of $\lceil \log(n) \rceil + 4\lceil log(m) \rceil + 4$ unit routes.

For a model with restricted communication, the same kind of method can be used but in such a model a shift of length l of power of 2 cost $O(\log(n/l))$. Hence the generation of all left shifts on the tree would cost: $\sum_{i=0}^{m/2} \log(n/i)$, that is $O(\log m \log n)$.

This algorithm has a double interest. First, it can be used like a basic string-matching resolver by more complex algorithms, exactly as the naive technique is used on the PRAM model to complete the search. Secondly, it lower down the resolution time on the hypercube model approaching closely the lower bound of the problem for this architecture. The value of this lower bound is still an open problem.

References

1. R.S. Boyer et J.S. Moore. *A fast string searching algorithm.* Comm of the ACM, 1977.
2. D. Breslauer et Z. Galil. *An optimal O(log log n) parallel string matching algorithm.* SIAM J. Comput, 1990.
3. D. Breslauer et Z. Galil. *An lower bound for parallel string matching.* In Proc. of the 23rd Ann ACM Symp on theory of computing, 1991.
4. KL. Chung. *A randomized parallel algorithm for string matching on hypercube.* Pattern Recognition, 1992.
5. M. Crochemore et D. Perrin. *Two-Way String-Matching.* Journal of the ACM, Juillet 91.
6. R. Cypher et C.G. Plaxton. *Deterministic sorting in nearly logarithmic time on the hypercube and related computers.* Proc of the 22nd Ann Symp on theory of computing, 1990.
7. Z. Galil. *Optimal parallel algorithms for string matching.* Inform. and Control, 1985.
8. Z. Galil. *A constant-time Optimal Parallel string-matching algorithm.* Proc of the 24th ACM symp. on Theory of Comput., 1992.
9. A. Gibbons et W. Rytter. *Efficient parallel algorithms.* Cambridge University Press, 1988.
10. R.M. Karp et M.O. Rabin. *Efficient randomized pattern matching algorithms.* IBM J. Res. Dev., 1987.
11. D.E. Knuth, J.H. Morris et U.R. Pratt. *Fast Pattern Matching in strings.* SIAM J. Comput, 1977.
12. D. Nassimi et S. Sahni. *Data broadcasting in SIMD computers.* IEEE Trans. on Comp, 1981.
13. S. Ranka et S. Sahni. *Hypercube Algorithms.* Springer-Verlag, 1990.
14. U. Vishkin. *Optimal parallel pattern matching in strings.* ICALP, 1985.
15. A. Wu. *Embedding of tree networks into hypercubes.* J. of Parallel and Distributed Computing, 1985.

Computation of Squares in a String [1]

(Preliminary Version)

S. Rao Kosaraju
Department of Computer Science
The Johns Hopkins University
Baltimore, Maryland 21218-2694

Abstract

We design a linear time algorithm for computing a square substring from each position of a given string over a finite alphabet. The algorithm exploits several subtle properties of suffix trees for strings.

1. Introduction

Given a string $X[1..n]$ we seek to compute for each position i the shortest square beginning at that position. We design a linear time algorithm for this problem. Main & Lorentz [7] and Crochemore [5] developed linear time algorithms for the special case of testing whether X contains a substring that is a square. Rabin [9] discovered an attractively simple randomized algorithm for the problem of computing the smallest position from which there is a square. His algorithm, thus, computes only a part of our information, and, in addition, runs in $O(n \ log \ n)$ steps. Apostolico & Preparata [1] and Main & Lorentz [6] designed $O(n \ log \ n)$ step algorithms for computing all the squares in a string. Neither approach can lead to a linear time algorithm for our problem.

Our linear time algorithm is developed in section 2. In the final version we show that the algorithm can be modified into one that computes all squares: after $O(n)$ preprocessing, given any position i, the modified algorithm outputs all the distinct squares starting at position i in linear time. (The number of steps is proportional to the number of distinct squares starting at position i.) A precise statement of this result is given in section 3.

2. Computation of Squares

A *square* is a string of the form yy with $| y | \geq 1$. Strings $'aa'$ and $'abab'$ are examples of squares. In the square computation problem, for any given string $X[1...n] = x_1 x_2 ... x_n$, we want to compute for every $1 \leq i \leq n - 1$ the minimum j such that $x_i x_{i+1} ... x_{i+j-1}$ is a square. If there is no such j, then the corresponding value is 0. The substring $x_i ... x_j$ of $X[1..n]$ is denoted by $X[i..j]$. The string $X[1..n]$ is *periodic* if there exists a $k \leq \frac{n}{2}$ such that $X[1..n]$ is a prefix of $(X[1..k])^{\lceil n/k \rceil}$. The corresponding k is a *period* of X. A square yy is a *prime square* if yy cannot be periodic with a period less than $| y |$. Note that $'aaaa'$ is a square but it is not a prime square.

The following two observations are well-known.

Observation 1: If a string is periodic with period k, then its prefix of length $2k$ is a square.

---

[1]Supported by NSF Grant #CCR9107293

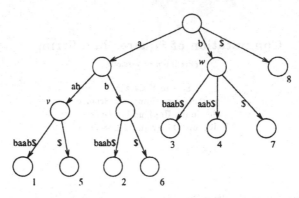

Figure 1: Suffix tree for *aabbaab*

Observation 2: For any string $X[1..n]$, if there exists an $i \leq \frac{n}{2}$ such that $X[1..n-i] = X[i+1..n]$, then, for any $2i \leq j \leq n$, $X[1..j]$ is periodic with period i.

Lemma 1: For any string $X[1..n]$ and any $k \geq 2$, if there exist an ℓ and ℓ distinct indices, $1 < i_1, i_2, ..., i_\ell \leq n-k+1$ such that $k\ell \geq 2n$ and $X[1..k] = X[i_1..i_1+k-1] = ... = X[i_\ell..i_\ell+k-1]$ then $X[1..k]$ is periodic with a period $\leq \frac{n}{\ell}$.

Proof: It is easy to establish that there exist j_1 and j_2 such that $i_{j_1} < i_{j_2}$ and $i_{j_2} - i_{j_1} \leq \frac{n}{\ell}$. Now by observation 2, $X[i_{j_1}..i_{j_1} + k - 1]$ is periodic with period $\leq \frac{n}{\ell}$. Hence $X[1..k]$ is periodic with period $\leq \frac{n}{\ell}$. (Note the $k \geq \frac{2n}{\ell}$.)

2.1 Suffix Trees:

The algorithm for the computation of squares relies upon suffix trees [1,3,8,10,12]. The suffix tree for a string X, denoted $\Sigma(X)$, is the compressed trie for all the suffixes of $X\$$, where $\$$ is a symbol not in the alphabet of X. The ith suffix of X and the corresponding leaf in the suffix tree are denoted by *suffix_i* and *leaf_i*, resp. The suffix tree for $X = aabbaab$ is shown in figure 1.

Even though the edges are labeled with substrings of $X\$$, the actual implementation will replace each such substring by a pair of pointers into $X\$$, so that the size of the data structure is $O(n)$.

For any string $X[1..n]$, we denote the suffix tree for $X[i..n]$ by $\Sigma(X, \geq i)$. When X is understood, we replace $\Sigma(X, \geq i)$ by $\Sigma(\geq i)$. For any node u, let σ_u be the concatenation of the edge labels on the path from the root to u. Let the *depth* of u be $| \sigma_u |$. Note that the depth is not the graph-theoretic depth. Thus in figure 1, $\sigma_v = aab$, and $depth(v)=3$. The parent of any node u is denoted by $p(u)$.

There are several $O(n)$ step algorithms for the computation of $\Sigma(X)$ [3,8,10,12], when the alphabet size is fixed. We base our square computation on Weiner's construction, the details of which are not necessary to follow our algorithm. At a high level, Weiner's algorithm constructs $\Sigma(\geq i)$ from $\Sigma(\geq i + 1)$ by installing *suffix_i* at an appropriate point in $\Sigma(\geq i + 1)$. The installation point can be a node of $\Sigma(\geq i + 1)$ or can be obtained by splitting an edge of $\Sigma(\geq i + 1)$ as shown in Figures 2a and 2b. For the example of figure 1, if X is extended to the left by symbol $'b'$ (i.e. the string is *baabbaab*), Weiner's construction locates $(w, leaf_4)$ edge, splits it into two edges with labels $'aab'$ and $'\$'$ and installs the new leaf as a child of the new node created. Throughout we assume that the suffix tree algorithm maintains the depth of each node in the suffix tree.

The following two properties play an important role in the development of our algorithm. In fact, the second property was the backbone of the algorithm of [1].

Observation 3: The depth of $p(leaf_i)$ in $\Sigma(\geq i)$ is no more than $1 +$ depth of $p(leaf_{i+1})$ in $\Sigma(\geq i+1)$.

Observation 4: For any string X and for any i and k, X has a minimum length $2k$ square starting at position i if and only if k is the minimum value such that in $\Sigma(X)$ the least common ancestor of leaves i and $i\text{-}k\text{+}1$ has a depth of at least k.

2.2 Algorithm Development

In this subsection we show how to augment the suffix tree construction with the computation of additional information that would facilitate the extraction of square information. At every stage of the suffix tree construction, at every node u we store the length of the shortest prefix of σ_u that is a square. (Value 0 indicates the non-existence of any prefix square). Let us denote this value as $square(u)$. In figure 1, $square(v) = 2$ since $\sigma_v = aab$ and its prefix $'aa'$ is a square. At the end of the computation, in $\Sigma(X)$ we can simply read off the square values at the leaves $1, 2, 3, ..., n-1$ in order as the output.

While computing the suffix tree (with this square information), we handle the suffixes in groups. Let us assume that at some stage, after inserting the previous group of suffixes, we have $\Sigma(\geq i+1)$. In the next phase we insert suffixes $i, i\text{-}1,...,i\text{-}k$, for a suitable value of k to be specified later on, successively while maintaining the square information. Now we discuss how to update the square information when each suffix in this group of suffixes is inserted.

The installation of $suffix_i$ into $\Sigma(\geq i+1)$ is shown schematically in figure 2.

It is easy to verify that the square values at the nodes of $\Sigma(\geq i+1)$ will remain invariant. When an additional node, node v, is created by splitting an edge, as in figure 2b, the $square(v)$ will be $square(w)$ if $square(w) \leq depth(v)$, otherwise it is 0. Whether an extra node is created or not, if the resulting square value at $p(leaf_i)$ in $\Sigma(\geq i)$ is not 0, then that value will also be the value at $leaf_i$. In such a case, the insertion of $suffix_i$ completes the current phase; i.e. $k=0$.

Now we describe the computation of $square(leaf_i)$ when $square(p(leaf_i)) = 0$. This is the most interesting part of the algorithm. Let the depth of $p(leaf_i)$ be Δ (i.e. $\mid \sigma_{p(leaf_i)} \mid = \Delta$). In addition to the main suffix tree, we construct an auxiliary suffix tree. The auxiliary suffix tree constructed is the suffix tree for the substring $X[i..i+3\Delta-1]$; i.e. for a substring of length 3Δ starting at position i. For clarity of presentation, in $X[i..i+3\Delta-1]$ and the auxiliary suffix tree, we carry over the suffix indexing from X. Thus $x_i...x_{i+3\Delta-1}\$$, the suffix from position 1 of $X[i..i+3\Delta-1]$, is $suffix_i$ - not $suffix_1$. Also, we refer to the main and the auxiliary suffix trees by Σ and Σ', resp.

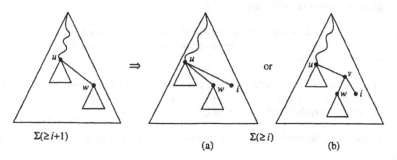

Figure 2: Modifying $\Sigma(\geq i+1)$ into $\Sigma(\geq i)$

In Σ', let α be the highest ancestor (least depth) of $leaf_i$ with depth $\geq \frac{\Delta}{2}$. We argue below that there are $O(1)$ leaves in the subtree of α and that in $O(1)$ steps we can compute $square(leaf_i)$ in Σ by processing these leaves.

After this computation, we have $\Sigma(\geq i)$, with square information. and $\Sigma'(\geq i)$ (i.e. $\Sigma(X[i..i + 3\Delta - 1]))$. Then we process $suffix_{i-1}$ by computing the main $\Sigma(\geq i - 1)$. Whether the square information at $leaf_{i-1}$ becomes available or not, we also update $\Sigma'(\geq i)$ to $\Sigma'(\geq i - 1)$ by applying Weiner's construction and inserting $x_{i-1}x_i...x_{i+3\Delta-1}$ into $\Sigma'(\geq i)$. In case the square information at $leaf_{i-1}$ of Σ was not available. we compute it using Σ' as before. We repeat this process and insert suffixes i-1, i-2,..., i-k until k becomes Δ or we encounter the first $suffix_{i-k-1}$ such that the depth of the $p(leaf_{i-k-1})$ in $\Sigma(\geq i - k - 1)$ becomes less than $\frac{\Delta}{2}$. Suffixes i, i-1, ..., i-k form the group of suffixes inserted in the current phase. We argue below that the overall algorithm requires $O(n)$ steps.

Lemma 2: During the current phase when any $suffix_{i-j}$, $0 \leq j \leq k$. gets inserted into Σ. depth of $p(leaf_{i-j})$ is in the range $[\frac{\Delta}{2}, \Delta + j]$.

Proof: By definition. the depth of $p(leaf_i)$ is Δ when $leaf_i$ gets inserted. By observation 3. on successive insertions the depth cannot increase by more than 1. The lower limit follows from the criterion for terminating the current phase.

Lemma 3: During the current phase when any $suffix_{i-j}$, $0 \leq j \leq k$. gets inserted into Σ, let the square value at $p(leaf_{i-j}) = 0$. At that instant, in Σ' let α be the highest ancestor of $leaf_{i-j}$ having depth $\geq \frac{\Delta}{2}$. Then the number of leaves in the subtree of α is no more than 17.

Proof: Since the square value at $p(leaf_{i-j})$ is 0, $suffix_{i-j}$ has no prefix of length $\leq \frac{\Delta}{2}$ that is a square. The length of the string considered for constructing Σ' (the substring $X[i - j..i + 3\Delta - 1]$) is $3\Delta + j$ which is $\leq 4\Delta$. By lemma 1, if there are more than 16 leaves, excluding $leaf_{i-j}$, in the subtree of α then σ_α is periodic with a period $< \frac{\Delta}{4}$. (Note that $k \geq \frac{\Delta}{2}$, $\ell = 16$, and $n = 4\Delta$.) Hence by observation 1, a prefix of σ_α of length $< \frac{\Delta}{2}$ is a square. Hence a prefix of $suffix_{i-j}$ of length $< \frac{\Delta}{2}$ is a square - a contradiction.

Lemma 4: During the current phase when any $suffix_{i-j}$, $0 \leq j \leq k$. gets inserted into Σ. let the square value at $p(leaf_{i-j}) = 0$. Then by making use of Σ' the square value at $leaf_{i-j}$ in Σ can be computed in $O(1)$ steps.

Proof: As in lemma 3, $suffix_{i-j}$ has no prefix of length $\leq \frac{\Delta}{2}$ that is a square. By lemma 2, the depth of $p(leaf_{i-j})$ in Σ is $\leq \Delta + j$. Hence we can easily show that $suffix_{i-j}$ has no prefix of length $> 2(\Delta + j)$ that is a square. Since $2(\Delta + j) \leq 3\Delta + j$ for any $0 \leq j \leq k \leq \Delta$, any prefix of $suffix_{i-j}$ that is a square is a prefix of $suffix_{i-j}$ in Σ'. (Note that $suffix_{i-j}$ in Σ' is $x_{i-j}x_{i-j+1}...x_{i+3\Delta-1}$$.) Since the subtree of α contains $O(1)$ leaves, by applying observation 4 to these leaves, in $O(1)$ steps we can compute the square value at $leaf_{i-j}$.

By lemma 4, and by making use of the fact that if the number of suffixes handled in the current phase is less than Δ then the reduction of the insertion depth during the current phase is at least $\frac{\Delta}{2}$, we can establish the following result.

Lemma 5: Let a phase end with $suffix_{i+1}$. Then the total number of steps needed for processing suffixes n+1, n,....,i+1 is no more than $c_1(n - i + 1) + c_2\Delta'$, where c_1 and c_2 are suitably large constants and Δ' is the depth of $p(leaf_{i+1})$ in $\Sigma(\geq i + 1)$.

As a consequence of this result, when all the phases are completed, we have the following main theorem.

Theorem 1: An $O(n)$ step algorithm which for any string X, of length n. computes the minimum square from each position i can be constructed.

3. Generalizations

So far we have restricted ourselves to computing the shortest square from each position of the given string. Now we generalize the problem to that of computing all the prime squares starting from each position.

Let the number of prime squares of X starting at position i be denoted $\pi_X(i)$. Let the total number of prime squares of X (i.e. $\sum_i \pi_X(i)$) be $\Pi(X)$. It is known that every string of length n has $O(n \log n)$ prime squares, and there are strings of length n having $\Omega(n \log n)$ prime squares [4]. Several $O(n \log n)$ step algorithms for computing all the prime squares also exist [1,6]. In the final version we show that the key ideas of this paper can be extended to the computation of all the prime squares of a string. The resulting algorithm will have a time complexity of $O(n + \Pi(X))$. None of the existing algorithms appears to be modifiable into this output-sensitive form. Note that our algorithm is a linear time algorithm with respect to the sum of the input and output sizes. In fact, we will design an algorithm which can preprocess any string X of length n in $O(n)$ steps, and subsequently given any i can output all the prime squares of X starting at position i in $O(\pi_X(i))$ steps.

References

[1] A. Apostolico and F. P. Preparata. Optimal off-line detection of repetitions in a string, *Theoretical Computer Science*, pages 297-315, 1983.

[2] W. L. Chang and E. L. Lawler. Approximate string matching in sublinear expected time. Proc. of *31st IEEE FOCS*, pages 116-124, 1990.

[3] M. T. Chen and J. Seiferas. Efficient and elegant subword tree construction. In A. Apostolico and Z. Galil, editors, *Combinatorial Algorithms on Words*, pages 97-107. Springer-Verlag, 1985.

[4] M. Crochemore. An optimal algorithm for computing the repetitions in a word, Inf. Processing Letters 12, pages 244-250, 1981.

[5] M. Crochemore. Transducers and repetitions, *Theoretical Computer Science*, pages 63-86, 1986.

[6] M. G. Main and R. J. Lorentz. An $O(n \log n)$ algorithm for finding all repetitions in a string, *J. of Algorithms*, pages 422-432, 1984.

[7] M. G. Main and R. L. Lorentz. Linear time recognition of squarefree strings. In A. Apostolico and Z. Galil, editors, *Combinatorial Algorithms on Words*, pages 271-278, Springer-Verlag, 1985.

[8] E. M. McCreight. A space-economical suffix tree construction algorithm. *J. of ACM*, pages 262-272, 1976.

[9] M. O. Rabin. Discovering repetitions in strings. In A. Apostolico and Z. Galil. editors, *Combinatorial Algorithms on Words*, pages 279-288, Springer-Verlag, 1985.

[10] J. Seiferas. Subword trees. Class Notes, February 1977.

[11] A. O. Slisenko. Detection of periodicities and string-matching in real time. *J. of Soviet Mathematics*, pages 1316-1386, 1983.

[12] P. Weiner. Linear pattern matching algorithms. *Proc. of 14th IEEE Symp. on Switching & Automata Theory*, pages 1-11, 1973.

Minimization of Sequential Transducers

Mehryar Mohri

Institut Gaspard Monge *
Université Marne la Vallée
e-mail address: mohri@univ-mlv.fr

Abstract. We present an algorithm for minimizing sequential transducers. This algorithm is shown to be efficient, since in the case of acyclic transducers it operates in $O(|E| + |V| + (E| - |V| + |F|) \cdot (|P_{max}| + 1)$ steps, where E is the set of edges of the given transducer, V the set of its vertices, F the set of final states, and P_{max} the longest of the greatest common prefixes of the output paths leaving each state of the transducer. It can be applied to a larger class of transducers which includes subsequential transducers.

1 Introduction

Finite automata and transducers are used in many efficient programs. They allow to produce in a very easy way lexical analyzers for complex languages. In some applications as in Natural Language Processing the involved finite-state machines can contain several hundreds of thousands of states. Reducing the size of these graphs without losing their recognition properties is then crucial.

This problem has been solved in the case of deterministic automata since any deterministic automaton has an equivalent automaton with the minimal number of states which classic algorithms help to compute from it in an easy way ([1]).

For sequential transducers, the existence of minimal transducers has already been shown in the case of group transducers (see [3]), but no algorithm has been proposed for their construction. Here we shall characterize minimal sequential transducers by means of an equivalence relation and present algorithms which allow to compute, given a sequential transducer, an equivalent minimal sequential transducer. These algorithms can be easily extended to the case of subsequential transducers.

The minimization algorithm for sequential transducers involves the computation of the *prefix of a non-deterministic automaton*. We present this algorithm first, as it is independent of the concept of sequential transducer. We then give a characterization of minimal sequential transducers and describe the entire algorithm allowing to obtain these transducers.

---

* I thank Maxime Crochemore and Dominique Perrin for interesting and helpful comments.

2 Prefix of a non-deterministic automaton

2.1 Definition

Let $G = (V, i, F, A^*, \delta)$ be a non-deterministic automaton where
- V is the set of states or vertices;
- $i \in V$ the initial state;
- $F \subseteq V$ the set of final states;
- A a finite alphabet;
- δ the state transition function which maps $V \times A^*$ to the set of subsets of V.[2]

We denote by
- G^T the transpose of G, namely the automaton obtained from G by reversing each transition;
- $Trans[u]$ the set of transitions leaving $u \in V$;
- $Trans^T[u]$ the set of transitions entering $u \in V$;
- $t.v$ the vertex reached by (resp. source of) t and $t.l$ its label, for any transition t in $Trans[u]$ (resp. in $Trans^T[u]$), $u \in V$;
- $out - degree[u]$ the number of edges leaving $u \in V$;
- $in - degree[u]$ the number of edges entering $u \in V$;
- E the set of edges of G.

In the following, we shall consider automata such that from any state there exists at least one path leading to a final state. We denote by $x \wedge y$ the greatest common prefix of x and y in A^*, and by ϵ the empty word of A^*. Let P be the function mapping V to A^* defined by the following:

if $u \in F$ $P(u) = \epsilon$,
else $\quad P(u) =$ greatest common prefix of the labels of all paths leading from u to F.

P is well defined as for any u in V there exists at least one path leading from u to F. This path is finite, thus the greatest common prefix of the labels of all paths leading from $u \in V - F$ to F is well defined and in A^*. P can be equivalently defined by the following recursive definition:

$$\forall u \in F \qquad P(u) = \epsilon;$$
$$\forall u \in V - F \quad P(u) = \bigwedge_{t \in Trans[u]} t.l \, P(t.v) \, .$$

Given a non-deterministic automaton G, we define $p(G)$, *the prefix of G*, as the non-deterministic automaton which has the same set of vertices and edges as G, the same initial state and final states, and, which only differs from G by the labels of its transitions in the following way: for any edge $e \in E$ with starting

---

[2] Notice that the input alphabet of G is A^*. Hence, labels of transitions can be words.

state u and destination state[3] v,

$$label_{p(G)}(e) = label_G(e)P(v) \qquad if \ \ u = i,$$
$$label_{p(G)}(e) = [P(u)]^{-1}label_G(e)P(v) \qquad else,$$

where $label_{p(G)}(e)$ is the label of the edge e in $p(G)$ and $label_G(e)$ its label in G. $p(G)$ is well defined as $P(u)$ is by definition a prefix of $label_G(e)P(v)$. It is easy to show that $p(G)$ recognizes the same language as G. $p(G)$ is obtained from G by *pushing* as much as possible labels from final states towards the initial state.

2.2 Computation

The definition of the function P suggests a recursive algorithm to compute $p(G)$ from G. This means to proceed by calculating P for all states of the adjacent list of $u \in V$ before calculating $P(u)$. However, in general, there does not exist a linear ordering of all states of G such that if the adjacent list of u contains v, then v appears before u in the ordering. Two states can indeed belong to a cycle. Hence, P cannot be computed in such a way.

In case G is a dag (directed acyclic graph) such an ordering exists. The reverse ordering of a *topological sort* of a dag meets this condition ([7], [4]).[4] Therefore, in case G is acyclic, we can consider states in such an order and compute for each of them the greatest common prefix corresponding to the recursive definition of P above. In this way, the greatest common prefix computation is performed at most once for each state of G. The computation of the automaton $p(G)$ from G can be performed in a similar way by considering the states of G in the same ordering.

In case G is not acyclic, we consider SCC's (strongly connected components) of G. There exists a linear ordering of SCC's such that if the adjacent list of a state in a SCC scc_1 contains a state of another SCC scc_2, then scc_2 appears before scc_1 in the ordering: the reverse ordering of a topological sort of the dag G^{SCC}, the *component graph* of G[5].

In order to compute $p(G)$ from G, we can gradually modify the transitions of G by considering its SCC's according to this ordering. Each time a SCC scc is considered, all the transitions leaving states of scc are transformed into those

---

[3] Notice that there can be several edges with starting state u and destination state v in G. These edges can even bear the same labels.

[4] This ordering can be obtained in linear time $O(|V| + |E|)$ since it corresponds to the increasing ordering of the finishing times in a depth-first search.

[5] Recall that the component graph of G is the dag which contains one state for each SCC of G, and which contains a transition $(u, a, v) \in V \times A^* \times V$ if there exists a transition from the SCC of G corresponding to u, to the SCC of G corresponding to v. It can be obtained in linear time $O(|E| + |V|)$ like strongly connected components of G ([1], [7], [4]).

of $p(G)$, and all the transitions from another SCC entering a state u of scc are modified in a way such that:

$$\forall t \in Trans^T[u], \quad t.l = label_G(t)\, P(u).$$

Thanks to the choice of the ordering, these transformations are operated only once for each SCC. Once all SCC's of G have been considered one obviously obtains $p(G)$.

Now we need to indicate how these transformations are performed. Suppose that all the SCC's visited before scc have been correctly modified[6]. Then, according to the definition of the function P, the following system of equations:

$$\forall u \in scc,\, X_u = \left(\bigwedge_{\substack{t\in Trans[u] \\ t.v \in scc}} t.l\, X_{t.v} \right) \wedge \left(\bigwedge_{\substack{t\in Trans[u] \\ t.v \notin scc}} t.l \right),$$

has a unique solution corresponding to the greatest common prefixes of each state of scc ($\forall u \in scc, X_u = P(u)$). In order to solve this system, we can proceed in the following way:

1- For each u in scc, we compute π_u, the greatest common prefix of all its leaving transitions:

$$\pi_u \leftarrow \left(\bigwedge_{\substack{t\in Trans[u] \\ t.v \in scc}} t.l\, X_{t.v} \right) \wedge \left(\bigwedge_{\substack{t\in Trans[u] \\ t.v \notin scc}} t.l \right) \quad if\ u \notin F,$$

$$\pi_u \leftarrow \epsilon \qquad\qquad\qquad\qquad\qquad\qquad\qquad else;$$

2- If $\pi_u \neq \epsilon$, we can make a change of variables: $Y_u \leftarrow \pi_u\, X_u$.

This second step is equivalent to storing the value π_u and solving the system modified by the following operations:

2'- $\forall t \in Trans[u], \quad t.l \leftarrow \pi_u^{-1}\, t.l$
$\quad\ \forall t \in Trans^T[u], \quad t.l \leftarrow t.l\, \pi_u.$

We can limit the number of times these two operations are performed by storing in a array N the number of empty labels leaving each state u of scc. As long as $N[u] \neq 0$, there is no use performing these operations as the value of π_u is ϵ.

---

[6] This is necessarily true for the first SCC considered as no transition leaves it to join a distinct SCC.

Also, if $N[u] = 0$ right after the computation of π_u, then π_u will remain equal to ϵ, as changes of variables will only affect suffixes of the transitions leaving u. We can store this information using an array F, in order to avoid performing step 1 in such situations or when u is a final state. We use a queue Q containing the set of states u with $N[u] = F[u] = 0$ for which the two operations above need to be performed, and an additional array INQ indicating for each state u whether it is in Q.

We start the above operations by initializing N and F to 0 for all states in scc, and by enqueuing in Q an arbitrarily chosen state u of scc. Each time the transition of a state v of $Trans^T[u]$ is modified, v is added to Q if $N[v] = F[u] = 0$. The property of SCC's and the initialization of N and F assure that each state of scc will be enqueued at least once. Steps 1 and 2' are operated until $Q = \emptyset$. This necessarily happens as, except for the first time, step 1 is performed for a state u if $N[u] = 0$. After the computation of the greatest common prefix we have or $N[u] = 0$ and then u will never be enqueued again, or $N[u] \neq 0$ and then a new non empty factor π_u of $P(u)$ has been identified. Thus, each state u is enqueued at most $(|P(u)| + 2)$ times in Q, and after at most $(|P_{max}| + 2)$ steps we have $Q = \emptyset$.

PREFIX_COMPUTATION(G)
1 for each $u \in V(G^{SCC})$ ▷ considered in order of increasing finishing times of
                               a DFS of G^{SCC}
2 do for each $v \in SCC[u]$
3   do $N[v] \leftarrow INQ[v] \leftarrow F[v] \leftarrow 0$
4      $Q \leftarrow v$ ▷ v arbitrarily chosen in $SCC[u]$
5      $INQ[v] \leftarrow 1$
6   while $Q \neq \emptyset$
7   do $v \leftarrow head[Q]$
8   DEQUEUE(Q)
9          $INQ[v] \leftarrow 0$
10         $p \leftarrow GCP(G, v)$
11   for each $t \in Trans^T[v]$
12   do if $(p \neq \epsilon)$
13   then if $(t.v \in SCC[v]$ and $N[t.v] > 0$ and $t.l = \epsilon$ and
                       $F[t.v] = 0)$
14   then $N[t.v] \leftarrow N[t.v] - 1$
15                 $t.l \leftarrow t.l \, p$
16   if $(N[t.v] = 0$ and $INQ[t.v] = 0$ and $F[t.v] = 0)$
17   then ENQUEUE($Q, t.v$)
18                 $INQ[t.v] = 1$

Once $Q = \emptyset$, it is easy to notice that the system of equations has a trivial solution: $\forall u \in scc, X_u = \epsilon$. As noticed above, it has a unique solution. Therefore, the system is resolved. Concatenating the factors π_u involved in the changes of variables corresponding to the state u gives the value of $P(u)$. The set of operations 2' are thus equivalent to multiplying the label of each transition joining the states u and v, $(v \in scc)$, at right by $P(v)$ and at left by $[P(u)]^{-1}$ if u is in scc. This shows that the transformations described above do modify the transitions leaving or entering states of scc as desired. Thus, the above pseudocode gives an algorithm computing $p(G)$ from G.

In this algorithm, $V(G^{SCC})$ represents the set of states of the component graph of G, and, for each u in $V(G^{SCC})$, $SCC[u]$ stands for the strongly connected component corresponding to u. The function $GCP(G, u)$ called in the algorithm is such that it returns p the greatest common prefix of all the transitions leaving u ($p = \epsilon$ if $u \in F$), replaces each of these transitions by dividing them at left by p, counts and stores in $N[u]$ the number of empty transitions, and, if $N[u] = 0$ after the computation of the greatest common prefix or if u is a final state gives $F[u]$ the value 1.

2.3 Complexity

Notice that the computation of the greatest common prefix of n ($n > 1$) words requires at most $(|p| + 1).(n - 1)$ comparisons, where p is the result of this computation. Indeed, this operation consists in comparing the letters of the first word to those of the $(n - 1)$ others until a mismatch or an end of word occurs. The same comparisons allow to obtain the division at left by p and the number of empty transitions. In case only one transition leaves v, the computation of the greatest common prefix can be assumed to be in $O(1)$. Hence, the cost of a call of the function GCP for a state v ($\in V - F$) is in $O((|p|+1)(out-degree(v)-1)+1)$ where p is the greatest common prefix of the transitions leaving v.

As noticed above, p is a factor of the greatest common prefix of v. Thus, the loop of lines 6-18 is performed at most $(|P(u)| + 2)$ times. The cost of each iteration of the loop of lines 11-18 can be considered as being in $O(1)$ as it requires a constant number of comparisons[7]. This loop is iterated $in - degree(v)$ times inside the loop of lines 11-18 when state v is considered at line 7. All other operations (initialization, computation of the $SCC's$ and of G^{SCC}, and the definition of the reverse topological sort of G^{SCC}) can be done in $O(|V|+|E|)$. Therefore, the total cost of the algorithm above is in

$$O(|V| + |E| + \sum_{u \in V'} (out - degree(u) - 1).|P(u)| + \sum_{u \in V} in - degree(u).|P(u)|),$$

where $V' = \{u \in V \mid out - degree(u) > 1\}$,
hence in:

$$O(|V| + |E|(|P_{max}| + 1)),$$

where P_{max} is the longest of the greatest common prefixes of the states of G. In

---

[7] The test $t.v \in SCC[v]$ of line 13 can be performed in constant time if we assume that we have designed an array indicating for each state v in G its corresponding state in G^{SCC}. This can be done in linear time $O(|V| + |E|)$ from the $SCC's$ of G.

case G is acyclic, each SCC is reduced to one state. Therefore, the loop of lines 7-18 is performed once for each state of G. The cost of the algorithm is then in

$$O(|V| + |E| + \sum_{u \in V'} (out - degree(u) - 1).|P(u)| + \sum_{u \in V} in - degree(u)),$$

hence in:

$$O(|V| + |E| + (|E| - (|V| - |F|)).|P_{max}|)$$

At worse, if all labels of the automaton are identical for instance, $|P_{max}|$ can reach the length of the longest path from the initial state to a final state without going through cycles. The following figure represents an acyclic automaton with identical labels we shall assume equal to an element $a \in A$, in which all states except the final one have an $out - degree$ of 2. Here, we have: $|P_{max}| = |V|/2$.

Fig. 1. Automaton G.

At each state, the number of comparisons needed for the computation of the greatest common prefix in the algorithm described above is proportional to $(d + 1)$, where d is its distance to the final state. It can be proved easily that PREFIX_COMPUTATION runs in $O(|V|^2)$ in this case. However, in most practical cases, $|P_{max}|$ is very small compared to $|V|$, and the algorithm can be considered to be very efficient. In the following section, we shall indicate an interesting application of this algorithm.

3 Minimized sequential transducers

3.1 Definitions

A sequential transducer (ST) T is a 7-tuple $(V, i, F, A, B^*, \delta, \sigma)$ where:
- V is the set of states;
- $i \in V$ the initial state;
- $F \subseteq V$ the set of final states;
- A and B finite sets corresponding respectively to the input and output alphabets of the transducer;
- δ the state transition function which maps $V \times A$ to V;
- σ the output function which maps $V \times A$ to B^*.

The functions δ and σ can be extended to map $V \times A^*$, by the following recursive relations:

$$\forall s \in V, \forall w \in A^*, \forall a \in A, \delta(s, \epsilon) = s, \delta(s, wa) = \delta(\delta(s, w), a);$$
$$\sigma(s, \epsilon) = \epsilon, \sigma(s, wa) = \sigma(s, w)\sigma(\delta(s, w), a).$$

In the following, we shall consider sequential transducers such that every state is reachable from the initial state, and such that for any state there exists a path leading to a final state. A sequential function f is a function which can be represented by a ST. Namely, if f is represented by $T = (V, i, F, A, B^*, \delta, \sigma)$, then for any $w \in A^*$ such that $\delta(i, w) \in F$, $f(w) = \sigma(i, w)$. We denote by $Dom(f)$ the set of words w for which f is defined.

3.2 Theorems and computation

For any sequential function f one can define the following relation on A^*:

$$\forall(u, v) \in A^*, \ u R_f v \iff \exists(u', v') \in B^* \times B^* /$$
$$\forall w \in A^*, \ uw \in Dom(f) \Leftrightarrow vw \in Dom(f),$$
$$\text{and,} \quad uw \in Dom(f) \Rightarrow u'^{-1}f(uw) = v'^{-1}f(vw).$$

It is easy to show that R_f is an equivalence relation. The following lemma shows that if there exists a ST T computing f with a number of states equal to the number of equivalence classes of R_f, then T is a minimal transducer computing f.

Lemma 1. *If f is a sequential function, R_f has a finite number of equivalence classes. This number is inferior or equal to the number of states of any ST computing f.*

Proof. Let $T = (V, i, F, A, B^*, \delta, \sigma)$ be a ST computing f. Choosing $u' = \sigma(i, u)$ and $v' = \sigma(i, v)$ in the above relation allows to show easily that:

$$\forall(u, v) \in (A^*)^2, \ \delta(i, u) = \delta(i, v) \Rightarrow u R_f v.$$

$\delta(i, u) = \delta(i, v)$ also defines an equivalence relation on A^*. Hence, the number of equivalence classes of this relation, namely the number of states of T, is superior or equal to the number of classes of R_f. This proves the lemma.

In the following, we define for any sequential function f a ST whose number of states is equal to the number of equivalence classes of R_f.

Theorem 1 *For any sequential function f, there exists a minimal ST computing it. Its number of states is equal to the number of equivalence classes of R_f.*

Proof. Let f be a sequential function. Let g be the function mapping A^* to B^* defined as follows:

$$\forall u \in A^*, g(u) = \bigwedge_{\substack{w \in A^* \\ uw \in Dom(f)}} f(uw).$$

We can define a transducer $T = (V, i, F, A, B^*, \delta, \sigma)$ by the following relations[8]:
- $V = \{\overline{u} | u \in A^*\}$;
- $i = \overline{\epsilon}$;
- $F = \{\overline{u} | u \in A^* \cap Dom(f)\}$
- $\forall u \in A^*, \forall a \in A, \delta(\overline{u}, a) = \overline{ua}$;
- $\forall u \in A^*, \forall a \in A, \sigma(\overline{u}, a) = [g(u)]^{-1}g(ua)$.

The definitions of V and F are correct, as, according to the previous lemma the number of equivalence states of R_f is finite. To show that δ and σ are correctly defined we need to prove that their definitions do not depend on the choice of the element u in $\overline{u}$. This is clearly true in the case of δ, as, if u and v belong to the same class, then ua and va are also equivalent for the relation R_f. The expression defining $\sigma(\overline{u}, a)$ is well-formed since, by definition[9]:

$$\forall w \in A^* / \ uw \in Dom(f), \quad g(u) \leq_p f(uw)$$
$$\Longrightarrow \forall w \in A^* / u(aw) \in Dom(f), g(u) \leq_p f(u(aw)) \Rightarrow g(u) \leq_p g(ua).$$

If u and v are equivalent, according to the definition of R_f we have:

$$\exists(u', v') \in B^* \times B^* / \forall w \in A^*,$$
$$uw \in Dom(f) \Leftrightarrow vw \in Dom(f),$$
$$\text{and, } uw \in Dom(f) \Rightarrow u'^{-1}f(uw) = v'^{-1}f(vw),$$

and: $\quad u(aw) \in Dom(f) \Leftrightarrow v(aw) \in Dom(f),$
$\quad\quad$ and, $u(aw) \in Dom(f) \Rightarrow u'^{-1}f(u(aw)) = v'^{-1}f(v(aw)).$

Considering the greatest common prefix of each member of the above identities leads to:

$$u'^{-1}g(u) = v'^{-1}g(v) \text{ and } u'^{-1}g(ua) = v'^{-1}g(va),$$
hence: $[g(u)]^{-1}g(ua) = [u'v'^{-1}g(v)]^{-1}u'v'^{-1}g(va) = [g(v)]^{-1}g(va).$

Therefore, the definition of the output function is correct. The set of words recognized by the left automaton of T is exactly $Dom(f)$:

$$\forall u \in A^*, u \in Dom(f) \Leftrightarrow \overline{u} \in F \Leftrightarrow \delta(i, u) \in F.$$

Also, notice that:

$$\forall(a, b) \in (A^*)^2, \sigma(\overline{\epsilon}, ab) = \sigma(\overline{\epsilon}, a)\sigma(\overline{a}, b) = [g(\epsilon)]^{-1}g(a)[g(a)]^{-1}g(ab) = g(ab).$$

---

[8] We denote by $\overline{u}$ the equivalence class of u. In order to shorten this presentation we shall assume in the following without reducing the generality of the theorem that $g(\epsilon) = \epsilon$.

[9] The notation $u \leq_p v$ means that u is a prefix of v.

A recursive application of these identities leads to: $\forall u \in A^*$, $\sigma(\bar{\epsilon}, u) = g(u)$.
Now, if $u \in Dom(f)$:

$$g(u) = \bigwedge_{\substack{w \in A^* \\ uw \in Dom(f)}} f(uw) \leq_p f(u).$$

Since f is sequential, we also have:

$$f(u) \leq_p g(u),$$

hence:

$$u \in Dom(f) \Rightarrow \sigma(\bar{\epsilon}, u) = g(u) = f(u).$$

Thus, T is a ST representing f which has the minimal number of states. This proves the theorem.

Considered as an $A \times B'$-automaton, where $B' \subseteq B^*$ is the set of its output labels, a ST $T = (V, i, F, A, B^*, \delta, \sigma)$ can be minimized. However, the corresponding algorithm ([1]) does not necessarily lead to a minimal transducer. We here prove that once the algorithm described in the previous section has been applied to the output automaton of T, namely to the automaton which has the same states and edges as T and whose labels are the output labels of T, the minimization of a ST in the sense of automata leads to the minimal transducer as defined above.

Given a ST $T = (V, i, F, A, B^*, \delta, \sigma)$, the application of the PRE-FIX_COMPUTATION algorithm to the output automaton of T has no effect on the states of T, nor on its transition function. Only its output function σ is changed. We can denote by $T_2 = (V, i, F, A, B^*, \delta, \sigma_2)$ the resulting transducer. Let P be the function which maps V to B^* defined by the following recursive relations:

$$\forall s \in F, \qquad P(s) = \epsilon;$$
$$\forall s \in V - F, P(s) = \bigwedge_{a \in A} \sigma(s, a) \, P(\delta(s, a)).$$

$P(s)$ is thus the greatest common prefix of the outputs of all words accepted by the left automaton of T when read from the state s. In order to simplify this presentation we shall assume that $P(i) = \epsilon$ [10]. According to the previous section, σ_2 is defined by:

$$\forall s \in V, \quad \sigma_2(s, a) = [P(s)]^{-1} \, \sigma(s, a) \, P(\delta(s, a)),$$
and, $\forall u \in A^*$, $\sigma_2(i, u) = \sigma(i, u) \, P(\delta(i, u))$. [11]

If $\delta(i, u) \in F$, we have: $\sigma_2(i, u) = \sigma(i, u)$. Therefore, T and T_2 compute the same sequential function. Let $T_3 = (V_3, i_3, F_3, A, B^*, \delta_3, \sigma_3)$ be the ST obtained from T_2 by minimization in the sense of automata. This operation can be performed

---

[10] This is equivalent to the assumption made above: $g(\epsilon) = \epsilon$.

[11] Notice that this is equivalent to: for any u in A^*, $\sigma_2(i, u) = g(u)$.

by merging the equivalent states of T_2 considered as an $A \times B'$-automaton, where $B' \subseteq B^*$ is the set of all output labels of T_2. Two states s_1 and s_2 of T_2 are equivalent in this sense iff:

$\forall w = w_1...w_n \in A^*$,
$\delta(s_1, w) \in F \Leftrightarrow \delta(s_2, w) \in F$ and $(w_0 = \epsilon)$:
$\delta(s_1, w) \in F \Rightarrow \forall i \in [0, n-1], \sigma_2(\delta(s_1, w_0...w_i), w_{i+1}) = \sigma_2(\delta(s_2, w_0...w_i), w_{i+1})$.

Let f be the sequential function computed by T_1 and T_2. The following lemma helps to prove that T_3 is the minimal ST computing f as defined in the previous section.

Lemma 2. *For any (u, v) in $(A^*)^2$, if $\overline{u} = \overline{v}$, then the states $\delta(i, u)$ and $\delta(i, v)$ are equivalent.*

Proof. Let (u, v) be in $(A^*)^2$, $s_1 = \delta(i, u)$ and $s_2 = \delta(i, v)$. It is easy to show that $\overline{u} = \overline{v}$ implies that:

$$\forall w \in A^*, \delta(s_1, w) \in F \Rightarrow \sigma_2(s_1, w) = \sigma_2(s_2, w).$$

Since for any i in $[0, n-1]$, $u \, R_f \, v$ implies $uw_0...w_i \, R_f \, vw_0...w_i$, for any $w = w_1...w_n \in A^*$ such that $\delta(s_1, w) \in F$, we also have:

$$\forall i \in [0, n-1], \, \sigma_2(\delta(s_1, w_0...w_i), w_{i+1}...w_n) = \sigma_2(\delta(s_2, w_0...w_i), w_{i+1}...w_n),$$

which is equivalent to:

$$\forall i \in [0, n-1], \, \sigma_2(\delta(s_1, w_0...w_i), w_{i+1}) = \sigma_2(\delta(s_2, w_0...w_i), w_{i+1}).$$

Therefore, s_1 and s_2 are equivalent states. This ends the proof of the lemma.

Now, since T_3 is obtained by merging the equivalent states of T_2 the lemma is equivalent to:

$$\forall(u, v) \in A^*, u R_f v \Rightarrow \delta_3(i_3, u) = \delta_3(i_3, v).$$

This implies that the number of states of T_3 is inferior or equal to the number of equivalence classes of R_f. As T_3 is a ST which computes the same function as T_2, by lemma 1 we prove that these two numbers are equal. T_3 is a minimal ST computing f, its states can be identified with the equivalence classes of R_f, and it is easy to show that it is exactly the minimal transducer as defined in the previous section. This proves the following theorem.

Theorem 2 *Given a ST T, a minimal ST computing the same function as T can be obtained by applying the PREFIX_COMPUTATION algorithm to the output automaton of T, and the minimization in the sense of automata to the resulting transducer. This minimal ST is the one defined in the previous section.*

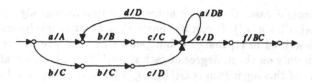

Fig. 2. Transducer T.

Figures 2-4 illustrate the minimization of sequential transducers in a particular case. Consider the ST T represented in figure 2. This transducer is minimal in the sense of automata. Still, it can be minimized following the process described above.

The application of the PREFIX_COMPUTATION algorithm leads to the transducer T_2 (figure 3) which computes the same function. Only outputs differ from those of T.

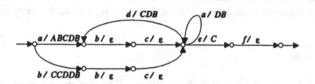

Fig. 3. Transducer T_2.

This ST is not minimal in the sense of automata. The corresponding minimization leads to the reduced transducer represented in figure 4 which is the minimal ST as defined previously.

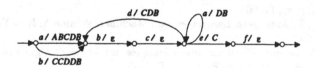

Fig. 4. Transducer T_3.

3.3 Complexity

In the case of acyclic transducers one may use a specific minimization algorithm for automata ([6]) which runs in linear time. Therefore, in this case, the whole process of minimization of a ST T can be done in $O(|V| + |E| + (|E| - (|V| - |F|)).|P_{max}|)$ steps, where P_{max} is the longest of the greatest common prefixes of the output paths leaving each state of T.

In the general case, the classic automata minimization algorithm (see [1]) runs in $O(|A|.|V|.\log|V|)$. It can be shown that a better implementation of the algorithm described in [1] makes it independent of the size of the alphabet. It then depends only on the in-degree of each state. Thus, a better evaluation of the running time of this algorithm is $O(|E|.\log|V|)$. And, the general minimization of sequential transducers runs in $O(|V| + |E|.(\log|V| + |P_{max}|))$.

4 Conclusion

The algorithm described above can obviously be also applied to subsequential transducers[12]. Indeed, any subsequential transducer $T = (V, i, F, A, B^*, \delta, \sigma, \varphi)$ can be transformed into a ST simply by adding a new special symbol \$ to the alphabet A, a new state to V being the single final state of T, and by replacing φ by transitions to this final state labeled by \$ and the value of φ. Several implementations of this algorithm in order to minimize the phonetic subsequential transducer of French, namely the transducer which associates with words of French their phonemic pronunciations, have proved it to be very efficient.

References

1. Aho, Alfred; John Hopcroft; Jeffrey Ullman. The design and analysis of computer algorithms. Reading, Mass.: Addison Wesley, (1974).
2. Berstel, Jean. Transductions and Context-Free Languages. Stuttgart: B.G.Teubner, (1979).
3. Choffrut, Christian.Contributions à l'étude de quelques familles remarquables de fonctions rationnelles. Université Paris 7, thèse de doctorat d'Etat, Paris: LITP, (1978).
4. Cormen, Thomas H.; Charles E. Leiserson; Ronald L. Rivest. Introduction to Algorithms. 2nd edition, Cambridge, Mass.: The MIT Press, New York: MacGraw-Hill Book Company, (1990).
5. Eilenberg, S.. Automata, Languages, and Machines. Volume A, New York: Academic Press, (1974).
6. Revuz, Dominique. Dictionnaires et lexiques, méthodes et algorithmes. Université Paris 7, thèse de doctorat, Paris: LITP, (1991).
7. Sedgewick, Bob. Algorithms. 2nd edition, Reading, Mass.: Addison Wesley, (1988).

---

[12] Subsequential transducers can be represented by an 8-tuple $(V, i, F, A, B^*, \delta, \sigma, \varphi)$, where $(V, i, F, A, B^*, \delta, \sigma)$ is a ST and φ a final function which maps F to B^* ([2]).

SHORTEST COMMON SUPERSTRINGS
FOR STRINGS OF RANDOM LETTERS

Kenneth S. Alexander[1]

Department of Mathematics
University of Southern California
Los Angeles, CA 90089-1113

Abstract. Given a finite collection of strings of letters from a fixed alphabet, it is of interest, in the contexts of data compression and DNA sequencing, to find the length of the shortest string which contains each of the given strings as a consecutive substring. In order to analyze the average behavior of the optimal superstring length, substrings with a specified collection of lengths are considered with the letters selected independently at random. An asymptotic expression, as the collection of lengths becomes large, is obtained for the savings from compression, that is, the difference between the uncompressed (concatenated) length and the optimal superstring length.

1 Introduction and Statement of Results

Given a finite alphabet $A = \{a_1, .., a_d\}$ and a collection $\{s_1, .., s_n\}$ of strings of letters with lengths $l_1, .., l_n$, one may wish to find a single string s which contains each s^i as a substring. (By a substring we always mean a consecutive one, i.e. $x_i x_{i+1}..x_j$ is a substring of $x_1..x_n$ for each $1 \leq i \leq j \leq n$. The substring is called *non-initial* if $i \geq 2$.) In the context of data transmission, compression can be achieved by transmitting only s and a label for the starting point of each s^i in s. Algorithms for such compression have been an object of considerable study in computer science; see [3], [5], [6] and the references therein. In the context of DNA sequencing, the search for such s may represent the idea that the s^i actually are overlapping substrings of an unknown s; see [4] and [5].

A string s, as above, of minimal length is called a *shortest common superstring* of $\{s^1, .., s^n\}$. The question we wish to discuss is, if the lengths $l_1, .., l_n$ are specified, and the letters of each s^i are then chosen independently according to some law μ on A, what will be the typical length of the shortest common superstring? In particular, how does it compare to the uncompressed, *i.e.* concatenated, length $l_1 + .. + l_n$? We will describe some heuristics and sketch the proof of a theorem; the full proof can be found in [1].

In the context of data compression, variants of our model are potentially of interest in the analysis of average behavior of algorithms. For DNA sequencing the independence assumption for distinct strings s^i is unrealistic, but the general problem of finding an "optimal superstring" for a set of random strings is

---
[1] Research supported by NSF grant DMS-9206139

important for the reconstruction of DNA sequences from fragments. In short, while our model does not describe a specific system of practical interest, it is a natural starting point in a wider class of models of significant interest.

Because the string lengths, and the chosen strings, are not necessarily distinct, it is convenient to represent each by a multiset, *i.e.* a set in which each element has a multiplicity. Suppose that for each $n \geq 1, \mathcal{L}_n$ is a multiset of positive integers representing the lengths of the strings to be selected, and $\mathcal{S}_n$ is the corresponding multiset of strings with independent random letters. The basic assumptions throughout will be:

(A1) $|\mathcal{L}_n| \to \infty$ as $n \to \infty$,
(A2) $\max \mathcal{L}_n \to \infty$ as $n \to \infty$.

Here $|C|$ denotes the cardinality of a multiset C, counted with multiplicities. If (A2) fails, then the string lengths remain bounded, at least on a subsequence, as $n \to \infty$, which makes the problem relatively trivial, so (A2) is not a strong assumption. Without (A1), one would not expect to be able to make a deterministic approximation for the length saved by compression.

Let

$$p_i := \mu(a_i)$$

and denote the entropy of μ by

$$H_\mu := -\sum_{i=1}^{d} p_i \log p_i.$$

It will be necessary to divide $\mathcal{S}_n$, and correspondingly $\mathcal{L}_n$, into short and long strings and string lengths. Let

$$t_n(k) := \sum_{l \in \mathcal{L}_n : l > k} (l - k)$$

be the number of possible starting points for a length-k string as an non-initial substring of a longer string in $\mathcal{S}_n$. Here and in all sums indexed by multisets, each summand is summed with its multiplicity. Let

$$k_n := \min\{k \in \mathcal{L}_n : t_n(k) < \exp(k H_\mu)\}.$$

Note that by (A2),

$$k_n \to \infty \text{ as } n \to \infty. \tag{1}$$

Let $|s|$ denote the length of a string s. We will call a string s, or its length,

short if $|s| < k_n$,
long if $|s| > k_n$.

For "nice" choices of $\mathcal{L}_n$ one can also consider s with $|s| = k_n$ to be long, but complications develop if there are too many such borderline strings, specifically if

$$|\{s \in \mathcal{S}_n : |s| = k_n\}| > \exp(k_n H_\mu)$$

by a large margin, so we must do the following: let $\mathcal{K}_n$ be a deterministically chosen subset of $\{s \in \mathcal{S}_n : |s| = k_n\}$ with

$$\mathcal{K}_n = \{s \in \mathcal{S}_n : |s| = k_n\} \quad \text{if } |\{s \in \mathcal{S}_n : |s| = k_n\}| \leq \exp(k_n H_\mu)$$
$$|\mathcal{K}_n| = \lfloor \exp(k_n H_\mu) \rfloor \quad \text{if } |\{s \in \mathcal{S}_n : |s| = k_n\}| > \exp(k_n H_\mu).$$

Here $\lfloor . \rfloor$ denotes the integer part, and by "deterministically chosen" we mean chosen without regard to the values of the random strings. We then say s is

short if $|s| = k_n$ and $s \notin \mathcal{K}_n$,
long if $|s| = k_n$ and $s \in \mathcal{K}_n$.

Let $\mathcal{S}_n^{(s)}$ and $\mathcal{S}_n^{(l)}$ denote the multisets of short and long strings respectively in $\mathcal{S}_n$, and $\mathcal{L}_n^{(s)}$ and $\mathcal{L}_n^{(l)}$ similarly for the lengths. Note that $\mathcal{S}_n^{(l)}$ is always nonempty.

Remark. 1 We can motivate our choice of short/long boundary with the following heuristic. Suppose $s \in \mathcal{S}_n$ is a string of length k. Will s, we may ask, likely be a non-initial substring of some longer string $s' \in \mathcal{S}_n$? The number of possible starting points for s in some s' is $|s'| - k$. With very high probability, s is a string for which the probability of occurrence is approximately $\exp(-H_\mu k)$. Therefore we can expect to find such an s' provided the expected number of occurrences of s inside an s', which is roughly $t_n(k) \exp(-H_\mu k)$, is at least of order 1. The short/long boundary is the largest k for which this is true. Thus short strings are those which are likely to be non-initial substrings of some longer string in $\mathcal{S}_n$. ∎

Let L_n denote the length of the shortest common superstring of $\mathcal{S}_n$. Let

$$V_n := \left(\sum_{l \in \mathcal{L}_n} l \right) - L_n$$

be the savings created by compression, and

$$j_n := |\mathcal{L}_n^{(l)}|$$
$$\tau_n([a, b]) := \sum_{l \in \mathcal{L}_n : a \leq l \leq b} l, \quad \text{for } a \leq b,$$
$$v_n := j_n (\log j_n)/H_\mu + \sum_{l \in \mathcal{L}_n^{(s)}} l.$$

We can now state our main result.

Theorem 1. *Suppose (A1) and (A2) hold. If*

$$\lim_{\epsilon \to 0} \lim_{n \to \infty} \sup \tau_n([(1-\epsilon)k_n, (1+\epsilon)k_n])/v_n = 0, \tag{2}$$

then

$$E|V_n/v_n - 1| \to 0 \quad as \ n \to \infty. \tag{3}$$

The need for a technical hypothesis such as (2), which says not too many strings are near the short/long borderline, is illustrated by the following example.

Example 1. Suppose $\mathcal{L}_n$ consists of the value 1 with multiplicity n, the value n with multiplicity 1, and a third value l_n with multiplicity 1. If n is large then l_n can be chosen so that the probability that the length-n string is a substring of the length-l_n string is between, say, .49 and .51; $\log l_n$ will be of order n. Then V_n is either near n or near $2n$, with probability near $1/2$ each, so there is no choice of v_n such that $E|V_n/v_n - 1| \to 0$, though (A1) and (A2) hold. ∎

One can obtain upper and lower bounds for V_n without using (2), then show that (2) is a sufficient condition for these bounds to be close to one another, and thereby obtain (3). Other sufficient conditions can be easily formulated; the main point is that the savings V_n should not be overly affected by the values of a few strings with lengths near the short/long boundary.

Remark. 2 In view of Remark 1, a heuristic for approximating V_n by v_n is as follows. With high probability, nearly every short string in $\mathcal{L}_n$ is contained in some long string, so the full length of nearly all short strings is saved by compression, while one can order the long strings in $\mathcal{L}_n$ so that each overlaps with its neighbors by roughly $(\log j_n)/H_\mu$.

This heuristic ignores the possibility of containment of long strings inside yet longer strings, yet it leads to a correct approximation v_n for V_n, at least under (2). This suggests that an algorithm which checks for such containment might not gain much over one that checks only for overlapping ends. ∎

Remark. 3 Let $q := \sum_{i=1}^{d} p_i^2$ be the probability that two independently selected letters match. An alternate heuristic for the choice of a largest k in Remark 1 is that the probability of matching s with a particular non-initial substring of another string s' is q^k, so the maximal k should be such that the expected number $t_n(k)q^k$ of substrings for which such a match occurs is of order 1. Except in the uniform case, q is larger than $\exp(-H_\mu)$ (see e.g. [2]), so the two heuristics are not compatible. Theorem 1 shows that the heuristic in Remark 1 is the correct one. The fault with the alternate heuristic is that if one considers k-letter substrings of two other strings s' and s'' (possibly $s' = s''$, in which case we consider nonoverlapping substrings), the events that a match with the fixed s occurs for s' and for s'' are highly correlated. The correlation occurs because a match of s with a substring of s' increases the probability that s is a string of relatively high probability (greater than $\exp(-H_\mu k)$) which makes it easier for s to also match the substring of s''. This means making the expected number of substrings for which a match occurs be one is not nearly the same as making the probability of at least one matching substring be non-negligible. Such a principle is also observed in other contexts involving matching; see [2]. ∎

Remark. 4 Instead of assuming that the letters in each word are chosen independently, we may consider the case that the words in S_n are truncated independent realizations of an ergodic A-valued stationary process with positive entropy, such as an irreducible aperiodic A-valued Markov chain in its stationary distribution. In other words, we have independent realizations $X^i = (\ldots, X^i_{-1}, X^i_0, X^i_1, \ldots), i \leq |\mathcal{L}_n|$, of a stationary process $X = (\ldots, X_{-1}, X_0, X_1, \ldots)$, and the ith word, of length l_i, in S_n is $X^i_1 \cdots X^i_{l_i}$. To obtain the analog of Theorem 1, one needs as an additional hypothesis some sort of mixing condition, to ensure roughly the following. Given several widely spaced locations in some long string, and given a short string, consider the event for each location that the short string is a substring of the long one starting at that location; these events should be nearly independent. More precisely, define $\mathcal{F}_{\leq n} := \sigma(\ldots, X_{n-1}, X_n)$ and $\mathcal{F}_{\geq n} := \sigma(X_n, X_{n+1}, \ldots)$. Then Theorem 1 remains valid under the additional assumption that for some $d_n = \exp(o(k_n))$,

$$\lim_{n \to \infty} \inf \inf \{ P(C|B)/P(C):$$
$$B \in \mathcal{F}_{\leq 0}, C \in \mathcal{F}_{\geq d_n}, P(C) \geq \exp(-2k_n H_\mu) \} > 0. \tag{4}$$

The proof remains the same, except for (2.18) and (2.19) of [1]. It is easily checked using (1) that (4) holds for every irreducible aperiodic A-valued Markov chain. ∎

Theorem 1 gives an approximate expression v_n for the savings V_n from compression. It would perhaps be more desirable to have an approximate expression for the compressed length L_n. This can of course be obtained directly from Theorem 1 when compression does not save nearly the full concatenated length, that is, when $v_n/(\sum_{l \in \mathcal{L}_n} l)$ is not near 1. When $v_n/(\sum_{l \in \mathcal{L}_n} l)$ is near 1, or equivalently $L_n = o(\sum_{l \in \mathcal{L}_n} l)$, the situation is more complicated. Consecutive strings from S_n in the shortest common superstring generally overlap almost completely, and one would have to estimate the total of the lengths of the small nonoverlapping parts.

The achievable savings from compression are most dramatic, naturally, when the entropy is small, as the following example illustrates.

Example 2. Suppose S_n contains 4 words of each length from 1 to 100. The concatenated length is then 20,200.

If the alphabet $A = \{0, 1\}, \mu(0) = .95$, and $\mu(1) = .05$, then $H_\mu = .1985, k_n = 45, j_n = 224$, and $(\log j_n)/H_\mu = 27.3$ so $v_n = 10,067$. Thus the shortest common superstring is about 50% of the length of the concatenated superstring. Strings of length 44 or less are typically substrings of longer strings, while strings of length 45 or more have an optimum overlap of around 27 letters.

If instead $\mu(0) = \mu(1) = .5$, then $H_\mu = .6931, k_n = 14, j_n = 348$, and $(\log j_n)/H_\mu = 8.4$, so $v_n = 3302$. The shortest common superstring is then about 84% of the length of the concatenated superstring.

If $A = \{a, c, t, g\}$ and $\mu(a) = \mu(c) = \mu(t) = \mu(g) = .25$, then $H_\mu = 1.3863, k_n = 8, j_n = 372$, and $(\log j_n)/H_\mu = 4.3$, so $v_n = 1700$. The shortest common superstring is then about 92% of the length of the concatenated superstring. ∎

It is possible to apply Theorem 1 to situations where $\mathcal{L}_n$ itself is random, independent of the choice of all letters. If (A1), (A2) and (2) all hold in probability, then Theorem 1 ensures that $V_n/v_n \to 1$ in probability.

We write $a_n \sim b_n$ to denote $a_n/b_n \to 1$ as $n \to \infty$.

Example 3. Suppose $\mathcal{L}_n$ consists of m_n independent geometric$(1/n)$ random values, with $m_n \to \infty$. It is readily shown that $t_n(k) \sim nm_n \exp(-k/n)$ in probability, uniformly in $k \le c \log nm_n$, for all $c > 0$, and then that $k_n \sim (\log nm_n)/H_\mu$ in probability. If $m_n \sim e^{\alpha n}$ for some $\alpha > 0$, it follows easily that $v_n \sim nm_n(1 - e^{-\alpha/H_\mu})$ in probability, as compared to the concatenated length of approximately nm_n. Thus the shortest common superstring has length $L_n \sim nm_n e^{-\alpha/H_\mu}$ in probability.

By contrast, if $\mathcal{L}_n$ consists of $m_n \sim e^{\alpha n}$ strings of nonrandom length n, it is easily checked that $L_n \sim nm_n(1 - \alpha/H_\mu)$ if $\alpha < H_\mu$, and $L_n = o(nm_n)$ if $\alpha \ge H_\mu$. ∎

2 Proof Sketch

In this section we sketch the proof of Theorem 1. Let $-1 < \epsilon < 1$; we first define $k_n(\epsilon), \mathcal{L}_n^{(l)}(\epsilon), j_n(\epsilon), v_n(\epsilon)$, etc., corresponding to $k_n, \mathcal{L}_n^{(l)}, j_n, v_n$, etc., roughly by replacing H_μ with $(1 + \epsilon)H_\mu$ in all definitions. It is easily shown using (2) that

$$\limsup_{\epsilon \to 0} \limsup_{n \to \infty} |v_n(\epsilon) - v_n(0)|/v_n(0) = 0,$$

so to prove Theorem 1 it is sufficient to show that for every $-1 < \epsilon < 0$,

$$E\left[V_n - v_n(\epsilon)\right]^+ = o(v_n(\epsilon)) \text{ as } n \to \infty \tag{5}$$

and for every $0 < \epsilon < 1$,

$$E\left[V_n - v_n(\epsilon)\right]^- = o(v_n(\epsilon)) \text{ as } n \to \infty \tag{6}$$

where x^+ and x^- denote $\max(x, 0)$ and $\max(-x, 0)$ respectively.

We begin with (5); here we use $\epsilon < 0$. Let $W^{l,n}(\epsilon)$ be a shortest common superstring for the set $\mathcal{S}_n^{(l)}(\epsilon)$ of long strings. Before forming $W^{l,n}(\epsilon)$ we can discard any string in $\mathcal{S}_n^{(l)}(\epsilon)$ which is a non-initial substring of some other string in $\mathcal{S}_n^{(l)}(\epsilon)$. We can view $W^{l,n}(\epsilon)$ as containing the undiscarded long strings in some order, with some overlap between the beginning of each undiscarded random long string W and the end of the string formed by its predecessors. We call the string that forms this overlap the *overlap string* of W. The compression savings for long strings only, denoted $V_n^{(l)}(\epsilon)$, is then bounded above by the total length of the discarded strings plus the total length of the overlap strings. Let $u_n(\epsilon) := \lfloor (1 - \epsilon) \log j_n(\epsilon)/H_\mu \rfloor + 1$, where $\lfloor \cdot \rfloor$ denotes the integer part. It is a standard fact from ergodic theory that for $\alpha < \exp(-H_\mu) < \beta$, there are constants $c_m(\alpha, \beta, \mu)$ such that

$$P\left(\{s \in A^k : P(s) < \alpha^k\}\right) \leq c_1 \exp(-c_2 k)$$

and

$$P\left(\{s \in A^k : P(s) > \beta^k\}\right) \leq c_1 \exp(-c_2 k).$$

In considering possible overlap strings of length k, this enables one to restrict attention to strings s for which the probability $P(s)$ is approximately $\exp(-H_\mu k)$. Let I_W denote the length of the overlap string of a given discarded long random string W. Then I_W is bounded by the maximum overlap length M_W between the beginning of W and the end of some other long string. Conditionally on the first k letters of W forming a particular string s, the probability that $M_W = k$ is at most $j_n(\epsilon)P(s)$, which decreases exponentially in $k - u_n(\epsilon)$ provided $P(s)$ is approximately $\exp(-H_\mu k)$. Using these facts, it is not hard to show that $E[I_W - u_n(\epsilon)]^+ = O(1)$. Summing over the at most $j_n(\epsilon)$ choices of W then yields

$$E\left[\sum_W I_W - j_n(\epsilon)u_n(\epsilon)\right]^+ = O(j_n(\epsilon)) = o(v_n(\epsilon)) \text{ as } n \to \infty. \tag{7}$$

For a long random string W of length k, given $W = s$, the length of W multiplied by the probability that W is discarded decreases exponentially in $k - u_n(\epsilon)$, provided $P(s)$ is approximately $\exp(-H_\mu k)$. From this it follows readily that the expected total length of all discarded long strings is $O(j_n(\epsilon))$. Combining this with (7) proves (5).

We next turn to (6); here we use $\epsilon > 0$. It is not hard to show that a given random short string is very likely to appear as a non-initial substring of one of the random long strings; in fact the expected sum of the lengths of the short strings which do not so appear is $o(v_n(\epsilon))$. Therefore to prove (6) it suffices to show that the compression savings for only long strings satisfies

$$EV_n^{(l)}(\epsilon) \geq (1 - o(1))j_n(\epsilon)u_n(\epsilon) \text{ as } n \to \infty. \tag{8}$$

Let us call a random string X a *partner* of a random string W if the final $u_n(\epsilon)$ letters of W coincide with the initial $u_n(\epsilon)$ letters of X. We can form a common superstring of $\mathcal{S}_n^{(l)}(\epsilon)$ by choosing an initial string in $\mathcal{S}_n^{(l)}(\epsilon)$, then choosing a successor $\Psi(W)$ for each string $W \in \mathcal{S}_n^{(l)}(\epsilon)$, then overlapping the final $u_n(\epsilon)$ letters of W with the initial $u_n(\epsilon)$ letters of $\Psi(W)$ whenever $\Psi(W)$ is a partner of W; this yields a common superstring provided the function Ψ is a permutation of $S_n^{(l)}(\epsilon)$ which consists of a single cycle. To obtain (8) Ψ must be chosen in such a way that most of the random strings have a partner string as successor. To obtain Ψ we begin by constructing a permutation φ which consists of a small number of cycles. To do this, first fix a value r which is a small power of $j_n(\epsilon)$. Then divide the long random strings into r equal-sized groups arbitrarily; for convenience we assume here that $j_n(\epsilon)$ is a multiple of r. For a given string s with length $u_n(\epsilon)$ and probability $P(s)$ near $\exp(-H_\mu u_n(\epsilon))$, for each m the number of random strings in group m ending with s and then number of random strings in group $m+1$ beginning with s are with high probability both large (thanks to the factor of $1 - \epsilon$ in the definition of $u_n(\epsilon)$) and approximately equal. Therefore one can choose a successor $\varphi(W)$ which is a partner of W in group $m+1$, for all but a small fraction of strings W in group m, choosing distinct successors for distinct random strings. (Here the number m or $m + 1$ of the group should be taken *mod r*.) The few "leftover strings" which are not assigned a partner as successor in this manner are then assigned arbitrary successors, with the restriction that φ be kept one-to-one, since φ must be a permutation. Any cycle of φ which contains no leftover random strings must have length which is a multiple of r; this is the purpose of the division into r groups. It follows that if there are Y_n leftover random strings, the number of cycles is φ is at most $Y_n + j_n(\epsilon)/r$. We then obtain Ψ by breaking each cycle of φ at an arbitrary point, then stringing the broken cycles together into a single cycle. The number of random strings W for which $\Psi(W)$ is not a partner of W is at most Y_n plus the number of cycles in φ, hence is at most $2Y_n + j_n(\epsilon)/r$. The expected value of this upper bound is $o(j_n(\epsilon))$ provided $j_n(\epsilon) \to \infty$; the case of bounded $j_n(\epsilon)$ is easily handled separately. From this we obtain (8), and (6) follows, completing the proof.

The author would like to thank M. Waterman and G. Benson for helpful discussions; M. Waterman in particular suggested considering problems of this type.

References

1. Alexander, K.S. Shortest common superstrings of random strings. (1993) Preprint.

2. Arratia, R. and Waterman, M.S. Critical phenomena in sequence matching. Ann. Probability 13 (1985) 1236-1249.

3. Blum, A., Jiang, T., Li, M., Tromp, J. and Yannakakis, M. Linear approximation of shortest superstrings. Proc. 23rd ACM Symp. on Theory of Computing, (1991) 328-336.

4. Peltola, H., Söderlund, H., Tarhio, J. and Ukkonen, E. Algorithms for some string matching problems arising in molecular genetics. In: Information Processing 83

(Proc. of the IFIP Congress 1983), R.E.A. Mason, ed. North-Holland, Amsterdam, (1983) 53-64.

5. Tarhio, J. and Ukkonen, E. A greedy approximation algorithm for constructing shortest common superstrings. Theor. Comp. Sci. 57 (1986) 131-145.

6. Turner, J. Approximation algorithms for the shortest common superstring problem. Information and Computation 83 (1989) 1-20.

Maximal common subsequences and minimal common supersequences

Robert W. Irving and Campbell B. Fraser*

Computing Science Department
University of Glasgow
Glasgow G12 8QQ
Scotland

Abstract. The problems of finding a longest common subsequence and a shortest common supersequence of a set of strings are well-known. They can be solved in polynomial time for two strings (in fact the problems are dual in this case), or for any fixed number of strings, by dynamic programming. But both problems are NP-hard in general for an arbitrary number k of strings. Here we study the related problems of finding a minimum-length maximal common subsequence and a maximum-length minimal common supersequence. We describe dynamic programming algorithms for the case of two strings (for which case the problems are no longer dual), which can be extended to any fixed number of strings. We also show that the minimum maximal common subsequence problem is NP-hard in general for k strings, and we prove a strong negative approximability result for this problem. The complexity of the maximum minimal common supersequence problem for general k remains open, though we conjecture that it too is NP-hard.

Key words: string algorithms, subsequence, supersequence, dynamic programming, NP-hard optimisation problems, approximation algorithms.

1   Introduction

A *subsequence* of a string α is any string that can be obtained from α by the deletion of zero or more symbols. A *supersequence* of α is any string that can be obtained from α by the insertion of zero or more symbols. Given a set S of k strings, a *common subsequence* of S is a string that is a subsequence of every string in S and a *common supersequence* of S is a string that is a supersequence of every string in S.

A common subsequence α is *maximal* if no supersequence of α is also a common subsequence of S — in other words, if α is not contained in any longer common subsequence of S. A *minimum maximal common subsequence (minmax common subsequence)* of S is a maximal common subsequence of shortest possible

---

* Supported by a postgraduate research studentship from the Science and Engineering Research Council

length. Clearly, a maximal common subsequence of *greatest* possible length is just a longest common subsequence, a concept that has been widely explored in the literature.

A common supersequence α is *minimal* if no subsequence of α is also a common supersequence of S — in other words, if α does not contain any shorter common supersequence of S. A *maximum minimal common supersequence (maxmin common supersequence)* of S is a minimal common supersequence of longest possible length. Clearly, a minimal common supersequence of *shortest* possible length is just a shortest common supersequence.

Example If $\alpha_1 = abc$, $\alpha_2 = bca$, the maximal common subsequences are a, bc and the unique minmax common subsequence is a, of length 1.

The minimal common supersequences are $bcabc$, $abca$, $bacbac$, and the unique maxmin common supersequence is $bacbac$, of length 6.

In this paper, we study the minimum maximal common subsequence problem (MinMaxCSub) and and the maximum minimal common supersequence problem (MaxMinCSup) from the complexity point of view. We show that, like the well-known longest common subsequence (LCS) and shortest common supersequence (SCS) problems, both of these new problems can be solved in polynomial time by dynamic programming for $k = 2$ (and, by extending the algorithms, for any fixed value of k). However, the dynamic programming algorithms are not quite so straightforward as those for the LCS and SCS problems, and have complexities $O(m^2n)$ and $O(mn(m+n))$ respectively for strings of lengths m and n. Note that the existence of polynomial-time algorithms for the MinMaxCSub and MaxMinCSup problems in the case of two strings is by no means obvious. Consider the problem of finding a maximum cardinality matching in a bipartite graph. This problem is well known to be solvable in polynomial time, whereas the problem of finding a minimum maximal bipartite matching is NP-hard [16].

We also show that, as is the case for the LCS problem, the MinMaxCSub problem is NP-hard when the number of strings k becomes a problem parameter. Furthermore, we prove a strong negative result regarding the likely existence of good polynomial-time approximation algorithms for MinMaxCSub in the case of general k. We conjecture that similar hardness results apply to the MaxMinCSup problem.

2 The MinMaxCSub problem for two strings

When restricted to the case of just two strings α and β of lengths m and n respectively, the classical LCS and SCS problems are easily solvable in $O(mn)$ time by dynamic programming. Indeed, in this case, the LCS and SCS problems are dual, in that $s = m + n - l$, where l and s are the lengths of a longest common subsequence and shortest common supersequence respectively. Much effort has gone in to finding refinements of dynamic programming and other approaches which lead to improvements in complexity in many cases. See, for example, [1, 5, 6, 7, 14, 15].

The following simple example serves to illustrate the fact that there is no obvious corresponding duality between the MinMaxCSub and MaxMinCSup problems in the case of two strings.

Example Let $\alpha = abc$, $\beta = dab$. Then the only maximal common subsequence is ab of length 2, while $dabc$, $abcdab$ and $abdcab$ are some of the minimal common supersequences, the latter two being maximum.

It is true, however, that if γ is a maximal common subsequence of length r of α and β, then forming an alignment of α and β in which the elements of γ are matched reveals a minimal common supersequence of α and β of length $m + n - r$. Hence, if l' is the length of a minmax common subsequence, and s' the length of a maxmin common supersequence, it follows that $s' \geq m + n - l'$. That the inequality can be strict is shown by the above example.

Hence the question arises as to whether either or both of the MinMaxCSub and MaxMinCSup problems can be solved in polynomial time, by dynamic programming or otherwise.

In this section we describe a polynomial-time algorithm to determine the length of a minmax common subsequence of two strings, and in the following section a polynomial-time algorithm to determine the length of a maxmin common supersequence of two strings. It turns out that these algorithms will determine the lengths of all maximal common subsequences and all minimal common supersequences respectively. They will also allow the construction of a minmax common subsequence, and indeed of all the maximal common subsequences (respectively a maxmin common supersequence, and all minimal common supersequences) of the two strings.

The algorithms use a dynamic programming approach based, as usual, on a table that relates the ith prefix $\alpha^i = \alpha[1\ldots i]$ of α and the jth prefix $\beta^j = \beta[1\ldots j]$ of β, for $i = 1,\ldots,m$, $j = 1,\ldots,n$, where m, n are the lengths of α, β respectively. However, as we shall see, for each i, j we need retain rather more information than merely the lengths of the maximal common subsequences, or of the minimal common supersequences, of α^i and β^j.

The MinMaxCSub Algorithm

Given a string α and a subsequence γ of α, we define

$sp(\alpha, \gamma)$ = length of the shortest prefix of α that is a supersequence of γ.

Given strings α, β of lengths m and n respectively, we define the set S_{ij}, for each $i = 1,\ldots,m$, $j = 1,\ldots,n$, by

$S_{ij} = \{(r,(x,y)) : \alpha^i$ and β^j have a maximal common subsequence γ of length r, and $sp(\alpha,\gamma) = x, sp(\beta,\gamma) = y\}$
with

$$S_{00} = \{(0,(0,0))\}.$$

For string α of length m, position i and symbol a, we define

$$next_\alpha(i, a) = \begin{cases} \min\{k : \alpha[k] = a, k > i\} & \text{if such a } k \text{ exists} \\ m + 1 & \text{otherwise.} \end{cases}$$

If α is a string and a a symbol of the alphabet, we denote by $\alpha + a$ the string obtained by appending a to α. Likewise, if the last character of α is a, we denote by $\alpha - a$ the string obtained by deleting the final a from α.

The algorithm for MinMaxCSub is based on a dynamic programming scheme for the sets S_{ij} defined above. So evaluation of S_{mn} reveals the length of the minmax common subsequence, but also finds the lengths of all maximal common subsequences of α and β (indeed of all maximal common subsequences of all pairs of prefixes of α and β). Furthermore, the use of suitable tracebacks in the array of S_{ij} values, can be used to generate, not only a minmax common subsequence, but all maximal common subsequences.

The basis of the dynamic programming scheme is contained in the following theorem:

Theorem 1 *(i) If $\alpha[i] = \beta[j] = a$ then*

$$S_{ij} = \{(r, (next_\alpha(x, a), next_\beta(y, a))) : (r - 1, (x, y)) \in S_{i-1,j-1}\}.$$

(ii) If $\alpha[i] \neq \beta[j]$ then

$$S_{ij} = \{(r, (x, j) \in S_{i-1,j}\} \cup \{(r, (i, y)) \in S_{i,j-1}\} \cup (S_{i-1,j} \cap S_{i,j-1}).$$

Proof (i) Suppose $(r - 1, (x, y)) \in S_{i-1,j-1}$ and that γ' is a maximal common subsequence of α^{i-1} and β^{j-1} of length $r-1$ with $sp(\alpha, \gamma') = x$ and $sp(\beta, \gamma') = y$. Then it is immediate that $\gamma = \gamma' + a$ is a maximal common subsequence of α^i and β^j, and that $sp(\alpha, \gamma) = next_\alpha(x, a)$, $sp(\beta, \gamma) = next_\beta(y, a)$.

On the other hand, suppose that γ is a maximal common subsequence of length r of α^i and β^j. Then the last symbol of γ is a, and $\gamma' = \gamma - a$ is certainly a common subsequence of α^{i-1} and β^{j-1}. If it were not maximal, then some supersequence δ of γ' would be a common subsequence of α^{i-1} and β^{j-1}, and therefore $\delta + a$, a supersequence of γ, would be a common subsequence of α^i and β^j, contradicting the maximality of γ. So $(r - 1, (sp(\alpha, \gamma'), sp(\beta, \gamma')) \in S_{i-1,j-1}$ and $(r, (sp(\alpha, \gamma), sp(\beta, \gamma)) \in S_{ij}$ with $sp(\alpha, \gamma) = next_\alpha(sp(\alpha, \gamma'), a)$ and $sp(\beta, \gamma) = next_\beta(sp(\beta, \gamma'), a)$.

(ii) Suppose $(r, (x, j)) \in S_{i-1,j}$, and that γ is a maximal common subsequence of α^{i-1} and β^j of length r with $sp(\alpha, \gamma) = x$, $sp(\beta, \gamma) = j$. Then γ is a common subsequence of α^i and β^j, and must be maximal since $\gamma + \alpha[i]$ cannot be a subsequence of β^j. A similar argument holds for $(r, (i, y)) \in S_{i,j-1}$. So $\{(r, (x, j) \in S_{i-1,j}\} \cup \{(r, (i, y)) \in S_{i,j-1}\} \subseteq S_{ij}$.

Further, if $(r, (x, y)) \in S_{i-1,j} \cap S_{i,j-1}$, then there is a string γ with $sp(\alpha, \gamma) = x < i$, $sp(\beta, \gamma) = y < j$, of length r, which is a maximal common subsequence of α^i and β^{j-1}, and of α^{i-1} and β^j. So γ must also be a maximal common subsequence of α^i and β^j. For any supersequence of γ that is a subsequence of α^i and β^j must either be a subsequence of α^i and β^{j-1}, or of α^{i-1} and β^j.

On the other hand, suppose that γ is a maximal common subsequence of length r of α^i and β^j.

case (iia) $sp(\alpha, \gamma) = i$. Then γ is a maximal common subsequence of α^i and β^{j-1}, and so $(r, (i, y)) \in S_{i,j-1}$ for some y.

case (iib) $sp(\beta, \gamma) = j$. Then γ is a maximal common subsequence of α^{i-1} and β^j, and so $(r, (x, j)) \in S_{i-1,j}$ for some x.

case (iic) $sp(\alpha, \gamma) < i, sp(\beta, \gamma) < j$. Then γ is both a maximal common subsequence of α^{i-1} and β^j, and of α^i and β^{j-1}. So $(r, (sp(\alpha, \gamma), sp(\beta, \gamma))) \in S_{i-1,j} \cap S_{i,j-1}$.

This completes the proof of the theorem. $\square$

Recovering a Minmax Common Subsequence

The recovery of a particular minmax common subsequence involves a standard type of traceback through the dynamic programming table from cell (m, n), during which the sequence is constructed in reverse order. To facilitate this traceback, each entry in position (i, j) in the table (for all i, j) should have associated with it, during the application of the dynamic programming scheme, one or more pointers indicating which particular element(s) in cells $(i - 1, j)$, $(i, j - 1)$ or $(i - 1, j - 1)$ led to the inclusion of that element in cell (i, j). For example, if $\alpha[i] = \beta[j] = a$, and $(r-1, (x, y)) \in S_{i-1,j-1}$ then $(r, (next_\alpha(x, a), next_\beta(y, a))$ is placed in cell (i, j) with a pointer to the element $(r-1, (x, y))$ in cell $(i-1, j-1)$.

With these pointers, any path from an element $(r, (x, y))$ in cell (m, n) to the element in cell $(0, 0)$ represents a maximal common subsequence of α and β of length r, namely the reversed sequence of matching symbols from the two strings corresponding to cells from which the path takes a diagonal step.

Analysis of the MinMaxCSub Algorithm

The number of cells in the dynamic programming table is essentially mn, so that if we could show that the number of entries in each cell was bounded by, say, $\min(m, n)$, and that the total amount of computation was bounded by a constant times the total number of table entries, then we would have a cubic time worst-case bound for the complexity of the algorithm. However, this turns out not to be the case, as the following example shows.

Example Consider two strings of length $n = p(p + 1)/2 + q$ over an alphabet $\Sigma = \{a_1, \ldots, a_n\}$, defined as follows

$$\alpha = \alpha_1 + \alpha_2 + \cdots + \alpha_p + a_{p(p+1)/2+1}, \ldots, a_n$$

$$\beta = \alpha_p + \alpha_{p-1} + \cdots + \alpha_1 + a_n, \ldots, a_{p(p+1)/2+1}$$

where $\alpha_1 = a_1$, $\alpha_2 = a_2 a_3$, $\ldots$, $\alpha_p = a_{(p-1)p/2+1} \cdots a_{p(p+1)/2}$, and $+$ denotes concatenation.

It is not hard to see that position (n, n) in the dynamic programming table contains the pq entries $(r, (x, y))$ for $r = 2, \ldots, p + 1$, $x = p(p + 1)/2 + 1, \ldots, n$, $y = n + 1 + p(p + 1)/2 - x$. With $q = \Theta(p^2)$, this gives $\Theta(n^{3/2})$ entries in the (n, n)th cell.

However, suppose that we wish to find only the length of a minmax common subsequence (and to construct such a sequence by traceback through the table). Then, if any particular cell in the table contains more than one entry $(r, (x, y))$ with the same (x, y) component, we may discard all but the one with the smallest r value. For if a maximal common subsequence γ has a prefix γ' such that $sp(\alpha, \gamma') = x$ and $sp(\beta, \gamma') = y$, then to make γ as short as possible, γ' must be chosen as short as possible.

Also, if the entries $(r, (x, y))$ in the (i, j)th cell are listed in increasing order of x, then they must clearly also be in decreasing order of y, and therefore, since $x \leq i$, $y \leq j$, the number of such entries with distinct (x, y) components cannot exceed $\min(i, j)$. Further, it is easy to see that by processing the lists of cell entries in this fixed order, the amount of work done in computing the contents of cell (i, j) is, in case (i) bounded by a constant times the number of entries in cell $(i - 1, j - 1)$, and in case (ii) bounded by a constant times the sum of the numbers of entries in cells $(i - 1, j)$ and $(i, j - 1)$. (In case (i), this assumes precomputation of the tables of *next* values, which can easily be achieved in $O(n|\Sigma|)$ time for a string of length n, where Σ is the alphabet.)

In conclusion, the length of a minmax common subsequence can be established by a suitably amended version of the above dynamic programming scheme in $O(m^2 n)$ time in the worst case, for strings of lengths m and n $(m \leq n)$. Furthermore, such a subsequence can also be constructed from the dynamic programming table without increasing that overall time bound. But it remains open whether the lengths of all maximal common subsequences can be established within that time bound. A trivial bound of $O(m^3 n)$ applies in that case, since the number of entries in each cell is certainly bounded by m^2.

3 The MaxMinCSup problem for two strings

The MaxMinCSup algorithm is not dissimilar in spirit to the MinMaxCSub algorithm, and there is a certain duality involving the terms in which the algorithm is expressed.

Given strings α and γ, we define

$lp(\alpha, \gamma) = $ length of the longest prefix of α that is a subsequence of γ.

Given strings α, β of lengths m and n respectively, we define the set T_{ij}, for each $i = 0, \ldots, m$, $j = 0, \ldots, n$, by

$T_{ij} = \{(r, (x, y)) : \text{there exists a minimal common supersequence } \gamma \text{ of } \alpha^i \text{ and } \beta^j$, of length r, such that $lp(\alpha, \gamma) = x, lp(\beta, \gamma) = y\}$.

Finally, for string α, position i and symbol a, we define

$$f_\alpha(i, a) = \begin{cases} i + 1 & \text{if } \alpha[i + 1] = a \\ i & \text{otherwise.} \end{cases}$$

The algorithm for MaxMinCSup is based on a dynamic programming scheme for the sets T_{ij} defined above. So evaluation of T_{mn} reveals the length of the maxmin common supersequence, but also finds the lengths of all minimal common supersequences of α and β (indeed of all minimal common supersequences of all pairs of prefixes of α and β). Furthermore, the use of suitable tracebacks in the array of T_{ij} values, can be used to generate, not only a maxmin common supersequence, but all minimal common supersequences.

The zero'th row and column of the T_{ij} table can be evaluated trivially, as follows:

$$T_{i0} = \{(i, (i, lp(\beta, \alpha^i)))\} \quad (1 \le i \le m)$$

and

$$T_{0j} = \{(i, (lp(\alpha, \beta^j), j))\} \quad (1 \le j \le n)$$

with

$$T_{00} = \{(0, (0, 0))\}.$$

The basis of the dynamic programming scheme is contained in the following theorem:

Theorem 2 *(i) If $\alpha[i] = \beta[j] = a$ then*

$$T_{ij} = \{(r, (f_\alpha(x, a), f_\beta(y, a)) : (r - 1, (x, y)) \in T_{i-1, j-1}\}$$

(ii) If $\alpha[i] = a \neq b = \beta[j]$ then

$$T_{ij} = \{(r, (f_\alpha(x, b), j)) : (r - 1, (x, j - 1)) \in T_{i, j-1}\}$$
$$\cup \{(r, (i, f_\beta(y, a)) : (r - 1, (i - 1, y)) \in T_{i-1, j}\}.$$

Proof (i) Suppose $(r - 1, (x, y)) \in T_{i-1, j-1}$ and that γ' is a minimal common supersequence of α^{i-1} and β^{j-1} of length $r - 1$ with $lp(\alpha, \gamma') = x$ and $lp(\beta, \gamma') = y$. Then it is immediate that $\gamma = \gamma' + a$ is a minimal common supersequence, of length r, of α^i and β^j, and that $lp(\alpha, \gamma) = f_\alpha(x, a)$ and $lp(\beta, \gamma) = f_\beta(y, a)$.

On the other hand, suppose that γ is a minimal common supersequence of length r of α^i and β^j. Then $\gamma[r] = a$, and $\gamma' = \gamma - a$ is certainly a common supersequence of α^{i-1} and β^{j-1}. If γ' were not minimal, then some subsequence δ of γ' would be a common supersequence of α^{i-1} and β^{j-1}, and therefore $\delta + a$, a subsequence of γ, would be a common supersequence of α^i and β^j, contradicting the minimality of γ. So $(r - 1, (x, y)) \in T_{i-1, j-1}$ with $x = lp(\alpha, \gamma')$, $y = lp(\beta, \gamma')$ and $lp(\alpha, \gamma) = f_\alpha(x, a)$, $lp(\beta, \gamma) = f_\beta(y, a)$.

(ii) Suppose $(r - 1, (i - 1, y)) \in T_{i-1, j}$, and that γ' is a minimal common supersequence of α^{i-1} and β^j of length $r - 1$ with $lp(\alpha, \gamma') = i - 1$, $lp(\beta, \gamma') = y$. (The argument is similar in the case $(r - 1, (x, j - 1)) \in T_{i, j-1}$.) Then $\gamma = \gamma' + a$ is a common supersequence of α^i and β^j with $lp(\alpha, \gamma) = i$ and $lp(\beta, \gamma) = f_\beta(y, a)$. Further, γ must be minimal. For suppose that a subsequence δ of γ is a common supersequence of α^i and β^j. If δ were a subsequence of γ', then γ' would not be a minimal common supersequence of α^{i-1} and β^j. So $\delta = \delta' + a$, where δ' is a subsequence of γ'. So δ' cannot be a common supersequence of α^{i-1} and β^j. If it is not a supersequence of α^{i-1} then $\delta' + a$ cannot be a supersequence of

α^i — a contradiction. If it is not a supersequence of β^j then, since $\delta' + a$ is a supersequence of β^j, we must have $\beta[j] = a$ — a contradiction.

On the other hand, suppose that γ is a minimal common supersequence of length r of α^i and β^j. Then $\gamma[r] = a$ or b.

case (iia) $\gamma[r] = a$. It is immediate that $lp(\alpha, \gamma) = i$, for otherwise $\gamma - a$ would be a common supersequence of α^i and β^j. So $\gamma' = \gamma - a$ is a minimal common supersequence of α^{i-1} and β^j with $lp(\alpha, \gamma') = i - 1$ and $lp(\beta, \gamma') = y$ for some y such that $lp(\beta, \gamma) = f_\beta(y, a)$.

case (iib) $\gamma[r] = b$. A similar argument shows that $\gamma' = \gamma - b$ is a minimal common supersequence of α^i and β^{j-1} with $lp(\beta, \gamma') = j - 1$ and $lp(\alpha, \gamma') = x$ for some x such that $lp(\alpha, \gamma) = f_\alpha(x, b)$.

This completes the proof of the theorem. $\square$

Recovering a Maxmin Common Supersequence

As in the case of a minmax common subsequence, the recovery of a particular maxmin common supersequence involves a traceback through the dynamic programming table from cell (m, n) to cell $(0, 0)$, during which the sequence is constructed in reverse order. To facilitate the traceback, each entry in position (i, j) in the table (for all i, j) should have associated with it, during the application of the dynamic programming algorithm, one or more pointers indicating which particular element(s) in cells $(i - 1, j)$, $(i, j - 1)$ or $(i - 1, j - 1)$ led to the inclusion of that element in cell (i, j). For example, if $\alpha[i] = a = \beta[j]$ and $(r - 1, (x, y)) \in T_{i-1, j-1}$, then $(r, (f_\alpha(x, a), f_\beta(y, a)))$ is placed in cell (i, j) with a pointer to the element $(r - 1, (x, y))$ in cell $(i - 1, j - 1)$.

With these pointers, any path from an element $(r, (x, y))$ in cell (m, n) to the element in cell $(0, 0)$ represents a minimal common supersequence of α and β of length r, namely the reversed sequence of symbols found by recording $\alpha[i]$ for a vertical or diagonal step from cell (i, j) and $\beta[j]$ for a horizontal step from cell (i, j).

Analysis of the MaxMinCSup Algorithm

As in the case of the MinMaxCSub algorithm, we can establish a cubic time bound for the restricted version of the MaxMinCSup algorithm that is designed to find the length of a maxmin common supersequence, and to construct such a common supersequence from the dynamic programming table. The trick again is the observation that, for this purpose, whenever $(r, (x, y))$ elements in the same cell have the same (x, y) component, only one need be retained, namely that with the largest r value. For if a minimal common supersequence γ has a prefix γ' such that $lp(\alpha, \gamma') = x$ and $lp(\beta, \gamma') = y$, then to make γ as long as possible, γ' should be chosen as long as possible.

By this means we can restrict the number of elements in the (i, j)th cell to at most $i + j$, recalling that each such entry $(r, (x, y))$ has either $x = i$ or $y = j$. This leads to a worst-case time bound of $O(mn(m + n))$ for this version of the algorithm. Again, it is not clear whether the lengths of all minimal common supersequences can be found in time better than $O(mn(m + n)^2)$ in the worst case, this arising from the obvious upper bound of $(m + n)^2$ on the number of elements in each cell of the table.

4 The MinMaxCSub problem for k strings

It is well-known that the problem of finding an LCS of k strings is NP-hard, even in a number of special cases [10, 12, 13]. Recently, Jiang and Li [9] described a polynomial-time transformation from the Maximum Clique problem for graphs to the LCS problem with the property that the strings constructed have a common subsequence of length r if and only if the original graph has a clique of size r. If the given graph has k vertices then the derived LCS instance comprises $2k$ strings. Now, it has recently been established by Arora et al [2] that, if P $\neq$ NP, then there cannot exist a polynomial-time approximation algorithm for the Maximum Clique problem with a performance guarantee of k^δ, for some $\delta > 0$. Hence, Jiang and Li were able to conclude that, unless P = NP, there cannot exist a polynomial-time approximation algorithm for LCS with a performance guarantee of k^δ, for some $\delta > 0$.

As we shall see in Theorem 3, the transformation, from Independent Set, given by Maier [10] to prove the NP-completeness for LCS also serves as a transformation from the Minimum Independent Dominating Set problem to MinMaxCSub. The former problem is also NP-hard [3], and was shown by Irving [8] not to have a polynomial-time approximation algorithm with a constant performance guarantee (if P $\neq$ NP). Halldórsson [4] has recently strengthened this result to show that, if P $\neq$ NP, then, for no $\delta < 1$, can there exist a polynomial-time approximation algorithm with performance guarantee k^δ. Maier's transformation has the property that the strings constructed have an LCS of length r if and only if the original graph has an independent set of size r. If the given graph has k edges then the derived LCS instance has $k + 1$ strings. It will therefore follow from the transformation, not only that MinMaxCSub is NP-hard, but also that this same strongly negative approximability result applies to the MinMaxCSub problem.

Theorem 3 *(i) The MinMaxCSub problem is NP-hard.*
(ii) If P $\neq$ NP, then, for no $\delta < 1$, can there exist a polynomial-time approximation algorithm for MinMaxCSub on k strings with performance guarantee k^δ.

Proof (i) Let $G = (V, E)$, t, with $V = \{v_1, \ldots, v_n\}$ and $E = \{e_1, e_2, \ldots, e_m\}$, be an arbitrary instance of (the decision version of) the Minimum Independent Dominating Set problem. We construct an instance of MinMaxCSub as follows. Include in the set S of strings the string $\beta = v_1 v_2 \ldots v_n$. For each edge $e_i = \{v_p, v_q\}$ ($p < q$) include in the set S of strings the string α_i defined by

$$\alpha_i = v_1 v_2 \ldots v_{p-1} v_{p+1} \cdots v_n v_1 v_2 \ldots v_{q-1} v_{q+1} \cdots v_n$$

We claim that G has an independent dominating set of size t if and only if S has a maximal common subsequence of length t.

To prove this claim, we must show that (a) if G has an independent dominating set U of size t then S has a maximal common subsequence of length t; (b) if S has a maximal common subsequence of length t then G has an independent dominating set of size t.

To prove (a), assume that $U = \{v_{u_1}, v_{u_2}, \ldots, v_{u_t}\}$ is an independent dominating set of size t, where $1 \leq u_1 < u_2 < \ldots < u_t \leq n$.

It can easily be checked that the string $\alpha = v_{u_1} v_{u_2} \ldots v_{u_t}$ is a common subsequence of S. If some supersequence, α', of α, is a common subsequence of S then observe for a contradiction, that $\exists\, v_p$ in α' but not in α, which is connected to $v_q \in U$, in G by edge $e_j = \{v_p, v_q\}$. Assuming $p < q$ then the string $\alpha_j = v_1 v_2 \ldots v_{p-1} v_{p+1} \ldots v_n v_1 v_2 \ldots v_{q-1} v_{q+1} \ldots v_n$. For α' to be a subsequence of β, v_p must precede v_q in α'. But this prevents α' from being a subsequence of α_j. A similar contradiction is obtained if $p > q$ is assumed.

To prove (b), assume $\alpha = v_{u_1} v_{u_2} \ldots v_{u_t}$, of length t, is a maximal common subsequence of the strings in S. The first observation is that if v_{u_p} and v_{u_q} are two symbols in α and $p < q$ then $u_p < u_q$. For otherwise α could not be a subsequence of β. The elements of α must form an independent set, U, of size t, in G. To see this, observe for a contradiction that if two elements, v_{u_p} and v_{u_q} $(p < q)$, of α are connected in G by edge $e_j = \{v_{u_p}, v_{u_q}\}$ then the string $v_{u_p} v_{u_q}$, a subsequence of α, would not be a subsequence of α_j and hence α would not be a subsequence of α_j. If U is not maximal then $\exists\, U'$, an independent set of size $t' > t$, and $U \subset U'$. Observe for a contradiction that this would imply $\exists v_j \in U'$ and $v_j \notin U$. Then it is easy to see that the string $\alpha' = v_{u_1} \ldots v_{u_p} v_j v_{u_{p+1}} \ldots v_{u_t}$, where $u_p < j < u_{p+1}$, a supersequence of α would be a common subsequence of all the strings in S, contradicting the maximality of α. This concludes the proof of part (i).

The proof of part (ii) follows from the observation that the reduction is linear [11], and therefore preserves the approximability of the Minimum Independent Dominating Set, and from the result of Halldórsson [4] on the approximability of that problem. □

5 Conclusion and open problems

We have shown that, in the case of two strings (or indeed any fixed number of strings), a minimum maximal common subsequence and a maximum minimal common supersequence can be found in polynomial time by dynamic programming. However, for general k, we have shown that finding a minimum maximal common subsequence of k strings is NP-hard, and further, that, unless P = NP, the length of a minimum maximal common subsequence cannot be approximated, in polynomial time, within a factor of k^δ for any $\delta < 1$.

It is natural to conjecture that finding a maximum minimal common supersequence of k strings is also NP-hard, but we have not been able to find an appropriate transformation to establish this result. It also seems likely that approximating the maximum minimal common supersequence will be no easier than approximating the minimum maximal common subsequence, but we have no result of this kind.

References

1. A. Apostolico and C. Guerra. The longest common subsequence problem revisited. *Algorithmica*, 2:315–336, 1987.
2. S. Arora, C. Lund, R. Motwani, M. Sudan, and M. Szegedy. Proof verification and hardness of approximation problems. In *Proc. 33rd IEEE Symp. Found. Comp. Sci.*, pages 14–23, 1992.
3. M.R. Garey and D.S. Johnson. *Computers and Intractability.* Freeman, San Francisco, CA., 1979.
4. M.M. Halldórsson. Approximating the minimum maximal independence number. *Information Processing Letters*, 46:169–172, 1993.
5. D.S. Hirschberg. A linear space algorithm for computing maximal common subsequences. *Communications of the A.C.M.*, 18:341–343, 1975.
6. D.S. Hirschberg. Algorithms for the longest common subsequence problem. *Journal of the A.C.M.*, 24:664–675, 1977.
7. J.W. Hunt and T.G. Szymanski. A fast algorithm for computing longest common subsequences. *Communications of the A.C.M.*, 20:350–353, 1977.
8. R.W. Irving. On approximating the minimum independent dominating set. *Information Processing Letters*, 37:197–200, 1991.
9. T. Jiang and M. Li. On the approximation of shortest common supersequences and longest common subsequences. Submitted for publication, 1992.
10. D. Maier. The complexity of some problems on subsequences and supersequences. *Journal of the A.C.M.*, 25:322–336, 1978.
11. C.H. Papadimitriou and M. Yannakakis. Optimization, approximation, and complexity classes. *Journal of Computer and System Sciences*, 43:425–440, 1991.
12. K-J. Raiha and E. Ukkonen. The shortest common supersequence problem over binary alphabet is NP-complete. *Theoretical Computer Science*, 16:187–198, 1981.
13. V.G. Timkovskii. Complexity of common subsequence and supersequence problems and related problems. *English Translation from Kibernetika*, 5:1–13, 1989.
14. E. Ukkonen. Algorithms for approximate string matching. *Information and Control*, 64:100–118, 1985.
15. S. Wu, U. Manber, G. Myers, and W. Miller. An O(NP) sequence comparison algorithm. *Information Processing Letters*, 35:317–323, 1990.
16. M. Yannakakis and F. Gavril. Edge dominating sets in graphs. *SIAM J. Appl. Math.*, 38:364–372, 1980.

Dictionary-Matching on Unbounded Alphabets: Uniform Length Dictionaries

Dany Breslauer*

Istituto di Elaborazione della Informazione
Consiglio Nazionale delle Ricerche
Via S. Maria 46, Pisa 56126, Italy

Abstract

In the string-matching problem one is interested in all occurrences of a short pattern string in a longer text string. Dictionary-matching is a generalization of this problem where one is looking simultaneously for all occurrences of several patterns in a single text.

This paper presents an efficient on-line dictionary-matching algorithm for the case where the patterns have uniform length and the input alphabet is unbounded. A tight lower bound establishes that our approach is optimal if the only access the algorithm has to the input strings is by pairwise symbol comparisons.

In an immediate application, the new dictionary-matching algorithm can be used in a previously known higher-dimensional array-matching algorithm, improving the performance of this algorithm on unbounded alphabets. The resulting algorithm is currently the fastest known algorithm for k-dimensional array-matching on unbounded alphabets, for $k \geq 3$.

1 Introduction

Given a collection of pattern strings $\mathcal{D} = \{\mathcal{P}_1, \mathcal{P}_2, \cdots, \mathcal{P}_{|D|}\}$, one is interested in finding all occurrences of the patterns in a text string $\mathcal{T}$. This problem, which is called the *dictionary-matching* problem or the *multi-pattern string-matching* problem, is a generalization of the standard string-matching problem where only occurrences of a single pattern are sought. For a survey on pattern matching algorithms see Aho's paper [1].

The collection of patterns $\mathcal{D}$ will be called a *dictionary* and its size is denoted by $d = |D|$. The length of the text is $n = |T|$ and the length of each pattern is $|P_i|$. When there is only one pattern or if all patterns in the dictionary have the same length, we use m to denote this length. The number of distinct symbols in the dictionary, which is called the *dictionary alphabet size*, is denoted by σ_D. Clearly $\sigma_D \leq \Sigma_{i=1}^{d} |P_i|$.

The assumptions on the alphabet that the input symbols are chosen from has a crucial role in the design of efficient string algorithms. To solve the string-matching or the dictionary-matching problems one only has to be able to compare pairs of input symbols

---

*The author was partially supported by the European Research Consortium for Informatics and Mathematics postdoctoral fellowship. Part of this work was done while visiting at the Centrum voor Wiskunde en Informatica, Amsterdam, The Netherlands, and part while visiting at the Institut National de Recherche en Informatique et en Automatique. Rocquencourt, France.

in order to get $(=,\neq)$ answers. However, since alphabet symbols are encoded numerically on a computer, the symbols are naturally (arbitrarily) ordered. This order can be used to obtain more efficient algorithms that access the input strings by $(<,=,>)$-comparisons. Furthermore, sometimes it is convenient to assume that the alphabet symbols are small integers which are bounded by some function of the input size, and in most practical cases the alphabet symbols are integers from a constant range (i.e. *ASCII* encoded characters).

The string-matching problem has several linear time algorithms. For instance, the classical string-matching algorithm of Knuth, Morris and Pratt [26] takes $O(m)$ time to pre-process the pattern and then $O(n)$ time to find all occurrences of the pattern in the text. The naive approach to the dictionary-matching problem could try to match each pattern separately by using this algorithm, resulting in a dictionary-matching algorithm that has an $O(\Sigma_{i=1}^{d}|P_i|)$ time dictionary preprocessing step and an $O(dn)$ time text scanning step. Thus, the dictionary-matching problem can be solved in $O(\Sigma_{i=1}^{d}|P_i| + dn)$ time.

Aho and Corasick [2] generalized the Knuth-Morris-Pratt string-matching algorithm and showed that on a constant size alphabet, the dictionary-matching problem can be solved by constructing an automaton in $O(\Sigma_{i=1}^{d}|P_i|)$ time and then scanning the text in $O(n)$ time using the automaton. (The time bounds for dictionary-matching algorithms usually includes also an additive factor of T_{occ}, the total number of occurrences of the dictionary patterns in the text (clearly $T_{occ} \leq dn$). More precisely, the text scanning step takes $O(n + T_{occ})$ time if all occurrences of the patterns have to be reported. However, in many cases it is sufficient to report some representation of all the occurrences, such as the longest pattern starting or ending at each text position, and thus to save the $O(T_{occ})$ time factor.) This is a significant improvement since in practice the text can be very long and since the same static pattern dictionary is often searched in several text strings. On unbounded ordered alphabets the Aho-Corasick algorithm takes $O((\Sigma_{i=1}^{d}|P_i| + n)\log\min(d,\sigma_{\mathcal{D}}))$ time. (In this paper the $\log n$ function usually means $\max(1,\log_2 n)$.) It is interesting to note that the Aho-Corasick algorithm is optimal in the case of a constant size alphabet and almost optimal if the only access to the input strings is by $(=,\neq)$-comparisons. (If $n \geq (1 + \epsilon)\max_{i=1..d}|P_i|$, for some constant $\epsilon > 0$, then the text scanning step requires at least $\Omega(nd)$ $(=,\neq)$-comparisons. This bounds is achieved by the Aho-Corasick algorithm and also by the naive algorithm that matches each pattern separately. When the text is shorter, it is possible to modify the text scanning step of the Aho-Corasick algorithm to be optimal by ignoring patterns that are too long to occur in the text.)

Commentz-Walter [18, 19] and Crochemore et al. [20] gave other dictionary-matching algorithms that are based on ideas from the Boyer-Moore [16] string-matching algorithm. These algorithms achieve faster average running times by matching the patterns from their end towards their start. Recently, the dictionary-matching problem gained interest after the discovery of algorithms that can handle dictionaries that are dynamically changing without having to recompute the dictionary preprocessing information from scratch [6, 7, 8, 9, 10, 24].

The main contribution of this paper is a new approach to the dictionary-matching problem with uniform length patterns on ordered alphabets. Similarly to the Commentz-Walter [18, 19] algorithm, the new algorithm tries to match the dictionary patterns from their end towards their start. However, our motivation is entirely different. While the Commentz-Walter algorithm matches the patterns from their end to start hoping to skip parts of the text, our algorithm does so since this order of comparisons allows to replace a logarithmic multiplicative factor in the running time of the Aho-Corasick algorithm by an additive factor and thus, to amortize some comparisons against large segments of the text.

The new dictionary-matching algorithm takes $O((\frac{\log d}{m}+1)n)$ time for scanning the text after an $O(dm \log \sigma_D)$ time dictionary preprocessing step. A tight lower bound shows that the text scanning step is the fastest possible if the algorithm can access the symbols of the input strings only by pairwise $(<, =. >)$-comparisons. The suggested implementation of the algorithm in the standard random-access machine model [3] requires a large sparse table which is used for fast access to the preprocessing data. We suggest two alternatives to store this table. The first uses $O(md^2)$ space that does not need to be initialized, and therefore is time efficient. The second alternative uses a hashing scheme of Fredman, Komlos and Szemeredy [22] and requires only $O(md)$ space, at the cost of introducing randomization in the dictionary preprocessing step. The preprocessing that creates the hash table takes $O(md)$ time with high probability [21]. Note that the rest of the dictionary preprocessing step takes $O(md \log \sigma_D)$ time, which dominates the time it takes to create the hash table with very high probability.

In the k-dimensional array-matching problem the pattern and text are k-dimensional arrays of sizes m^k and n^k respectively. (The discussion below assumes that the pattern and the text are k-dimensional cubes. It trivially generalizes to any rectangular k-dimensional arrays.) Bird [15] and Baker [13] demonstrated that the k-dimensional array-matching problem can be solved by k iterations of a one-dimensional dictionary-matching algorithm. By using the Aho-Corasick algorithm they were able to obtain an $O(m^2 + n^2)$ time two-dimensional and an $O(m^k + n^k \log m)$ time k-dimensional algorithms on constant size alphabets. When the alphabet is larger. the time bounds become $O(m^k \log m + n^k \log m)$. Other algorithms for the two-dimensional and the k-dimensional problems were suggested by Karp and Rabin [25], Zhu and Takaoka [31] and by Baeza-Yates and Régnier [12].

Amir, Benson and Farach [5] designed a two-dimensional array-matching algorithm that does not resort to one-dimensional techniques. Their algorithm has an $O(n^2)$ time text processing step that uses only $(=, \neq)$-comparisons. However, their pattern processing step still uses one-dimensional techniques and takes $O(m^2 \log m)$ time using $(<, =, >)$-comparisons. Galil and Park [23] improved the preprocessing step by giving an $O(m^2)$ time algorithm that uses only $(=, \neq)$-comparisons.

We show that using the new dictionary-matching algorithm it is possible to implement the Bird-Baker approach with an $O(n^k)$ time text scanning step and an $O(m^k \log m)$ time dictionary preprocessing using $(<. =. >)$-comparisons. This algorithm improves on the best previous bounds for k-dimensional array-matching on unbounded alphabets. for $k \geq 3$.

2 Dictionary-Matching on Ordered Alphabets

This section first outlines the ideas behind the dictionary-matching algorithm for uniform length dictionaries on unbounded ordered alphabets. The discussion proceeds in the comparisons model and it is simple and elegant. The implementation details of the algorithm in the standard random-access machine model [3] are more involved and given in Section 2.3.

2.1 The new algorithm

Without loss of generality assume that the text is of length $2m - 1$. We say that there is a *potential occurrence* of a pattern $\mathcal{P}_i$ starting at text position t if such an occurrence has not been ruled out by the results of previous comparisons. Initially, there is a potential occurrence of each pattern in the dictionary starting at all text positions t. such that $1 \leq t \leq m$. Thus, there are a total of dm potential occurrences. See Figure 1.

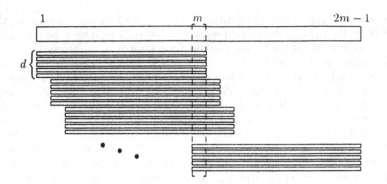

Figure 1: The dm potential occurrences in a text of length $2m - 1$.

The Aho-Corasick dictionary-matching algorithm tries to match the patterns from their start towards their end. It might take up to $\log d$ comparisons to determine that none of the patterns in the dictionary occurs starting at the first text position. Even when the Aho-Corasick algorithm reaches this conclusion it has eliminated only d potential occurrences. Up to $m \log d$ comparisons might be required to eliminate all the potential occurrences.

The new dictionary-matching algorithm will start matching the patterns from their end towards their start. The algorithm starts in a *backward phase* from text position m and proceeds towards the beginning of the text. Later, it will continue in a *forward phase* to match the remainder of the text starting from text position $m + 1$ and proceeding towards the end of the text. In both phases, the algorithm advances to the next text position only after it has obtained an equality answer to a comparison between the symbol in the current text position and some pattern symbol.

Consider all potential occurrences aligned with the text starting at the text positions where they might occur (Figure 1). The algorithm begins in a backward phase by comparing the symbol in the current text position, m, to the median of the alphabet symbols that appear in the potential occurrences in the column which is aligned with this text position.

If the comparison results in an equal answer, the algorithm "knows" which pattern symbol appears in the current text position and it will not compare it again. The algorithm eliminates all potential occurrences which do not agree with the result of the comparison and proceeds to the next text position.

After the algorithm moves to a smaller text position, there might be some potential occurrences which have not been eliminated, but do not intersect the column under the current text position. These are still valid potential occurrences and are considered inactive for the moment. These potential occurrences become active again when the algorithm starts the forward phase.

If the comparison results in an unequal answer, either less-than or greater-than, then the algorithm also eliminates all potential occurrences which do not agree with the result of the comparison. Since the compared dictionary symbol was the median, at least half of the active potential occurrences are eliminated. (Sometimes, by far more than the number of potential occurrences that might be eliminated in one step of the Aho-Corasick algorithm.) The algorithm then continues comparing the symbol at the current text position to the median of the alphabet symbols that appear in the surviving potential occurrences in the column which is aligned with this text position.

The algorithm proceeds this way in the backward phase until the beginning of the text or until all potential occurrences which start at the current text position and at smaller positions are eliminated. Then, the algorithm starts the forward phase and proceeds similarly. After the algorithm moves to next text position, there might also be some potential occurrences which have not been eliminated, but do not intersect the column under the current text position. Only now, if a potential occurrence does not intersects the column under the current text position, then this potential occurrence is an actual occurrence and it may be reported as such.

The number of potential occurrences at the beginning of the forward phase is obviously not larger than dm, their initial number. Clearly, the algorithm does not make more than $2m - 1$ comparisons that result in an equal answer and no more than $2 \log dm$ comparisons which result in an unequal answer.

Thus, the total number of comparisons is bounded by $O(\log d + m)$ and the $\log d$ multiplicative factor of the Aho-Corasick algorithm was converted to an additive factor that disappears if $\log d = O(m)$.

2.2 Tight bounds

We summarize the ideas presented above and prove the following tight bounds.

Theorem 2.1 *The dictionary matching problem with uniform length dictionaries on ordered alphabets takes $\Theta((\frac{\log d}{m} + 1)n)$ comparisons after preprocessing.*

The theorem follows from the next two lemmas.

Lemma 2.2 *The dictionary-matching problem with uniform length dictionaries on ordered alphabets can be solved using $O((\frac{\log d}{m} + 1)n)$ comparisons. The dictionary preprocessing that is required for the algorithm uses $O(dm \log \sigma_D)$ comparisons.*

Proof: Partition the text into $O(n/m)$ overlapping blocks of length $2m - 1$ (the last block might be shorter) in such a way that each occurrence of a pattern in the text is exactly in one such block. The blocks are processed one by one starting from the beginning of the text. We have shown that the number of comparisons made in each such block is bounded by $O(\log d + m)$ and thus, the number of comparisons required for the whole text is $O((\frac{\log d}{m} + 1)n)$.

The dictionary preprocessing has to compute the medians which are used in the text scanning step and to identify all equal pattern symbols. This can obviously be done in the comparison model by sorting all the pattern symbols using $O(dm \log \sigma_D)$ comparisons [3] to obtain complete information about their relative order. $\square$

An important property of the Aho-Corasick dictionary-matching algorithm is that it processes the text symbols from left to right and it does not go back to access previous symbols. This means that the text symbols can be processed immediately after they are given as input and there is no need to store them. Note, that an occurrence of a pattern starting at a certain text position can not be reported before the next m text symbols have been processed. We say that a dictionary-matching algorithm for uniform length dictionaries is *on-line* if the algorithm reports whether there are occurrences of dictionary patterns starting at text position t before examining symbols in any text positions that are larger than or equal to $t + m$. (Regardless of the time it spends processing any part of the text). The algorithm in Lemma 2.2 is clearly on-line.

The lower bound that we prove next only applies to the text scanning step. This lower bound holds for any algorithm that has access to the symbols of the input strings only by pairwise $(<, =, >)$-comparisons. We assumes that pairwise comparisons of text symbols are not permitted.

Lemma 2.3 *The dictionary-matching problem requires at least* $\Omega((\frac{\log d}{m} + 1)n)$ *comparisons.*

Proof: Assume that all the symbols of the pattern strings $\mathcal{P}_1, \cdots, \mathcal{P}_d$ are different and are ordered arbitrarily. Partition the text into $O(n/m)$ non overlapping consecutive blocks of length $2m - 1$ each (the last block might be shorter). The lower bound is proved for each block separately.

Initially, each pattern $\mathcal{P}_i$ can occur at text positions between positions 1 and m of each block. Thus, there are dm potential occurrences of patterns in a text block. Each comparison between a pattern symbol and a text symbol can be answered less than, equal to, or greater than, in such a way that at least a third of the potential occurrences survive. Therefore, at least $\log_3 dm$ comparisons are required until there is exactly one potential occurrence left.

The algorithm has to make at least m comparisons to verify that there is an actual occurrence. Thus, $\Omega(\max(\log dm, m)) = \Omega(\log d + m)$ comparisons are required in each block of length $2m - 1$. The total number of comparisons required in the whole text is $\Omega((\frac{\log d}{m} + 1)n)$. $\square$

2.3 Implementation details

This section shows how to implement the comparison model algorithm that was described above in the standard random access machine model [3]. The implementation uses suffix trees in a similar way to Amir and Farach [6] and Amir et al. [8].

2.3.1 Suffix trees

Suffix trees are a compressed form of *digital search trees* that are very useful in many algorithms on strings. The usual definition of a suffix tree is the following:

Definition 2.4 *Let $S[1..k]$ be a string whose last symbol $S[k] = $ '#' and '#' is a special alphabet symbol that does not appear anywhere else in S. The suffix tree T_S of $S[1..k]$ is a rooted tree with k leaves and $k - 1$ internal nodes such that:*

1. *Each edge is labeled with a non-empty substring of S.*

2. *No two sibling edges are labeled with substrings that start with the same symbol.*

3. *Each leaf is labeled with a distinct position of S.*

4. *The concatenation of the labels of the edges on the path from the root to a leaf labeled with position i is equal to the suffix $S[i..k]$.*

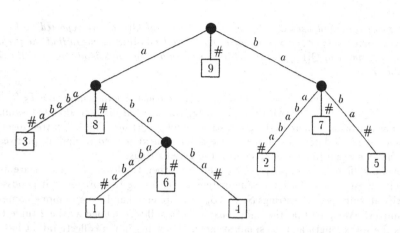

Figure 2: The suffix tree of $S[1..9] = $ 'abaababa#'. The edges are labeled with substrings of S and the leaves with the position in S at which the corresponding suffix starts.

An example of a suffix tree is given in Figure 2. Note that a suffix tree has no internal nodes with one child. Substrings of $S[1..k]$ can be represented by their starting and ending positions and therefore a suffix tree T_S can be stored in $O(k)$ space.

The important property of suffix trees that we use in this paper is that if two suffixes $S[i..k]$ and $S[j..k]$ have the same prefix, namely if $S[i..i + l - 1] = S[j..j + l - 1]$ and $S[i + l] \neq S[j + l]$, then the paths from the root to the leaves labeled with positions i and j share an initial segment. The concatenation of the labels of the edges on this initial segment is equal to $S[i..i + l - 1]$. The special alphabet symbol '#' that was assumed to be the last symbol of the string $S[1..k]$ is normally appended at the end of a given string to guarantee that the suffix tree has distinct leaves that correspond to each suffix. There exist several efficient algorithms to construct suffix trees [17, 27, 30].

Theorem 2.5 *Given a string $S[1..k]$, the suffix tree T_S can be built in $O(k \log \sigma_S)$ time using $(<, =, >)$-comparisons, where σ_S is the number of distinct symbols in S.*

We generalize the notion above to deal with several strings. Given a collection of strings $\{S_1, \cdots, S_p\}$, $S = S_1 '\#_1' \cdots S_p '\#_p'$ is obtained by concatenating the strings and special distinct alphabet symbols '$\#_1$'.....'$\#_p$' that do not appear anywhere in the strings. The suffix tree T_S identifies identical prefixes of suffixes of these strings and has distinct leaves that correspond to each suffix of $S_i '\#_i'$, for $i = 1, \cdots, p$. Note that the concatenation of the labels of edges from the root to a leaf is equal to a suffix of one of the strings S_i followed by the complete strings S_j, $j = i + 1, \cdots, p$, and the special alphabet symbols that are appended at the end of these strings. The special alphabet symbols '$\#_i$' are identified by their position within S and there is no need represent them as real alphabet symbols since they are not equal to any other symbol.

Lemma 2.6 *Given a collection of strings $\{S_1, \cdots, S_p\}$, we construct a data structure that is used to find all prefixes of a string Q that are equal to suffixes of some of the given strings.*

Let σ_S denote the number of distinct symbols in the strings $\{S_1, \cdots, S_p\}$ and let $l = \Sigma_{i=1}^p |S_i|$ denote the sum of their lengths. Then, the data structure is constructed in $O(l \log \sigma_S)$ *time and uses $O(l)$ space. The prefixes of $Q[1..q]$ are reported on-line in* $O(q + \log l)$ *time using $(<, =, >)$-comparisons, where by on-line we mean that the prefixes ending at position i of $Q[1..q]$ are reported before examining symbols at positions that are larger than i.*

Proof: Given a collection of strings $\{S_1, \cdots, S_p\}$, we construct the suffix tree T_S for the string $S = S_1\#_1 \cdots S_p\#_p$, where $\#_1, \ldots, \#_p$ are special distinct alphabet symbols that do not appear anywhere in the given strings and in $Q[1..q]$. This suffix tree together with some auxiliary data structures that we describe next is used to find all prefixes of $Q[1..q]$ that are equal to suffixes of some of the given strings.

If several suffixes of strings $S_1, \cdots, S_p$ are equal, then the leaves in T_S that are associated with these suffixes have the same parent node. Using this property, it is possible to identify all suffixes of the strings $S_1, \cdots, S_p$ that are equal and assign unique names to equal suffixes. We agree that the unique name for a suffix is defined as the 2-tuple that contains the suffix length and the smallest index of a string in the collection that has the same suffix. It is possible to find all the equal suffixes and assign the unique names to the suffixes by a simple traversal of the suffix tree T_S in $O(l)$ time.

Since we can not hope to report all the strings in the collection that have a given suffix within the required time bounds, we will report only the unique name that was assigned to each suffix that is found. Using this unique name, it is possible to find the rest of the strings that have the same suffix, if necessary.

The algorithm proceeds by traversing the suffix tree T_S starting from the root, while Q is given on-line a symbol at a time. The algorithm follows an edge from a node v to a node w in the suffix tree if the label on this edge is equal to the next symbols of Q. Since sibling edges in the suffix tree are labeled with substrings that start with different symbol, the algorithm can decide which edge to follow based only on the current symbol of Q. The algorithm then continues and checks that the complete label of that edge is equal to the next symbols of Q.

When the traversal reaches a node v of the suffix tree, the algorithm has to report if it found a suffix of any of the strings $S_1, \cdots, S_p$. If there is an an edge from v to a leaf, and the edge has a label that starts with one of the special symbols '$\#_i$', then a suffix of the string S_i was found. Note that there might be many strings that have this same suffix, and as agreed before, we report only the smallest index of a string that has this suffix (this index was precomputed and stored at the node v of the suffix tree).

It is also possible that the algorithm did not find an edge whose label matches the next symbols of Q. In this case the algorithm terminates since it is not possible that longer prefixes of Q are equal to suffixes of the strings $S_1, \cdots, S_p$. However, the algorithm might have found an edge whose label has a nonempty prefix that matches the next symbols of Q. If the label on this edge matches the next symbols of Q up to a special symbol '$\#_i$', then the algorithm reports also that this prefix of Q is equal to a suffix of S_i. Note that in this case there is only one suffix to be reported.

When the algorithm has to choose which edge to follow from a node, it can decide based only on the first symbols of the edge labels. Normally, if $(<, =, >)$-comparisons are used, one can search for the current symbol of Q in the list of first symbols of the edge labels using binary search or binary search trees [3, 29]. This binary search is the cause of the $\log \sigma_S$ multiplicative factor in the running time of the suffix tree construction algorithm and also in the running time of the Aho-Corasick dictionary-matching algorithm. Note, that since the special symbols '$\#_i$' do not appear in Q, they do not have to be considered in this binary search.

However, a binary search does not give us the bounds we are trying to achieve. We rather proceed in the spirit of the algorithm given in Section 2.1 and create a data structure that will be used to guide the search for the edge that the algorithm follows more efficiently.

Given a suffix tree T_S, we define the weight of a node v as the number of leaves in the subtree rooted at v. The weights of all nodes can be computed by a simple traversal of the suffix tree in $O(l)$ time.

In order to find the edge whose label starts with the current symbol of Q, we use a weighted search. Similarly, to the binary search, the edges at a node v are ordered by the first symbols on their labels. However, instead of choosing the median of the remaining possible edges, we choose the weighted median where each edge has the weight of the node it is leading to. This way, instead of eliminating half of the possible edges in case of a mismatch, we eliminate half of the weight. Since the weight of the current node that the algorithm considers only decreases when the algorithm finds the right edge and follows it to the next node, we are guaranteed that there are no more than $\log l$ mismatches throughout the algorithm. Since there are up to q comparisons that result in equal answers and all other steps take constant time, the total time spent is bounded by $O(q + \log l)$.

To implement the weighted search efficiently, we precompute a search tree for each node of the suffix tree T_S. The search trees of all the nodes can be computed in $O(l \log \sigma_S)$ time and stored in $O(l)$ space using standard data structures. □

The same data structure can also be used to find all suffixes of a string $Q[1..q]$ that are equal to prefixes of some strings $\{S_1, \cdots, S_p\}$ by considering the reversed strings.

2.3.2 The implementation

Recall that the dictionary patterns have a uniform length m, and assume without loss of generality that the text T is of length $2m - 1$. The backward and forward phases are implemented independently and symmetrically. Then, their results are combined to report the occurrences of the patterns in the text.

The dictionary preprocessing builds the data structure described in Lemma 2.6 to find all prefixes of the dictionary patterns that occur at the end of $T[1..m]$ and all suffixes of the dictionary patterns that occur at the beginning of $T[m+1..2m-1]$. The occurrences of patterns in the text are reported by checking if the concatenation of some of the prefixes that were found with some of the suffixes is equal to some patterns in the dictionary.

To check efficiently if the string obtained by the concatenation of the pattern prefix $\mathcal{P}_k[1..i]$ with the pattern suffix $\mathcal{P}_l[i+1..m]$ is equal to some dictionary pattern, the dictionary preprocessing step computes a table $presuf_i(k, l)$ the gives the index of a dictionary pattern that is equal to prefix-suffix concatenation if such a pattern exists. If there are several patterns in the dictionary that are equal, we agree that we report only a representative which is the smallest index of a pattern that was found. If there is a need to report the indices of all patterns an additional list data structure is created to identify all equal patterns.

The $presuf_i(k, l)$ table is defined for $i = 1, \cdots, m$ and $k, l = 1, \cdots, d$ and it seems to require $O(md^2)$ space. However, since the data structure in Lemma 2.6 reports a unique representative for each pattern prefix or suffix, and since there are only d strings in the dictionary, the table $presuf_i(\cdot, \cdot)$, has only d relevant entries, for each $i = 1, \cdots, m$. Therefore, the whole table has $O(md)$ relevant entries (note that this table need not be initialized). Using the hashing scheme of Fredman, Komlos and Szemeredy [22] it is possible to represent this table in $O(md)$ space with a constant time access, after an initialization step that is done in the dictionary preprocessing step and takes $O(md)$

time with high probability [21]. Alternatively. given a constant $\epsilon > 0$. this table can be represented in $O(md^{1+\epsilon})$ space with an $O(1/\epsilon)$ access time similarly to Apostolico et al. [11].

Lemma 2.7 *There exists a dictionary-matching algorithm for uniform length dictionaries with d patterns of length m and a text of length $2m - 1$ that takes $O(\log d + m)$ time using $(<, =, >)$-comparisons. The dictionary preprocessing takes $O(md \log \sigma_D)$ time and uses $O(md^2)$ space.*

Proof: The pattern preprocessing first creates the data structure that was described in Lemma 2.6 for the forward and backward phases. These data structures are created in $O(md \log \sigma_D)$ time and use $O(md)$ space. The $presuf_i(k, l)$ table is created in $O(md)$ time and uses $O(md^2)$ space.

The text scanning step uses these data structures to find the prefixes and suffixes of dictionary patterns in the text in $O(\log d + m)$ time. It then reports the occurrences of the dictionary patterns using the $presuf_i(k.l)$ table in $O(m)$ time. □

The lemma above is used repeatedly when the text is longer. The following theorem summarizes the properties of the new dictionary matching algorithm.

Theorem 2.8 *There exists an on-line dictionary-matching algorithm for uniform length dictionaries with d patterns of length m and a text of length n that takes $O((\frac{\log d}{m} + 1)n)$ time using $(<, =, >)$-comparisons.*

3 Higher Dimensional Array Matching

Bird [15] and Baker [13] independently discovered a higher-dimensional array-matching algorithm that uses a one-dimensional dictionary-matching algorithm as a subroutine. Their original algorithm used the Aho-Corasick [2] algorithm that was available at the time, but any dictionary-matching algorithm can be used. It particular, by using the dictionary-matching algorithm that was described in Section 2.3 it is possible to improve the running time of the Bird-Baker array-matching algorithm on unbounded ordered alphabets.

Bird and Baker reduced the k-dimensional array matching problem into k iterations of a one-dimensional dictionary matching problems. We explain how their algorithms works in two-dimensions.

Assume that the pattern is given as $\mathcal{P}[1..m. 1..m]$ and the text is $T[1..n. 1..n]$. Note that each row or column of a two-dimensional array can be regarded as a one-dimensional string.

1. Denote the pattern rows $\mathcal{P}_i = \mathcal{P}[i. 1..m]$. The algorithm starts by solving a dictionary matching problem that finds all occurrences of the uniform length pattern rows $\mathcal{P}_i$, $i = 1..m$, in each row of the text.

   In fact, since some of the pattern rows might be identical. the algorithm first finds a unique representative for each set of equal pattern rows and searches only for these representatives. Namely. $\mathcal{P}_{u(i)}$ is the representative for pattern row $\mathcal{P}_i$, such that $\mathcal{P}_{u(i)} = \mathcal{P}_i$ and $u(i) = u(j)$. for all pattern rows $\mathcal{P}_i = \mathcal{P}_j$. The pattern is clearly represented as:

$$\mathcal{P}[1..m, 1..m] = \begin{array}{l} \mathcal{P}_{u(1)} \\ \mathcal{P}_{u(2)} \\ \cdot \\ \cdot \\ \mathcal{P}_{u(m)}. \end{array}$$

The output of the dictionary matching problem is an array $T_1[1..n, 1..n]$ such that $T_1[j, k] = u(i)$ if an occurrence of $\mathcal{P}_{u(i)}$ was discovered in the text row starting at $T[j, k]$, and $T_1[j, k] = 0$ if no occurrence of a pattern row starts at $T[j, k]$. More formally, $T_1[j, k] = u(i)$ if and only if $T[j, k + l - 1] = \mathcal{P}[u(i), l]$, for $l = 1..m$.

The unique representatives $\mathcal{P}_{u(i)}$ can be found while the dictionary of the pattern rows $\mathcal{P}_i$, $i = 1..m$, is created. The dictionary preprocessing takes $O(m^2 \log m)$ time and the text scanning step takes $O(n^2)$ time.

2. Clearly, there is an occurrence of the pattern $\mathcal{P}[1..m, 1..m]$ starting at $T[j, k]$ if and only if $T_1[j + l - 1, k] = u(l)$, for $l = 1..m$.

   Therefore, the occurrences of the pattern can be discovered by using a string matching algorithm to search for occurrences of the sequence $u(1)u(2) \cdots u(m)$ in each column of $T_1[1..n, 1..n]$.

   These occurrences can be found by using the Knuth-Morris-Pratt string-matching algorithm with $O(m)$ time pattern preprocessing and $O(n^2)$ time text scanning.

Thus, the overall time complexity of the Bird-Baker approach using the new dictionary-matching algorithm is $O(n^2 + m^2 \log m)$. The extension to k-dimensions is straightforward. It requires $O(n^k)$ time for the text scanning step and $O(m^k \log m)$ time for the pattern preprocessing. It is important to note the following two facts:

1. The last iteration of the Bird-Baker algorithm searches only for a single pattern and can use the Knuth-Morris-Pratt string matching algorithm.

2. The matching problems solved after the first iteration of the Bird-Baker algorithm do not involve the input alphabet. Therefore, even if the input alphabet is of constant size, the alphabet size in the later iterations might be larger.

3.1 Remarks

1. In the discussion above we assume that the input arrays are of a fixed dimension k. This allows the *Big-O* notation to hide multiplicative factors that depend on k.

   In fact, all published higher-dimensional array-matching algorithms that we are familiar with, including the Bird-Baker algorithm, require in one form or another k iterations over the text array. Therefore, the time complexity bounds of these algorithms include a multiplicative factor of k. A second multiplicative factor is usually hidden in the fact that indexing a k-dimensional array required $O(k)$ time. However, in many cases it is possible to avoid this dependence on k by paying a careful attention to the access pattern and by using the one-dimensional memory representation of arrays.

   The bounds of our algorithm sometimes hide a few more multiplicative k factors.

2. The discussion above clearly generalizes to rectangular k-dimensional array. However, to obtain good time bounds one has to pay attention to the direction, or order of coordinates, in which the arrays are processed.

   For example, suppose that the pattern is an $m_1 \times m_2$ two-dimensional array where $m_1 \ll m_2$ and the text is an $n \times n$ square. The Bird-Baker approach solves a dictionary-matching problem followed by a string matching problem. The time complexity of the dictionary-matching problem depends on the direction in which the algorithm proceeds: $O((\frac{\log m_1}{m_2} + 1)n^2)$ time for m_1 pattern strings of length m_2 each, or alternatively, $O((\frac{\log m_2}{m_1} + 1)n^2)$ time for m_2 pattern strings of length m_1 each. Clearly, the first direction is better, yielding an $O(n^2)$ algorithm.

   The same principle holds in higher dimensions. If the coordinates are ordered in such a way that the pattern is an $m_1 \times \cdots \times m_k$ array with $m_1 \leq \cdots \leq m_k$ and the text is an $n_1 \times \cdots \times n_k$ array, then the text scanning step takes $O(n_1 \times \cdots \times n_k)$ time and the pattern preprocessing step takes $O(m_1 \times \cdots \times m_k \times \log m_k)$ time.

3. It is interesting to point out that there exist a trivial comparison model algorithm for k-dimensional array-matching using $(2^{k-1} + 1)m^k + 2n^k$ $(=, \neq)$-comparisons, including preprocessing. In the standard model, linear bounds are currently achieved only in one and two-dimensions. The complicated details of the two-dimensional algorithm of Galil and Park [23] might suggest that better understanding of periodicity (overlap) properties of higher dimensional arrays [4, 14, 28] would be required before linear time higher dimensional array-matching algorithms are available in the standard model.

4   Acknowledgments

The author wishes to thank Alberto Apostolico, Wojciech Szpankowski and Laura Toniolo for several discussions and comments on this work, Yossi Matias for his advise on the hashing literature and Zvi Galil for his help in obtaining copies of some bibliography items.

References

[1] A.V. Aho. Algorithms for Finding Patterns in Strings. In J. van Leeuwen, editor, *Handbook of Theoretical Computer Science*. pages 257–300. Elsevier Science Publishers B. V., Amsterdam, the Netherlands. 1990.

[2] A.V. Aho and M.J. Corasick. Efficient string matching: An aid to bibliographic search. *Comm. of the ACM*. 18(6):333–340, 1975.

[3] A.V. Aho, J.E. Hopcroft. and J.D. Ullman. *The Design and Analysis of Computer Algorithms*. Addison-Wesley, Reading, MA., 1974.

[4] A. Amir and G. Benson. Two-dimensional periodicity and its applications. In *Proc. 3rd ACM-SIAM Symp. on Discrete Algorithms*, pages 440–452, 1992.

[5] A. Amir, G. Benson, and M. Farach. Alphabet-Independent Two-Dimensional Matching. In *Proc. 24th ACM Symp. on Theory of Computing*, pages 59–68, 1992.

[6] A. Amir and M. Farach. Adaptive dictionary matching. In *Proc. 32th IEEE Symp. on Foundations of Computer Science*, pages 760–766, 1991.

[7] A. Amir and M. Farach. Two-dimensional dictionary matching. *Inform. Process. Lett.*, 44:223–239, 1992.

[8] A. Amir, M. Farach, Z. Galil, R. Giancarlo, and K. Park. Dynamic Dictionary Matching. Manuscript, 1992.

[9] A. Amir, M. Farach, R.M. Idury, J.A. La Poutré, and A.A. Schäffer. Improved Dynamic Dictionary-Matching. In *Proc. 4nd ACM-SIAM Symp. on Discrete Algorithms*, pages 392–401, 1993.

[10] A. Amir, M. Farach, and Y. Matias. Efficient Randomized Dictionary-Matching Algorithms. In *Proc. 3rd Symp. on Combinatorial Pattern Matching*, number 644 in Lecture Notes in Computer Science, pages 262–275. Springer-Verlag, Berlin, Germany, 1992.

[11] A. Apostolico, C. Iliopoulos, G.M. Landau, B. Schieber, and U. Vishkin. Parallel construction of a suffix tree with applications. *Algorithmica*, 3:347–365, 1988.

[12] R. Baeze-Yates and M. Régnier. Fast two-dimensional pattern matching. *Inform. Process. Lett.*, 45:51–57, 1993.

[13] T.P. Baker. A Technique for Extending Rapid Exact-Match String Matching to Arrays of More than One Dimension. *SIAM J. Comput.*, 7(4):533–541, 1978.

[14] G.E. Benson. *Two-Dimensional Periodicity and Matching Algorithms*. PhD thesis, Dept. of Computer Science, University of Maryland, 1992.

[15] R.S. Bird. Two Dimensional Pattern Matching. *Inform. Process. Lett.*, 6(5):168–170, 1977.

[16] R.S. Boyer and J.S. Moore. A fast string searching algorithm. *Comm. of the ACM*, 20:762–772, 1977.

[17] M.T. Chen and J. Seiferas. Efficient and elegant subword-tree construction. In A. Apostolico and Z. Galil, editors, *Combinatorial Algorithms on Words*, volume 12 of *NATO ASI Series F*, pages 97–107. Springer-Verlag, Berlin, Germany, 1984.

[18] B. Commentz-Walter. A string matching algorithm fast on the average. In *Proc. 6th International Colloquium on Automata, Languages, and Programming*, Lecture Notes in Computer Science, pages 118–132. Springer-Verlag, Berlin, Germany, 1979.

[19] B. Commentz-Walter. A string matching algorithm fast on the average. Technical Report 79.09.007, IBM Wissenschaftliches Zentrum, Heidelberg, Germany, 1979.

[20] M. Crochemore, A. Czumaj, L. Gasieniec, S. Jarominek, T. Lecroq, W. Plandowski, and W. Rytter. Fast Practical Multi-Pattern Matching. Technical Report 93-3, Institut Gaspard Monge, Université de Marne la Vallée, Marne la Vallée, France, 1993.

[21] M. Dietzfelbinger, J. Gil, Y. Matias, and N. Pippenger. Polynomial Hash Functions Are Reliable. In *Proc. 19th International Colloquium on Automata, Languages, and Programming*, number 623 in Lecture Notes in Computer Science. Springer-Verlag, Berlin, Germany, 1992.

[22] M.L. Fredman, J. Komlos, and E. Szemeredi. Storing a sparse table with $O(1)$ worst case access time. *J. Assoc. Comput. Mach.*, 31(3):538–544, 1984.

[23] Z. Galil and K. Park. Truly Alphabet-Independent Two-Dimensional Pattern Matching. In *Proc. 33th IEEE Symp. on Foundations of Computer Science*, pages 247–256, 1992.

[24] R.M. Idury and A.A. Schäffer. Dynamic Dictionary-Matching with Failure Functions. In *Proc. 3rd Symp. on Combinatorial Pattern Matching*, number 644 in Lecture Notes in Computer Science, pages 276–287. Springer-Verlag, Berlin, Germany, 1992.

[25] R.M. Karp and M.O. Rabin. Efficient randomized pattern matching algorithms. *IBM J. Res. Develop.*, 31(2):249–260, 1987.

[26] D.E. Knuth, J.H. Morris, and V.R. Pratt. Fast pattern matching in strings. *SIAM J. Comput.*, 6:322–350, 1977.

[27] E.M. McCreight. A space economical suffix tree construction algorithm. *J. Assoc. Comput. Mach.*, 23:262–272, 1976.

[28] M. Régnier and L. Rostami. A Unifying Look at d-dimensional Periodicities and Space Coverings. In *Proc. 4rd Symp. on Combinatorial Pattern Matching*, number 684 in Lecture Notes in Computer Science, pages 215–227. Berlin, Germany, 1993. Springer-Verlag.

[29] R.E. Tarjan. *Data Structures and Network Algorithms*. SIAM, Philadelphia, PA., 1985.

[30] P. Weiner. Linear pattern matching algorithms. In *Proc. 14th Symposium on Switching and Automata Theory*, pages 1–11, 1973.

[31] R.F. Zhu and T. Takaoka. A Technique for Two-Dimensional Pattern Matching. *Comm. of the ACM*, 32(9):1110–1120, 1989.

Proximity Matching Using Fixed-Queries Trees

Ricardo Baeza-Yates[1]*, Walter Cunto[2], Udi Manber[3]** and Sun Wu[4]

[1] Dpto. de Ciencias de la Computación, Universidad de Chile,
Blanco Encalada 2120, Santiago, Chile
[2] IBM Consulting Group, Aptdo. 64778 &
Dpto. de Computación y Tecnología de la Información,
Univ. Simón Bolívar, A.P. 68000, Caracas, Venezuela
[3] Dept. of Computer Science, University of Arizona, Tucson, AZ 85721, USA
[4] Dept. of Computer Science, National Chung-Cheng Univ.,
Ming-Shong, Chia-Yi, Taiwan

Abstract. We present a new data structure, called the fixed-queries tree, for the problem of finding all elements of a fixed set that are close, under some distance function, to a query element. Fixed-queries trees can be used for any distance function, not necessarily even a metric, as long as it satisfies the triangle inequality. We give an analysis of several performance parameters of fixed-queries trees and experimental results that support the analysis. Fixed-queries trees are particularly efficient for applications in which comparing two elements is expensive.

1 Introduction

Search structures such as hashing and trees are at the basis of many efficient computer science applications. But they usually support only exact queries. Finding things approximately, that is, allowing some errors in the query specifications, is much harder. The first question that a prominent biologist once asked one of the authors when finding that he is a computer scientist is whether it is possible to adapt binary search to allow approximate queries. In this paper we present a new data structure that makes some progress towards this end. It does not, by any measure, solve the problem, but it may improve the efficiency of approximate search for some applications.

We assume that we have a fixed set X, which we can preprocess and store in a data structure. The problem is to find all elements in the set that are close, under some distance function – we study in detail the Hamming and Levenshtein distance functions – to a query element. This problem has been extensively studied for distance functions based on a lexicographical order (closest neighbor or closest

---

* This work was partially supported by Grant 1930765 from Fondecyt.
** Supported in part by NSF grants CCR-9002351 and CCR-9301129, and by the Advanced Research Projects Agency under contract number DABT63-93-C-0052. The information contained in this paper does not necessarily reflect the position or the policy of the U.S. Government or other sponsors of this research. No official endorsement should be inferred.

point problem). We refer the reader to [SW90, Mur83] for techniques that assume linear orderings, Euclidean, and similar distance functions. The problem is much harder, however, for distance functions that are not related to a linear ordering.

The practical approach for finding all elements in a database close to a query element is typically to design quick algorithms that approximate the distance (e.g., BLAST [AGMML90] and FASTA [LP85] for biological sequences, and [BGKN90] for speech recognition). The approximated distance can be used to filter elements that seem too far away, but these algorithms need to be applied to all elements of the database. Their running time is therefore linear in the size of the database.

Another approach, for arbitrary distance functions, was given by Burkhard and Keller [BK73] in the context of database queries. They designed a tree structure, later called BK-trees, making heavy use of the triangle inequality. This idea was improved by Shapiro [Sha77] by using a set of trees and deriving stricter criteria for filtering objects. BK-trees were later compared [NK82] to a variation of k-d trees [FBF77], and found in general to have slower running time for closest neighbor searching. However, the comparison was done only for the Hamming distance and very small data sets. Further work [SDDR89] studied the effect of approximating the Leveshtein distance using simpler distance functions that are easier to compute and bound from above the original function. However, the reduction in the computation of the distance was traded by the number of extra comparisons needed. Shasha and Wang [SW90] extended BK-trees to any set of precomputed distances, using them in an optimal way. They compute an approximate distance map of the database to guide the search by using a Floyd-Warshall style algorithm of $O(n^3)$ running time. They study empirically the effect of the number of precomputed distances and the distribution of the distances for small sets (hundreds of objects), and compare their algorithm with BK-trees obtaining better performance, which improves with more precomputed distances. They also conjecture that star-like precomputed graphs are the best topology, and that explains why the multiple star topology of a BK-tree leads to good performance.

Different algorithmic approaches that lead to $o(N)$ expected performance both theoretically and in practice were developed recently by several people. They are quite different, although they all use the idea of filtering large regions of the database (or large text) so that more expensive techniques are used only for small regions. Myers [My94] designed an $O(N^\alpha)$ algorithm to match biological sequences. Although designed specifically for DNA and protein matching, his approach is applicable essentially to any distance function (although the complexity depends on it). Ukkonen [Uk92] presented another approach based on comparing k-grams. Gonnet et. al. [GCB92, BYG90] developed techniques, based on suffix arrays, to compare a sequence against a database or even the whole database against itself. Ukkonen [Uk93] independently developed similar techniques using suffix trees. Chang and Lawler [CL90] also gave an $o(N)$ algorithm, based on suffix trees, for this problem. Recently [BR+93], the monotonous bisector tree used in computational geometry plus the use of q-grams of [Uk92] as strings profiles was presented. However, their results show that the filtering achieved by using a triangle inequality distance bound based on q-grams were not very good, obtaining a reasonable performance only for very long keys (64 or more) and close proximity searchings.

In this paper we present yet another approach, similar to BK-trees, which achieves

$O(N^\alpha)$, with $\alpha < 1$, expected complexity for finding close matches. We make heavy use of the triangle inequality. We present a tree structure that can handle proximity queries for any distance function that satisfies the triangle inequality. The main goal of this structure is to minimize the number of element comparisons, as opposed to other data structure operations (such as tree traversal). This is especially important in applications where computing distances between two elements is a much more expensive operation than, say following pointers. Our tree structure differs from other tree structures in that the keys on each level of the tree are all the same, so we have just one key per level. In other words, which comparisons we make does not depend on the results of previous comparisons up the tree. Thus the name fixed-queries tree or FQ-tree. This may seem like a poor strategy (e.g., regular binary search trees will yield a linear search with this policy), but it has several advantages for our intended applications. First, it minimizes element comparisons, because traversing different parts of the tree does not require more comparisons. Second, it allows all comparisons to the keys to be done in one batch, enabling easy parallelism. Third, it simplifies the data structure and its analysis.

A major strength of FQ-trees, which also applies to BK-trees, is that the data structure and the search algorithms are independent of the distance function, as long as it satisfies the triangle inequality. If someone comes up with different distance measures or different algorithms to compute them, they will still be able to use our scheme without any other change to filter their data. The amount of filtering may depend on the distance distribution but not the algorithms. Our scheme allows one to use the expensive exact algorithms (e.g., dynamic programming) and still be able (at least for small distances) to search in a large database.

One of the main contributions of the paper is the analysis of FQ-trees. The analysis concentrates in particular on two classical distance functions: Hamming (mismatches only) and Levenshtein (also insertions and deletions). We show that FQ-trees provide logarithmic expected time for exact search, and sublinear expected time for proximity or neighborhood search, among other results. In particular we study the effect of the key length, the alphabet size, the maximal error allowed in a proximity search, and the bucket size. These results are corroborated by experimental results.

2 Fixed-Queries Trees

Let U be the set of all possible elements defined by the problem. For the rest of the paper we will assume that $X \subset U$ is a large set of elements, each of which is a large object by itself, and that U is much larger than X. In the analysis section we will concentrate on strings as elements, but they can be arbitrary. An FQ-tree differs from a regular search tree in two major respects. First, the keys are not numbers from an ordered set, they are members of U. Second, all internal nodes at the same level are associated with the same key. We will use keys to denote the elements that are compared while traversing the tree, and elements to denote members of X. Let *dist* be a distance function that satisfies the triangle inequality and assume that the set of all possible distances is a finite set $\{d_0, d_1, d_2, ...d_s\}$. (If this set is not finite, or if it is very large, we can discretize it to s different ranges of values.) An FQ-tree for a set of elements X and a distance function *dist* is a tree that satisfies the following four properties.

- All elements of X are associated with the leaves.
- If X contains only b elements, then the tree consists of one (leaf) node containing all elements (b is the *bucket* size).
- All internal nodes of depth r are associated with the same key k_r. The keys can be selected at random, they can be members of X, or they can be selected especially to optimize the tree; we assume that they are random.
- Every internal node v is a root of a valid FQ tree associated with a set $X_v \subset X$. An internal node v with key k_r has one subtree for every non-empty set X_i defined by

$$X_i = \{x \in X_v \text{ such that } dist(k_r, x) = d_i \geq 0\}.$$

Figure 1 shows an example of an FQ-tree.

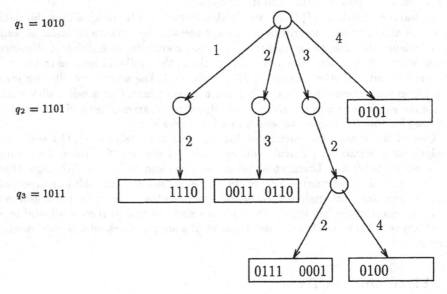

Fig. 1. Example of an FQ-tree with $b = 2$ using the Hamming distance.

A search for a query element q proceeds down the tree by computing at each level the distance between q and the key k_i for that level. If the search is an exact search, then at each level the child that corresponds to the distance between q and k_i is selected until a leaf is reached. A neighborhood or proximity search is more complicated. Suppose that we want to find all elements of X that are within distance D to q. We use the triangle inequality to filter nodes. If we are at an internal node v with associated key k_i and if $dist(q, k_i) = d_i$, then all children of v associated with distances d such that $d_i - D \leq d \leq d_i + D$ may contain elements with distance D to q. We search, recursively, all of them. This is where the advantage of having the same key at all nodes of the same level comes to play. Even though we may traverse many children, we make only one key comparison per level. Therefore, the main cost of a proximity search is related to the sum of the height of the tree (which indicates how many key comparisons we need to make), and the content of all buckets that are reached in the search (which are all unfiltered elements – those that are within distance $d_i - D$ to $d_i + D$ for all d_i). The overhead of traversing the data structure (e.g., finding the right children), is only secondary, because following pointers is much less expensive than comparing two complex objects.

3 Analysis of FQ-trees

In the analysis below, we make several simplifying assumptions, which makes the analysis imprecise. Nevertheless, we believe that the results reasonably approximate the true behavior of FQ trees and the empirical results support this belief.

We assume that the elements of X are strings over a finite alphabet Σ of size $|\Sigma| = \sigma$. We assume that the possible range of valid distances is level independent. This is clearly not true, because two elements at distance d to a given key are at most $2d$ apart, by the triangle inequality. We discuss the error made by the analysis at the end of the section.

Let p_k ($k \geq 0$) be the probability of two strings being at distance d_k for a given distance function for $0 \leq k \leq s$. We define $p_k = 0$ for $k > s$. This probability distribution will be modeled later for our two distance functions: Hamming and Levenshtein. We start with some terminology:

- n – the number of elements in the tree;
- b – the size of the bucket;
- $H(n)$ – the expected height of the tree (not counting the leaf level);
- $C(n)$ – the expected number of comparisons for an exact search;
- $I(n)$ – the average internal path length to a leaf;
- $N_D(n)$ – the expected number of internal nodes traversed in a proximity query of distance D;
- $B_D(n)$ – the expected number of leaf elements compared in a proximity query of distance D;
- $P_D(n)$ – the expected number of comparisons for a proximity query of distance D;

Under the assumptions above we have that

$$H(n) = 1 + \max_{k=0}^{s}(H(\lceil p_k n \rceil)), \quad H(i) = 0 \ (i \leq b)$$

where we approximate the expected number of nodes in the k-th subtree as $\lceil p_k n \rceil$. That is, the expected height is given by the expected largest subtree. Although this is not exactly true, an exact analysis is very difficult. Since the distributions we use are centered and concentrated in a small range, the approximations are good as shown by the experimental results.

Assuming without loss of generality that $p_k < 1$ for all k, the recurrence for $H(n)$ has the following solution

$$H(n) = \frac{\log(n)}{\log(1/\max_k(p_k))} + O(\log b)$$

The recurrence for $I(n)$ is

$$I(n) = 1 + \sum_{k=0}^{s} p_k I(\lceil p_k n \rceil), \quad I(i) = 0 \; (i \leq b)$$

which gives

$$I(n) = \frac{\log(n)}{\sum_{k=0}^{s} p_k \log(1/p_k)} + O(\log b)$$

The recurrence for $C(n)$ is similar to that of $I(n)$, except that the boundary condition is changed to $C(i) = i$ for $i \leq b$. Clearly,

$$I(n) \leq C(n) \leq I(n) + b$$

so $C(n)$ has the same complexity as $I(n)$.

The recurrence for $N_D(n)$ is

$$N_D(n) = 1 + \sum_{k=0}^{s} p_k \sum_{j=\max(0,k-D)}^{\min(k+D,s)} N_D(\lceil p_j n \rceil), \quad N_D(i) = 0 \; (i \leq b)$$

where we are using the triangle inequality to prune some subtrees, searching only on the subtrees at distance $k - D$ to $k + D$ of the current distance k. This equation can be rewritten as

$$N_D(n) = 1 + \sum_{k \geq 0} \gamma_k(D) N_D(\lceil p_k n \rceil)$$

where $\gamma_k(D) = \sum_{i=\max(0,k-D)}^{\min(k+D,s)} p_i$. Using induction one can show that for a fixed value of D and constant b we have

$$N_D(n) = O(n^\alpha)$$

with $0 < \alpha < 1$. The value of α is obtained from the following transcendental equation

$$\sum_{k=0}^{s} \gamma_k(D) p_k^\alpha = 1$$

Specific values for α are given later for the Hamming and Levenshtein distances. Similarly, the recurrence for $B_D(n)$ is

$$B_D(n) = \sum_{k=0}^{s} p_k \sum_{j=\max(0,k-D)}^{\min(k+D,s)} B_D(\lceil p_j n \rceil), \quad B_D(i) = i \; (i \leq b)$$

A string can be also seen as a point in an m-dimensional space with integer coordinates in the set $\{1, 2, ..., |\Sigma|\}$. The Hamming metric can be computed in $O(m)$ time.

We assume here that all strings have the same length m and that each key symbol is drawn independently from the alphabet Σ. Let $q = 1 - 1/\sigma$ be the probability that two symbols differ. The probability that the distance between two elements a and b is k is given by the binomial distribution

$$p_k(m, \sigma) = Pr\{Ham(a, b) = k\} = \binom{m}{k} q^k (1 - q)^{m-k}$$

with $0 \leq k \leq m$. The maximum value of $p_k(m, \sigma)$, the mode, is given by k such that $p_k \geq p_{k-1}$ and $p_k > p_{k+1}$. Solving these equations we get $mq - q \leq k \leq mq - q + 1$. That is, k is very close to the expected value of the distribution which is mq. So, using $k = (\sigma - 1)m/\sigma$ we obtain the following upper bound when σ is finite and m is large

$$H(n) \leq 2(1 + O(1/\log m)) \log_m n$$

similarly to random m-ary tries [GBY91]. It is difficult to obtain other closed expressions given the complexity of the analysis. Using the formula for $N_D(n)$ we can compute the values for α, which gives the complexity of $P_D(n) = O(n^\alpha)$. Figure 2 shows the values for α in a proximity searching for different values of m and D with $\Sigma = 4$. We also include experimental values for α (dotted lines) obtained by using a least squares fit on the data.

Using the probability distribution p_k we can compute all the performance measures using the formulas of the previous section. Figure 3 shows analytical and experimental values for proximity searching ($P_D(n)$) in function of n for different values of D when $\sigma = 4$ and $m = 16$. Experimental values are given by the dotted lines (as in all the other graphs) and agree reasonable well with the analysis. We have used a logarithmic y-axis to be able to represent several values of D in the same graph (the same occurs in subsequent graphs).

Figures 4 and 5 shows the the effect of the key length and the alphabet size for proximity searching. When m increases, the number of comparisons decrease. This happens because the fan-out of the tree increases. On the other hand, when σ increases the number of comparisons does the same. This happens because the probability distribution of the possible distances concentrates near the average (approaching m) and the average fan-out decreases. Figure 6 shows the effect of the bucket size b on $H(n)$ and $I(n)$ for $m = 16$, $\Sigma = 4$, and $n = 90000$. Figure 7 shows the effect of the bucket size on $P_D(n)$ (solid lines) for $n = 90000$ with $m = 16$ and $\Sigma = 4$. Clearly, the optimal bucket size is $b = 1$. Using the formulas for $B_D(n)$ and $N_D(n)$ we can compute how much we are saving with respect to BK-trees by not having to do one comparison per internal node, which would give $B_D(n) + N_D(n)$ instead of $P_D(n)$. Figure 7 shows also this quantity (dashed lines), and that the improvement over BK-trees is more than 50% for small b. We also have experimental results for finding the closest match, and for that case, the number of comparisons after some point decreases with n. This is due to the fact that for larger n the probability of finding a close match increases, which prunes the tree search more rapidly.

which has the same complexity as $N_D(n)$ but different constant multiplicative factor. The total number of comparisons done in a proximity query, $P_D(n)$, in a FQ-tree is then bounded by

$$B_D(n) + I(n) \leq P_D(n) \leq B_D(n) + H(n)$$

Because $I(n)$ and $H(n)$ are logarithmic, $P_D(n) = O(B_D(n)) = O(n^\alpha)$. (In the original BK-tree, the terms $I(n)$ and $H(n)$ are similar but they have to be replaced by $N_D(n)$; although the complexity does not change, there is a significant reduction in the constant factor when using FQ-trees.)

The above analysis is an approximation because of several facts:

- In the recurrences we use the ceiling of the expected number of elements in every branch. This may increase the number of elements in a subtree by 1. In the recurrences for $H(n)$ and $I(n)$ the effect is minimal. For $N_D(n)$ the effect of this error increases with D and over estimates the real values. So, the formulas only are approximate. We can fix this error by computing exact values for small n, and we have done so for $N_D(n)$, although the results obtained are similar.

- Another source of error is the fact that any pair of elements which are at a given distance k to another key, have their distance bounded by $2k$ (by the triangle inequality). We can take this in account into the recurrences by adding a second parameter that carries the maximal possible distance (in other words, two close strings have common segments). For example, for $H(n)$ we get

$$H(n,r) = 1 + \max_{k=0}^{s}(H(\lceil p_k n \rceil, \min(2k, r))), \quad H(i,r) = 0 \ (i \leq b, \ r \geq 0)$$

where r denotes the maximal distance. The effect of this new parameter depends on the probability distribution and only if p_k, $k \leq s/2$ is significative. This is the case for the Hamming distance when $\sigma = 2$ and for the Leveshtein distance. For the Hamming distance, if $\sigma > 2$ the probability distribution p_k is very skewed and the results are the same.

In spite of the two problems mentioned above, the experimental results shown in the next section are very close to the analytical results showing that the final error is small. On the other hand, any exact analysis must consider all possible tree arrangements and/or multinomial distributions as in B-trees or m-ary trees [Mah92].

4 Distance Functions

We study the two most common distance functions for strings – the Hamming and Levenshtein distances. In both cases we use Σ as the alphabet of size σ.

4.1 Fixed Length Keys: Hamming Case

The Hamming distance for two strings of length m is defined as the number of symbols in identical positions that are different. For example, $Ham(string, strong)$ is 1. For two strings a and b of size m we have

$$0 \leq Ham(a, b) \leq m$$

The experimental results were obtained by using random keys and running between 10 and 50 experiments, depending on the complexity measure. The largest variation was obtained for the height (as expected). For example, the experimental value for $H(90000)$ with $\sigma = 4$, $m = 16$, and $b = 2$ was on average 9.80 with a standard deviation of 0.42. In other measures the deviation was less than 1%. For almost all the figures given, the experimental results are very close to the approximated analytical values.

4.2 Variable Length Keys: Levenshtein Case

The Levenshtein distance is defined as the minimal number of characters that we need to change, insert, or delete to transform one of the strings into the other string. For example, $Lev(string, song)$ is 3. For two strings a and b we have

$$0 \leq Lev(a, b) \leq \max(|a|, |b|)$$

The Levenshtein distance can be computed in time $O(|a| \times |b|)$ by using dynamic programming.

We model the variable length key by generating independently the length of each key. We use a positive Poisson distribution with parameter λ for the length. This distribution models the fact that shorter strings are more probable than very long strings. Let L_i be the probability of a string a having length $i > 0$, then

$$L_i = Pr\{|a| = i\} = \frac{\lambda^{i-1}}{(i-1)!}e^{-\lambda}$$

The distance between two strings a and b is computed by using the length difference of the two elements and for the common prefix of length $m = \min(|a|, |b|)$, because both strings are random, we use the Hamming distance. So, without losing generality we have

$$Lev(a, b) = ||a| - |b|| + Ham(a_1...a_m, b_1...b_m).$$

But we know already the probability distribution p_k for the Hamming distance. Thus, the probability of $Lev(x, y)$ being k is

$$P_k = \sum_{i \geq 1} L_i \left(\sum_{j=\max(1,i-k)}^{i} L_j p_{k-(i-j)}(j, \sigma) + \sum_{j=i+1}^{i+k} L_j p_{k-(j-i)}(i, \sigma) \right)$$

Because the distance can be unbounded, we define $P'_k = P_k$ for $k < s = 2\lambda$, and $P'_s = 1 - \sum_{k=0}^{s-1} P_k$. With this probability distribution P' we can compute all the performance measures defined in the previous section as in the Hamming case. For example, Figure 8 shows the complexity of proximity searching for some values of D and λ (average string length) for $\Sigma = 30$ (a simple model for lower case text).

Figure 9 shows the effect of n on proximity searching for $\lambda = 5$ and $\sigma = 30$. The results are similar to the Hamming case with λ replacing m. The main difference in this case is that the probability distribution is more centered than for the Hamming case. We are currently working on experimental results for this distance function.

5 Conclusions and Future Work

In this preliminary paper we introduced FQ trees and showed their potential as a data structure that supports fast approximate queries. The novelty of FQ trees is having one fixed key per level to filter the input. This idea is obviously not generally applicable because it leads to traversal of many nodes. However, as we showed, the number of key comparisons is decreased by using our data structure. More work is necessary to see how FQ-trees compare to other data structures for specific applications in which comparing two elements is the major cost of the search. Also, an exact analysis would be desirable.

We mention briefly here one important variant of FQ-trees that takes the tradeoff of reducing the number of key comparisons vs. increasing the number of traversed nodes even further. Instead of building the tree recursively and stopping when a node contains no more than b elements, we can insist that every leaf has no less than a certain depth. In other words, we may want to replace leaves with paths. When we arrive at a leaf that contains even one element, there is no guarantee than this element fits the search criteria. We still need to compare the element directly to the query adding one more key comparison. But if we add more levels to a leaf, we increase the probability that the triangle inequality will filter out the element corresponding to that leaf. So, we will traverse more nodes, but at the end we will be left with a smaller set to compare directly to the query. And since we insist on one key per level, the total number of key comparisons will be significantly reduced.

A preliminary analysis shows that in expectation it is possible to achieve logarithmic number of key comparisons for a fixed small D. (The performance degrades significantly when D is increased.) Preliminary experimental data supports this analysis. It is too early, however, to predict whether this variant will be practical.

There are several other ways to improve and extend the basic idea of FQ-trees:

- The key selection, which we did at random, can be adapted to the data. For example, if the data is known to be divided into clusters, then one key per cluster is a good choice.
- Elements can be partitioned into smaller objects to allow local similarity. For example, for sequence comparisons we can divide each sequence into smaller sequences and treat each smaller part as an element. We will do the same for the query and search each part separately.
- Further study is needed to optimize the use of secondary memory and to improve the bucket utilization. It may be possible to have partial split procedures that improve storage utilization similarly to multiway trees [BYC92].

Acknowledgements

We would like to thank the helpful comments of the referees.

References

[AGMML90] Altschul S.F., W. Gish, W. Miller, E. W. Myers, and D. J. Lipman, "Basic local alignment search tool," *J. Molecular Biology* 15 (1990), 403–410.

[BYC92] Baeza-Yates, R.A. and Cunto, W., "Unbalanced Multiway Trees Improved by Partial Expansions", *Acta Informatica*, **29** (5), 1992, 443–460.

[BYG90] Baeza-Yates, R.A. and Gonnet, G.H., "All-against-all Sequence Matching", Dept. of Computer Science, Universidad de Chile, 1990.

[BGKN90] Bahl L. R., P. S. Gopalakrishnan, D. S. Kanevsky, and D.S. Nahamoo, "A fast admissible method for identifying a short list of candidate words," IBM tech report RC 15874 (June 1990).

[BR+93] Bugnion, E. and Roos, T. and Shi, F. and Widmayer, P. and Widmer, F. "A Spatial Index for Approximate Multiple String Matching", 1st South American Workshop on String Processing, Belo Horizonte, Sept 1993, 43–54.

[BK73] Burkhard, W.A. and Keller, R.M. "Some Approaches to Best-Match File Searching", *Communications of the ACM* **16** (4), April 1973, 230-236.

[CL90] Chang W.L., and E.L. Lawler, "Approximate matching in sublinear expected time," *Proc. of the 31st IEEE Symp. on Foundations of Computer Science* (1990) 116–124.

[FBF77] Friedman, J.H. and Bentley, J.L. and Finkel, R.A. "An Algorithm to find best matches in logarithmic expected time", *ACM Trans. on Math. Software* 3(3), 1977.

[GBY91] Gonnet, G.H. and Baeza-Yates, R. Handbook of Algorithms and Data Structures, Addison-Wesley, second edition, 1991.

[GCB92] Gonnet, G.H., M.A. Cohen, and S.A. Benner, "Exhaustive matching of the entire protein sequence database," *Science* **256**, 1443.

[LP85] Lipman D. J., and W.R. Pearson, "Rapid and sensitive protein similarity searches," *Science* **227** (1985), 1435-1441.

[Mah92] Mahmoud, H. Evolution of Random Search Trees, John Wiley, New York, 1992,

[Mur83] Murtagh, F. "A Survey of Recent Advances in Hierarchical Clustering Algorithms", *IEEE Computer* **26** (4), 1983, 354–359.

[My92] Myers, E. "Algorithmic Advances for Searching Biosequence Databases," *Proceedings of the International Symposium on Computational Methods in Genome Research* (Heidelberg, 1992), to appear.

[My94] Myers, E. "A Sublinear Algorithm for Approximate Keyword Matching," *Algorithmica*, in press.

[NK82] Nevalainen, O. and Katajainen, J. "Experiments with a Closest Point Algorithm in Hamming Space", *Angewandte Informatik* 5, 1982, 277-281.

[SDDR89] Santana, O. and Diaz, M. and Duque, J.D. and Rodriguez, J.C. "Increasing radius search schemes for the most similar strings on the Burkhard-Keller tree", International Workshop on Computer Aided Systems Theory, EUROCAST'89, 1989.

[Sha77] Shapiro, M. "The Choice of Reference Points in Best-Match File Searching", *Communications of the ACM* **20** (5), May 1977, 339-343.

[SW90] Shasha, D. and Wang, T-L. "New Techniques for Best-Match Retrieval", *ACM Transactions on Information Systems* 8, 1990, 140-158.

[Uk92] Ukkonen, E., "Approximate string matching with q-grams and maximal matches," *Theoretical Computer Science* (1992), 191–212.

[Uk93] Ukkonen, E., "Approximate string-matching over suffix trees," *4th Annual Combinatorial Pattern Matching Symp.*, Padova, Italy (June 1993), 228–242.

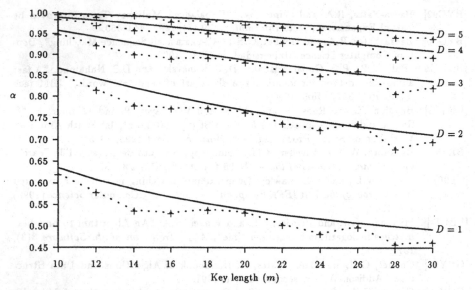

Fig. 2. Complexity of proximity searching depending on D and m (Hamming). Experimental results are shown with dotted lines and + symbols.

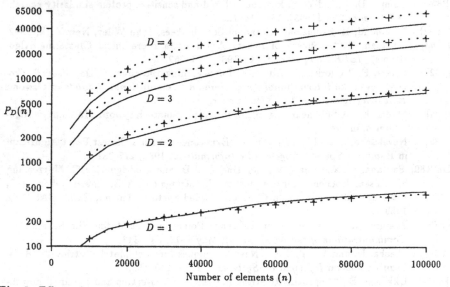

Fig. 3. Effect of n on proximity searching depending on D for $m = 16$, $\sigma = 4$ and $b = 1$ (Hamming). Experimental results are shown with dotted lines and + symbols.

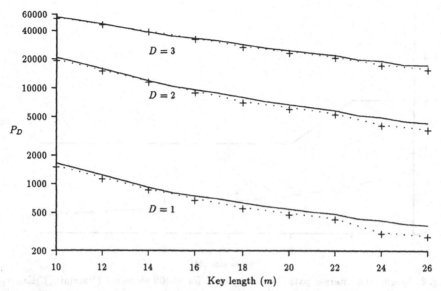

Fig. 4. Effect of key length on proximity searching depending on D for $\sigma = 4$, $b = 3$ and 90000 elements (Hamming). Experimental results are shown with dotted lines.

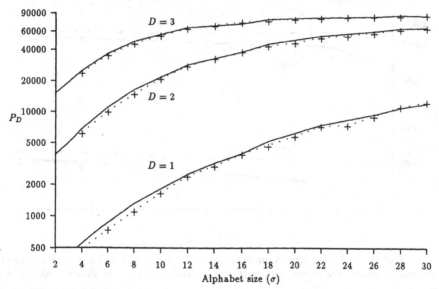

Fig. 5. Effect of alphabet size on proximity searching depending on D for $m = 20$, $b = 3$ and 90000 elements (Hamming). Experimental results are shown with dotted lines.

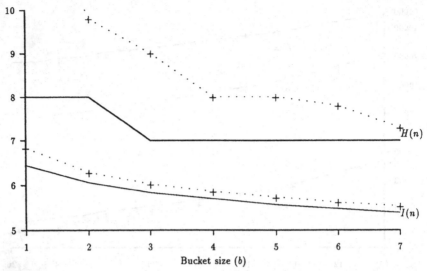

Fig. 6. Height and internal path depending on b for 90000 elements (Hamming). Experimental results are shown with dotted lines and $+$ symbols.

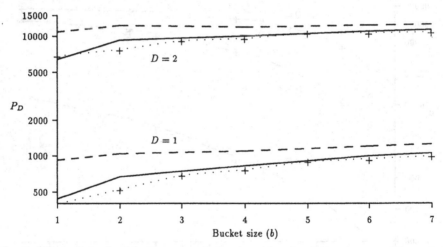

Fig. 7. Effect of the bucket size on proximity searching for 90000 elements (Hamming). Experimental results are shown with dotted lines and BK-trees are shown with dashed lines.

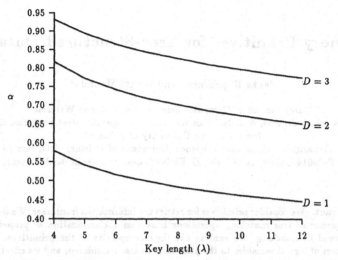

Fig. 8. Complexity of proximity searching depending on D and λ (Leveshtein).

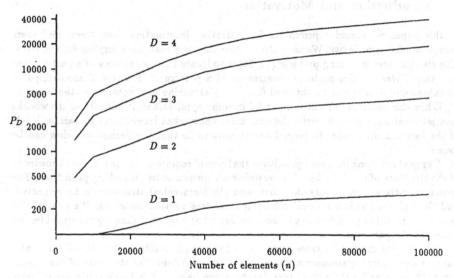

Fig. 9. Effect of n on proximity searching depending on D for $\lambda = 5$, $\sigma = 32$ and $b = 1$ (Leveshtein).

Query Primitives for Tree-Structured Data*

Pekka Kilpeläinen[1] and Heikki Mannila[2]

[1] Department of Computer Science, University of Waterloo
Waterloo, Ontario, N2L 3GI Canada, e-mail: pkilpela@watsol.uwaterloo.ca
(on leave from University of Helsinki)
[2] Department of Computer Science, University of Helsinki, P.O. Box 26
SF-00014 University of Helsinki, Finland, e-mail: mannila@cs.helsinki.fi

Abstract. We consider primitives for retrieving information from trees. We define a sequence of tree matching operations based on a classification of properties preserved in matching. We analyse the time complexity of the primitives. The addition of logical variables to the primitives is also considered, and its effects on the complexities is studied.

1 Introduction and Motivation

In this paper we consider primitives for retrieving information from trees, and their computational complexity. We use pattern matching as a basis for querying data in trees. The classical tree matching problem [14, 24] is to locate the occurrences of a pattern tree in a target tree. In this problem occurrences of a pattern P in a tree T are defined to be subtrees of T that can be obtained from P by attaching new subtrees to the leaves of P. While this matching primitive is useful in some applications, it has serious drawbacks as a retrieval operation. Namely, the user must know what labels occur at various levels of the target tree in order to formulate the pattern so that the desired matches can be found.

Our goal is to look for query primitives that would require from the user less knowledge of the structure of the data. For this, we define a sequence of tree matching primitives. The looser primitives require only the labels and the hierarchical structure to be respected, and the tightest one is a special case of classical tree pattern matching. We analyze the complexities of the primitives and also consider what happens when logical variables are added to the patterns.

Our specific interest in trees comes from their applicability in structured text databases. Context-free grammars are often used for describing the structure of text documents [3, 5, 9], and a structured text database can be considered as a collection of parse trees [12, 17]. Thus the grammars correspond to schemas, and parse trees correspond to instances.

We could use tree matching for instance to locate information of linguistic interest. For example, the following queries could be reasonable on trees like the parse tree shown in Figure 1.

1. Locate those sentences that contain a verb phrase including verb "holds", noun "cat", and some adverb, in this order.

---

* Research supported by the Academy of Finland.

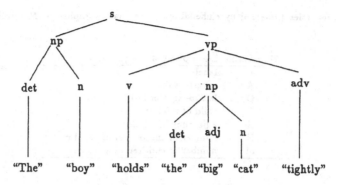

Fig. 1. The parse tree of a sentence

2. Extract the words that appear in a sentence both as a verb and a noun.

The primitives of this paper allow this kind of queries to be expressed easily.

We define a hierarchy of tree matching primitives by considering what properties are preserved by the embeddings that take the pattern to the target tree. The classification can serve as a starting point in relating the complexities of the different tree matching problems to each other. The complexities of the primitives range from NP-complete to linear. Generally, the more restricted problems are easier to solve. No non-trivial lower bound complexities are known for the problems. Some of the primitives and results presented here have been known before, but some have only been reported in the PhD thesis of the first author [16]. Also, the general framework is new.

The rest of the paper is organized as follows. In Section 2 we define the various tree matching primitives. In Section 3 we present a fairly complete complexity analysis of the primitives. Section 4 presents the extension of the tree matching problems where logical variables are allowed in the patterns. In Section 5 we summarize the complexity results for the problems with logical variables. Section 6 is a short conclusion.

2 Tree Inclusion Problems

In this section we introduce the collection of *tree inclusion problems*. The trees that we consider are rooted and ordered, and consist of labeled nodes.

Let P and T be trees. An injective function f from the nodes of P to the nodes of T is an *embedding* of P in T. An embedding f *preserves* a binary property φ between nodes, if for all nodes u and v of P we have that $\varphi(u, v)$ holds in P if and only if $\varphi(f(u), f(v))$ holds in T. Table 1 gives the list of properties that we consider.

Given a set S of properties to be preserved, an *S-embedding* is an embedding that preserves the properties in S. Given a pattern tree P and a target tree T, the *decision*

---

[3] An embedding preserves the adjacency of siblings if it maps pattern nodes $v_1, \ldots, v_k$ to k adjacent siblings in the target whenever nodes $v_1, \ldots, v_k$ are siblings. To be precise, this is a k-ary property.

Table 1. Properties preserved by embeddings. Note that C implies A, M implies N, and N implies S.

Abbreviation	Description of the property
L	labels
A	ancestorship
O	left-to-right ordering
C	child-of relation
S	adjacency of siblings[3]
N	number of children of internal nodes
M	number of children of all nodes

problem of S-matching is to find whether there exists an S-embedding of P in T. For retrieval purposes the answers to decision problems are not so illuminating; thus the *retrieval problem of S-matching* is to find (the roots of) all subtrees U of T such that there is an S-embedding of P in U. These subtrees are the S-*occurrences of* P *in* T. If a tree U is an S-occurrence of P we say that (the root node of) P S-*matches* at the root of U. An S-embedding of P in U is generally also an S-embedding of P in any tree which has U as a subtree. Sometimes we are interested only in the minimal subtrees that are occurrences of the pattern. The primitives that we consider are defined in Table 2.

Table 2. Different tree matching operations and the properties they preserve.

Abbreviation	Description	Preserved properties
UTI	unordered tree inclusion	LA
OTI	ordered tree inclusion	LAO
UPI	unordered path inclusion	LC
OPI	ordered path inclusion	LCO
URI	unordered region inclusion	LCS
ORI	ordered region inclusion	LCSO
UCI	unordered child inclusion	LCN
OCI	ordered child inclusion	LCNO
UST	unordered subtree	LCM
OST	ordered subtree problem	LCMO

For labeled structures the preservation of labels is essential of any pattern matching operation. For trees, the preservation of the hierarchical relationship between the nodes of the pattern is naturally important. Therefore all the primitives we consider preserve at least labels (L) and ancestorship (A). The matching problem where only these properties are required to be preserved is called *unordered tree inclusion (UTI)*; if furthermore also the left-to-right ordering (O) of the pattern nodes is required to be preserved, the problem is called *ordered tree inclusion (OTI)*. Unordered tree inclusion, or {LA}-matching, is exemplified by pattern P and target T in Figure 2. Note that pattern P would {LAO}-match only at the root of T.

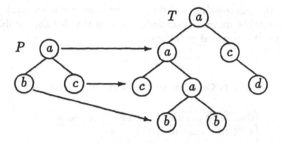

Fig. 2. An {LA}-embedding of P in its minimal {LA}-occurrence in T.

In (un)ordered tree inclusion, only the basic shape of the pattern tree has to occur in the target. Therefore the tree inclusion primitives are tolerant to a wide class of variations in the structure of the data.

Other variants of the inclusion problems are obtained by restricting the corresponding embeddings further. The *path inclusion* problems require that also the child-of relationship (C) is preserved. Thus this primitive is considerably tighter than the tree inclusion primitives. Note that the path inclusion problems are variants of the subgraph isomorphism problem restricted to trees.

Some of the problems described in Table 2 may seem less important as query primitives, but are included as intermediate steps between more interesting problems. This is the case for the *region inclusion* problems URI and ORI, where in addition to the requirements of path inclusion, adjacent siblings in the pattern may be matched only at adjacent siblings in the target.

An {N}-embedding maps *internal* nodes with k children to nodes with k children. The problem of locating the target subtrees in which there is an {LCN}-embedding of the pattern is called *unordered child inclusion (UCI)*. Restricting to {LCNO}-embeddings gives the problem of *ordered child inclusion (OCI)*, which is known as classical tree pattern matching [14].

Finally, if we restrict the {LCN}-embeddings to map *every* node to a node with a similar number of children, we get in the unordered case the *unordered subtree problem (UST)*. In UST we search for the subtrees of the target that are isomorphic to the pattern. In the corresponding *ordered subtree problem (OST)* we search for identical subtrees. These problems are related to recognizing common substructures, and thereby to data compressing.

3   Algorithms and Complexities

We have collected to this section results of the computational complexity of the proposed primitives. We denote the number of nodes in the pattern by m and the number of nodes in the target by n.

Table 3 shows the complexities of the various tree inclusion problems. The NP-completeness in the table means NP-completeness of the decision problem. The other bounds are given for the retrieval problems.

Table 3. Complexities of tree inclusion problems

Unordered problems	Ordered problems
UTI NP-complete	OTI $O(mn)$
UPI $O(m^{1.5}n)$	OPI $O(mn)$
URI $O(m^2 n)$	ORI $O(mn)$
UCI $O(m^{1.5}n)$	OCI $O(n\sqrt{m}\,polylog(m))$
UST $O(n)$	OST $O(n)$

The NP-completeness proof for UTI is by a reduction from satisfiability. The proof shows that the problem is NP-complete already for patterns of depth 2 and targets of depth 3. The difficulty of UTI is due to choosing the correct order for the pattern subtrees. If the out-degree of the pattern nodes is bounded by a constant, the problem can be solved in polynomial time [18].

Recently a proof of the NP-completeness of UTI with *unlabeled* trees was published in [22]. That proof is a reduction from EXACT COVER BY 3-SETS. It shows that the problem is NP-complete even if we restrict to targets whose depth is bounded or whose branching factor is bounded for other nodes except for the root.

The polynomiality of the other primitives can be shown by using variations of the following algorithm schema for solving S-matching problems. At the core of the schema we need to decide whether a pattern node v matches at a target node w. The decision is parameterized by a *match predicate* ψ that expresses the required relation between w and the occurrences of the subtrees of v. For example, in the problem OTI a pattern P with immediate subtrees $P_1, \ldots, P_k$ can {LAO}-match at node w only if there are descendants $u_1, \ldots, u_k$ of w in left-to-right order such that no one of them is a descendant of another one, and that P_i {LAO}-matches at u_i for all $i = 1, \ldots, k$. The instantiations of the schema are dynamic programming algorithms, which compute in a bottom-up manner for each target node w the set $M(w)$ of pattern nodes that S-match at w.

Algorithm 3.1 *General tree matching schema.*
Input: *Pattern tree P and target tree T.*
Output: *Occurrences of P in T.*
Method:

for each *target node w in a bottom-up order* **do**
    $M(w) := \emptyset$;
    for each *pattern node v in a bottom-up order* **do**
        if *label$(v) = $ label(w)* **then**
            Let $v_1, \ldots, v_k$ be the children of v; (k = 0 if v is a leaf)
            if *there are descendants $u_1, \ldots, u_k$ of w*
                such that $\psi(w, u_1, \ldots, u_k)$ holds

$$\text{and } v_i \in M(u_i) \text{ for } i = 1, \ldots, k \text{ then}$$

            comment: v S-matches at w;

                $M(w) := M(w) \cup \{v\}$;

            fi;

        fi;

    od;

    if $root(P) \in M(w)$ then

        comment: P S-matches at w;

        output w;

    fi;

od;

Solving the match predicate ψ for the problems of unordered path inclusion UPI, unordered region inclusion URI and unordered child inclusion UCI involves computing a maximum matching in a bipartite graph. It is an old result that applying the maximum matching algorithm of Hopcroft and Karp [15] in Algorithm 3.1 leads to an $O(m^{3/2}n)$ time solution for UPI [23, 25]. Unordered child inclusion constrains pattern leaves to match at target leaves only. This requires only a minor modification to the algorithm, and the same time complexity applies to UCI.

The match predicate of unordered region inclusion URI requires the children $v_1, \ldots, v_k$ of pattern node v to match at a contiguous sequence of k adjacent sibling nodes. This is a restricted variant of bipartite matching, which can be solved for bipartite graphs $G = (X \cup Y, V)$ in time $O(|X|^2|Y|)$ [16]. Plugging this result in Algorithm 3.1 gives the time bound of $O(m^2 n)$ for unordered region inclusion.

Ordered tree inclusion OTI can be solved in $\Theta(mn)$ time by applying dynamic programming and restricting the search for {LAO}-embeddings to so called left embeddings [19, 18]. The $\Theta(mn)$ space requirement of this algorithm can be improved by an algorithm which does not maintain an $m \times n$ table but keeps track of the parts of the pattern that can be embedded, during a traversal of the target. The non-tabulating algorithm requires the same asymptotical worst-case time but only $O(m\,depth(T))$ space [16].

Let the children of target node w be $w_1, \ldots, w_l$. The match predicate $\psi(w, u_1, \ldots, u_k)$ for ordered S-problems with child-of relation preserved[4] requires that nodes $u_1, \ldots, u_k$ form an ordered subsequence of $w_1, \ldots, w_l$. A suitable sequence $u_1, \ldots, u_k$ for ordered path inclusion OPI can be found, if one exists, by scanning the children of w from left to right and using a counter for the number of children of v already matched. The algorithm for ordered region inclusion ORI has to find a contiguous sequence $w_j, \ldots, w_{j+k-1}$ of children of w, such that node v_i {LCSO}-matches at w_{j+i-1} for all $i = 1, \ldots, k$. This can be done by a method resembling trivial string matching in $O(kl)$ time, using the results computed in $M(w_1), \ldots, M(w_l)$. Plugging these methods of solving the match predicate in Algorithm 3.1 gives the $O(mn)$ solutions to ordered path inclusion OPI and ordered region inclusion ORI.

The asymptotically fastest algorithm presented so far for ordered child inclusion OCI, or classical tree pattern matching, works in time $O(n\sqrt{m}\,polylog(m))$. It is worth noting that obtaining this bound requires quite complicated methods [6]. The worst case complexity of practical OCI-algorithms falls to the $O(mn)$ category of the other ordered

---

[4] That is, {CO} $\subseteq S$.

tree inclusion problems [14]. The expected time complexity of the naive "traverse-and-compare" algorithm has been show to be linear [26].

Finally, the unordered and ordered subtree problems UST and OST can be solved in linear time. Only those subtrees of the target need to be checked that have as many nodes as the pattern. They can be easily located in $O(n)$ time. Because the total size of these subtrees is $O(n)$, the problems are solvable in linear time if we can show that each of them can be tested against pattern P in $O(m)$ time [7, 13]. For ordered subtree problem this is easy to do by simply comparing P and its candidate occurrence node-by-node against each other. Testing whether trees P and U are isomorphic can also be done in linear time [1, p. 84–86].

Problems OPI and ORI do not seem to be easily solvable in linear time, since string pattern matching with don't care symbols can be reduced to them in linear time [21, 27]. The tightest known upper bound for string matching with don't care symbols is $O(polylog(m)\,n)$ [8]. The obstacle to obtaining linear-time algorithms for the various matching primitives is that generally the occurrences may overlap, and therefore it is difficult to avoid considering the same target nodes multiple times. If we require the target to be *nonperiodic* in the sense that no label appears more than once on any root-to-leaf path, the problems of OTI, OPI and OCI can be solved in linear time [20]. The restriction is a strong one, but it seems that many text-databases can be modeled using nonrecursive grammars, and therefore represented as nonperiodic trees.

4   Adding Logical Variables

The tree inclusion problems discussed above can be used as primitives for locating the occurrences of the pattern in the target. Equality constraints for the substructures of the pattern occurrences can be stated by labeling pattern leaves with *logical variables*. Leaves labeled by variables match at arbitrary nodes with the following restriction. An embedding f of P in T is said to *respect variables* if whenever u and v are two nodes of P labeled by the same variable, f maps u and v to the roots of identical subtrees of T. We get the *tree inclusion problems with logical variables* from the above tree inclusion problems by requiring the corresponding embeddings to respect variables. Abbreviations for the problems with logical variables are formed by adding a suffix "-V" to the abbreviation of the corresponding tree inclusion problem. For example, abbreviation UPI-V stands for Unordered Path Inclusion with logical Variables.

As an example, consider the pattern P and the target T shown in Figure 3. Since permuting siblings does not change pattern P, its occurrences are the same in instance (P, T) of both the ordered and the unordered variants of each particular inclusion problem. A variable respecting {LA}-embedding of P in T can map the x-nodes of P either to the b-nodes or to the c-nodes of T. Variable respecting {LC}-embeddings of P in T map the root of P to the root of T and the leaves of P to the b-nodes of T. Finally, there are no variable respecting {LCS}-embeddings of P in T since T does not have two identical subtrees adjacent to each other.

5   Complexity of the Primitives with Logical Variables

Most of the inclusion problems with logical variables are computationally difficult. The intuitive reason for this is the local freedom in those problems to perform a suitable

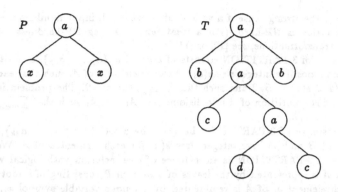

Fig. 3. An instance of tree inclusion problems with logical variables.

matching combined with the global requirement of consistency that embeddings respect variables. The same holds for string pattern matching if we allow repeating logical variables in the pattern. Angluin has shown in [2] that it is NP-complete to decide whether a given string can be obtained by substituting strings to the logical variables in a given string pattern.

Table 4 presents the complexity results for the inclusion problems with logical variables. Again, NP-completeness relates to the decision problems and the polynomial bounds refer to the retrieval problems.

Table 4. Complexities of tree inclusion problems with logical variables

Unordered problems	Ordered problems
UTI-V NP-complete	OTI-V NP-complete
UPI-V NP-complete	OPI-V NP-complete
URI-V NP-complete	ORI-V NP-complete
UCI-V NP-complete	OCI-V $O(mn)$
UST-V NP-complete	OST-V $O(n)$

The decision problems "Is there a variable respecting S-embedding of P in T?" are in NP, since checking whether a mapping satisfies the conditions can be done in polynomial time. If no variable appears twice in the pattern, the problems with logical variables are not harder than the corresponding tree inclusion problems. They can be solved by straightforward variants of the corresponding tree inclusion algorithms.

In the case of the unordered inclusion problems the difficulty is to find the correct ordering for the branches of the pattern. The proof that all unordered inclusion problems with logical variables are NP-hard is based on a pseudo-polynomial reduction from 3-PARTITION, which is NP-hard in the strong sense. Intuitively, a decision problem Π

is NP-hard in the strong sense if a variant of Π where all input numbers are expressed in unary notation is NP-hard. (For a treatment of strong NP-hardness and pseudo-polynomial transformations, see [10] or [11].)

An instance of 3-PARTITION consists of a set $A = \{a_1, \ldots, a_{3k}\}$, a positive integer bound B and a positive integer size $s(a)$ for each element a of A, such that each size $s(a)$ satisfies $B/4 < s(a) < B/2$ and such that $\sum_{a \in A} s(a) = kB$. The problem is to decide whether there is a partition of A into disjoint sets $A_1, \ldots, A_k$ such that $\sum_{a \in A_i} s(a) = B$ for each $i = 1, \ldots, k$.

Let an instance of 3-PARTITION be given by a set $A = \{a_1, \ldots, a_{3k}\}$, a positive integer bound B and positive integer sizes $s(a)$ for each element a of A. We represent the instance of 3-PARTITION as an instance of tree inclusion with logical variables as follows. Set A is represented by the leaves of a pattern P consisting of a root and of kB leaves. Each element a_i of A is represented by a unique variable symbol x_i, called the *label* of a_i. For each element a_i of A we label $s(a_i)$ leaves of P by the label x_i of a_i. The root of P is labeled by 0. A target T consisting of a root and kB leaves represents the partitioning of A. The leaves of T are divided into k groups of B nodes; the nodes in group i are labeled by i. The root of T is labeled by 0. (See Figure 4.)

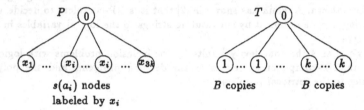

Fig. 4. Trees P and T corresponding to an instance of 3-PARTITION.

Lemma 1. *Let a set $A = \{a_1, \ldots, a_{3k}\}$ with a size $s(a)$ for each of its elements a and a bound B form an instance of 3-PARTITION. Let (P, T) be the representation of the instance of 3-PARTITION as an instance of the unordered subtree problem with logical variables. Then there is a variable respecting $\{LCM\}$-embedding of P in T if and only if there is a partition of A into disjoint sets $A_1, \ldots, A_k$ such that $\sum_{a \in A_i} s(a) = B$ for each $i = 1, \ldots, k$.*

□

In the above construction any variable respecting $\{LA\}$-embedding of P in T is also an $\{LCM\}$-embedding since both P and T consist of a root node and kB leaves. This shows that none of the unordered inclusion problems with logical variables is easier than 3-PARTITION.

The logical-variable variants of the ordered inclusion problems that may ignore parts of occurrences can be shown NP-complete by a reduction from 3SAT. An instance of 3SAT consists of a collection $C = \{c_1, \ldots, c_k\}$ of clauses, each containing three literals over a finite set of variables U. The problem is to decide whether there is a truth assignment for U that satisfies each clause in C.

The problem 3SAT can be reduced in polynomial time to OTI-V, OPI-V, and ORI-V as follows. Let an instance of 3SAT be given by the set of clauses $C = \{c_1, \ldots, c_k\}$, where $|c_i| = 3$ for all $i = 1, \ldots, k$. We represent C as the following instance (P, T) of ordered tree inclusion with logical variables. Pattern tree P consists of a root node labeled by 0 and of k immediate subtrees $P_1, \ldots, P_k$. Each pattern subtree P_i is a straightforward representation of clause c_i; for an example see Figure 5.

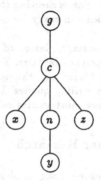

Fig. 5. The pattern subtree representing clause $\{x, \bar{y}, z\}$.

The target tree T consists also of a root node labeled by 0 and of k immediate subtrees $T_1, \ldots, T_k$. Target subtree T_i is a representation for the group of clauses consisting of the seven true instances of clause c_i. Each immediate subtree of T_i is obtained from the immediate subtree of pattern subtree P_i by substituting zeros and ones for its variables. An example is shown in Figure 6.

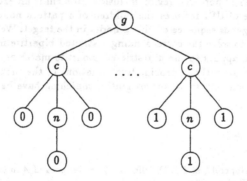

Fig. 6. The target subtree representing the true instances of clause $\{x, \bar{y}, z\}$.

Now there is a variable preserving {LAO}-embedding of P in T if and only if C is satisfiable. Further, there is an {LAO}-embedding of P in T if and only if there is an

{LCO}-embedding of P in T, which holds if and only if there is an {LCSO}-embedding of P in T. This shows that the problems of OTI-V, OPI-V, and ORI-V are NP-hard.

Related complexity results for term matching have been presented in [4] and in [28].

Ordered child inclusion with logical variables, also called *nonlinear tree pattern matching*, can be solved in $O(mn)$ time, using for example the algorithm of [24].

The ordered subtree problem with logical variables, OST-V, can be solved in $O(n)$ time. The algorithm and the argumentation are almost identical to the previous ones for the problem OST. The only complication is checking that repeated variables are matched against identical target leaves, which can easily be done in $O(m)$ time for each of the possible pattern occurrences.

Some, in practice probably important, classes of ordered matching problems with logical variables can be solved in polynomial time. For example, this is the case if the variables occur in suitably disjoint blocks, i.e., the occurrences of a variable do not influence any other variables [16]. In writing queries for text database applications this condition is often satisfied, and hence most queries can be evaluated efficiently.

6 Conclusions and Further Research

We have presented a sequence of tree matching problems, and analyzed the complexity of the problems. We also considered the case where patterns may contain logical variables, and showed that those problems are in general difficult.

The classification opens many interesting questions and directions for further work. Little is known about lower-bound complexities of the problems. An interesting general problem is to classify exactly those sets S of interesting tree properties such that the S-matching problem can be solved in polynomial or linear time. Results on graph minors could be useful here. A specific open question is whether ordered tree inclusion can be solved in linear time.

An observation from the classification is that usually the more restricted problems are easier to solve. The unordered region inclusion problem is an exception to this rule. The match predicate of URI requires the children of a pattern node to be matched, in any order, at a contiguous sequence of sibling nodes in the target. We have not found any straightforward way to solve the corresponding restricted bipartite matching problem in $O(m^{1.5}n)$ time which applies to the unrestricted bipartite matching.

A general treatment of the approximative variations of the problems would also be of interest. Some recent results on approximative matching have been published in [29, 30, 31].

References

1. A. V. Aho, J. E. Hopcroft, and J. D. Ullman. *The Design and Analysis of Computer Algorithms*. Addison-Wesley, 1974.
2. D. Angluin. Finding patterns common to a set of strings. *Journal of Computer and System Sciences*, 21:46–62, 1980.
3. F. Bancilhon and P. Richard. Managing texts and facts in a mixed data base environment. In G. Gardarin and E. Gelenbe, editors, *New Applications of Data Bases*, pages 87–107. Academic Press, 1984.

4. D. Benanav, D. Kapur, and P. Narendran. Complexity of matching problems. *Journal of Symbolic Computation*, 3(1&2):203–216, February/April 1987.

5. G. Coray, R. Ingold, and C. Vanoirbeek. Formatting structured documents: Batch versus interactive. In J. C. van Vliet, editor, *Text Processing and Document Manipulation*, pages 154–170. Cambridge University Press, 1986.

6. M. Dubiner, Z. Galil, and E. Magen. Faster tree pattern matching. In *Proc. of the Symposium on Foundations of Computer Science (FOCS'90)*, pages 145–150, 1990.

7. P. Dublish. Some comments on the subtree isomorphism problem for ordered trees. *Information Processing Letters*, 36:273–275, 1990.

8. M. J. Fischer and M. S. Paterson. String-matching and other products. In *Complexity of Computation*, pages 113–125. SIAM-AMS, 1974.

9. R. Furuta, V. Quint, and J. André. Interactively editing structured documents. *Electronic Publishing*, 1(1):19–44, 1988.

10. M. R. Garey and D. S. Johnson. "Strong" NP-completeness results: Motivation, examples and implications. *Journal of the ACM*, 25(3):499–508, July 1978.

11. M. R. Garey and D. S. Johnson. *Computers and Intractability*. W. H. Freeman and Company, 1979.

12. G. H. Gonnet and F. Wm. Tompa. Mind your grammar - a new approach to text databases. In *Proc. of the Conference on Very Large Data Bases (VLDB'87)*, pages 339–346, 1987.

13. R. Grossi. A note on the subtree isomorphism for ordered trees and related problems. *Information Processing Letters*, 39:81–84, 1991.

14. C. M. Hoffman and M. J. O'Donnell. Pattern matching in trees. *Journal of the ACM*, 29(1):68–95, January 1982.

15. J. E. Hopcroft and R. M. Karp. An $n^{5/2}$ algorithm for maximum matching in bipartite graphs. *SIAM Journal on Computing*, 2(4):225–231, December 1973.

16. P. Kilpeläinen. *Tree Matching Problems with Applications to Structured Text Databases*. PhD thesis, University of Helsinki, Dept. of Comp. Science, November 1992.

17. P. Kilpeläinen, G. Lindén, H. Mannila, and E. Nikunen. A structured document database system. In R. Furuta, editor, *EP90 – Proceedings of the International Conference on Electronic Publishing, Document Manipulation & Typography*, The Cambridge Series on Electronic Publishing. Cambridge University Press, 1990.

18. P. Kilpeläinen and H. Mannila. Ordered and unordered tree inclusion. Report A-1991-4, University of Helsinki, Dept. of Comp. Science, August 1991. To appear in *SIAM Journal on Computing*.

19. P. Kilpeläinen and H. Mannila. The tree inclusion problem. In S. Abramsky and T. S. E. Maibaum, editors, *TAPSOFT'91, Proc. of the International Joint Conference on the Theory and Practice of Software Development, Vol. 1: Colloquium on Trees in Algebra and Programming (CAAP'91)*, pages 202–214. Springer-Verlag, 1991.

20. P. Kilpeläinen and H. Mannila. Grammatical tree matching. In A. Apostolico, M. Crochemore, Z. Galil, and U. Manber, editors, *Proceedings of the Third Annual Symposium on Combinatorial Pattern Matching*, pages 162–174. Springer-Verlag, 1992.

21. S. R. Kosaraju. Efficient tree pattern matching. In *Proc. of the Symposium on Foundations of Computer Science (FOCS'89)*, pages 178–183, 1989.

22. J. Matoušek and R. Thomas. On the complexity of finding iso- and other morphisms for partial k-trees. *Discrete Mathematics*, 108(1–3):343–364, October 1992.

23. D. W. Matula. An algorithm for subtree identification. *SIAM Rev.*, 10:273–274, 1968. Abstract.

24. R. Ramesh and I. V. Ramakrishnan. Nonlinear pattern matching in trees. *Journal of the ACM*, 39(2):295–316, April 1992.

25. S. W. Reyner. An analysis of a good algorithm for the subtree problem. *SIAM Journal on Computing*, 6(4):730–732, December 1977.

26. J.-M. Steyaert and P. Flajolet. Patterns and pattern-matching in trees: An analysis. *Information and Control*, 58:19–58, 1983.

27. R. M. Verma. Strings, trees, and patterns. *Information Processing Letters*, 41:157–161, March 1992.

28. R. M. Verma and I. V. Ramakrishnan. Some complexity theoretic aspects of AC rewriting. In B. Monien and R. Cori, editors, *STACS89 - 6th Annual Symposium on Theoretical Aspects of Computer Science*, pages 407–420. Springer-Verlag, 1989.

29. K. Zhang and D. Shasha. Simple fast algorithms for the editing distance between trees and related problems. *SIAM Journal on Computing*, 18(6):1245–1262, December 1989.

30. K. Zhang, D. Shasha, and J. T.-L. Wang. Fast serial and parallel algorithms for approximate tree matching with VLDC's. In A. Apostolico, M. Crochemore, Z. Galil, and U. Manber, editors, *Proceedings of the Third Annual Symposium on Combinatorial Pattern Matching*, pages 151–161. Springer-Verlag, 1992.

31. K. Zhang, R. Statman, and D. Shasha. On the editing distance between unordered labeled trees. *Information Processing Letters*, 42(3):133–139, May 1992.

Multiple Matching of Parameterized Patterns

Ramana M. Idury *[1] and *Alejandro A. Schäffer* **[2]

[1] University of Southern California
[2] Rice University

Abstract. We extend Baker's theory of parameterized pattern matching [*Proc. 25th Annual STOC*, 1993, pp. 71–80] to algorithms that match multiple patterns in a text. We first consider the case where the patterns are fixed and preprocessed once, and then the case where the pattern set can change by insertions and deletions. Baker's algorithms are based on suffix trees, whereas ours are based on pattern matching automata.

Address for correspondence: Ramana M. Idury; Department of Mathematics; University of Southern California; Los Angeles, CA 90089-1113, USA.

1 Introduction

In classic string matching a *match* of pattern to text is either an exact match or a match with some small number of errors. Commands such as global-replace in editors such as emacs allow users to substitute all occurrences of string x with string y instantaneously. If there are many occurrences of x in the original text, the new text will contain y at each occurrence and it will not match the old text using the usual approximate string matching criteria.

Global replacements are a common way to reuse code with small modifications. However, code duplicated by copying and global replacements is often associated with bugs and plagiarism, so we would like to match the original code against the duplicated and edited copy using string matching techniques. What is needed is a criterion for matching that allows all occurrences of one string in the pattern to match occurrences of some other string in the text. To address this goal, Baker [6] defined a theory of parameterized pattern matching and used her theory to search for duplication in large programs.

In this paper we present algorithms for matching a dictionary D of parameterized patterns against an input text T. We give algorithms both for the case where the dictionary is static and can be preprocessed, and for the case where the dictionary changes over time by the insertion and deletion of individual patterns. We have completely implemented in C the algorithms for the more interesting dynamic case.

Baker's paper studies the problem of preprocessing a fixed text T, so as to quickly search for all occurrences of input pattern P in the text. The best solution to this

* idury@hto.usc.edu; Supported by NSF grant DMS-90-05833.
** schaffer@cs.rice.edu; Partially supported by NSF grant CCR-9010534.

problem in classic string matching preprocesses T into a data structure called a *suffix tree* [16, 13, 7]; Baker uses a variant of this data structure she calls a *p-suffix tree* to solve the parameterized version of the fixed text searching problem.

In this paper we are concerned with "dual" problems where the patterns are preprocessed and the text is treated as the input (variable). The most basic such problem is:

- **Fixed Pattern Matching(FPM):** Given a fixed pattern P of length p over alphabet Σ, preprocess P so as to be able to find all occurrences of P in a *query* text T of length t.

There are several solutions to FPM that preprocess the pattern in time $O(p)$ and search the text in time $O(t)$. In this paper we are particularly interested in the linear time algorithm of Knuth, Morris, and Pratt (KMP for short) [12] because their automata-theoretic approach can be generalized to solve some multiple pattern matching problems. One extension of FPM is to multiple patterns:

- **Multiple Pattern Matching(MPM):** Given a fixed set $D = P_1, P_2, \ldots, P_k$ of patterns over an alphabet Σ preprocess D so as to be able to search for all pattern occurrences in a query text T. The set D is sometimes called the *dictionary*.

Aho and Corasick (AC for short) [1] first solved MPM by generalizing the KMP automaton method to multiple patterns. The AC algorithm preprocesses the pattern dictionary in time $O(d \log \sigma)$ and searches a text in time $O(t \log \sigma + tocc)$, where d is the total size of all patterns, σ is the number of characters that occur in a pattern, and $tocc$ is the total number of occurrences reported.

Meyer [15] posed the problem of extending the MPM problem to allow the dictionary to increase by inserting single patterns. Amir and Farach [2] considered allowing the dictionary to change by both insertions and deletions, and got the first interesting time bounds. They defined the problem as follows:

- **Dynamic Dictionary Matching(DDM):** Preprocess and maintain a dictionary D of patterns under the operations *insert* a pattern, *delete* a pattern, and *search* for a query text T for all occurrences of patterns currently in the dictionary.

The DDM problem for one dimensional strings was studied further in [3, 11, 4] and we summarize the best bounds known for the case of an alphabet of arbitrary size:

Preprocessing: $O(d \log \sigma)$;
Insertion/Deletion: $O(p \log d)$, where p is the length of the pattern;
Text Scanning: $O((t + tocc) \log d)$.

Other search/update time combinations are achievable [11]. Slightly better time bounds are achievable when the alphabet is finite [4]. The DDM algorithms in [2, 3] use suffix trees, the algorithm in [11] uses automata methods, and the best asymptotic algorithm in [4] combines features of the two approaches.

We now summarize Baker's definitions for parameterized pattern matching. The alphabet of plain characters is Σ. The alphabet of *parameter characters* is called Π, and is disjoint from Σ. The two alphabets Σ and Π are assumed not to contain any natural numbers. A string over $\Sigma \cup \Pi$ is called a *parameterized string* or *p-string*. In our examples, we will use $\Sigma = \{a, b, c\}$, $\Pi = \{x, y, z\}$; a sample p-string is $axybcxxzbcx$. Two p-strings x and y *p-match* if x can be transformed into y by applying a one-to-one function on $\Sigma \cup \Pi$ that is the identity on Σ. For example $yyzbcy$ p-matches $axybcxxzbcx$ at the right end. We can combine the notion of parameterized matches with the three problems above.

- **Parameterized Fixed Pattern Matching(PFPM):** Given a p-pattern P of length p over $\Sigma \cup \Pi$, preprocess P so as to be able to find all p-matches of P in a *query* text T of length t.

- **Parameterized Multiple Pattern Matching(PMPM):** Given a set D of p-patterns $P_1, P_2, \ldots, P_k$ over alphabet $\Sigma \cup \Pi$, preprocess D so as to be able to search for all p-matches in a query text T.

- **Parameterized Dynamic Dictionary Matching(PDDM):** Preprocess and maintain a dictionary D of p-patterns under the operations *insert* a p-pattern, *delete* a p-pattern, and *search* for a query text T for all occurrences of patterns currently in the dictionary.

In Section 2, we show how to solve the PMPM problem within the same time bounds as the AC algorithm. An extension of the KMP automaton to solve the PFPM problem will follow as a special case. Baker (personal communication) also noted that it is possible to extend the KMP automaton to solve the PFPM problem. Amir, Farach, and Muthukrishnan [5] have recently provided some lower bounds for parameterized pattern matching.

A more interesting question is whether it is possible to extend the automata approach to solve the PDDM problem which combines the dynamic dictionary paradigm with the notion of parameterized matches. From a practical perspective, the dynamic dictionary paradigm applies naturally to the code duplication problem that motivated the definition of parameterized matching. The dictionary of patterns could be small pieces of Program 1 that we would want to match to Program 2. If we change what parts of Program 1 we want to match, or have multiple versions of Program 1, then we get an instance of the PDDM problem.

Our main result, given in Section 3, is that it is possible design an automaton algorithm to solve the PDDM problem. Our time bounds for the deletion and search are the same as for the DDM problem. Inserting a pattern takes time $O(p(\log^2 d + \log \sigma))$. One interesting feature of our PDDM algorithm is that we work with two dual representations of a parameterized pattern and one of the representations is computed only *implicitly* but never stored or printed.

2 Parameterized Multiple Pattern Matching

The AC algorithm scans the text one character at a time computing for each text prefix $t_1, \ldots, t_i$, the longest *prefix* of a pattern that matches a suffix of the text scanned so far. If the matching pattern prefix has a full pattern as a suffix, then all full patterns that are suffixes are output.

The AC algorithm uses an automaton whose states are precisely the pattern prefixes. We use a state and the corresponding pattern prefix interchangeably; we call the pattern prefix the *label* of its state. There are two partial functions, *goto* and *fail*, that represent the forward and backward automaton transitions. There is a third function *output*, such that for any state x, $output(x)$ is the list of all patterns that are suffixes of x.

For state x and character a, define $goto(x,a) = xa$, if state xa exists and let it be undefined if state xa does not exist. For the start state, ϵ, define $fail(\epsilon) = \epsilon$. For a nonempty state x, define $fail(x)$ to be the longest proper suffix of x that is a state in the AC automaton. The *goto* transitions form an outward tree whose root is ϵ; the *fail* transitions form an inward tree whose root is ϵ [1].

For any given choice of *state* and *symbol* we may have to take the *fail* transition repeatedly, but this shortens the length of the new *state*. The total time needed to scan a text T using this algorithm is $O(t(g + f))$, where $g + f$ is the time needed to make one evaluation of *goto* and *fail* [1].

Baker defined a procedure *prev* to convert a string S on $\Sigma \cup \Pi$ to a string of the same length on $\Sigma \cup \mathsf{N}$ that helps in finding p-matches. The procedure *prev* is defined from left to right on S. For each $c \in \Sigma$, *prev* maps the character to itself. For each $c \in \Pi$, *prev* maps the first occurrence of c to 0 and each successive occurrence of c to the number of characters since the previous occurrence. For example, $prev(axbybbyxyxay) = a0b0bb3622a3$. Baker defined a *p-suffix* of S to be a string obtained by applying *prev* to a suffix of S. Also $psuffix(S, i) = prev(S[i, \ldots n])$, where $n = |S|$. The utility of *prev* is that:

Lemma 1 [6]. *P-strings S_1 and S_2 p-match if and only if $prev(S_1) = prev(S_2)$.*

Lemma 2 [6]. *If P is a p-string pattern and T is a p-string text, then P p-matches T starting at $T[i]$ if and only if $prev(P)$ is a prefix of $psuffix(T, i)$.*

Example 1. Suppose $P = axbxy$ and $T = byaybyxcc$. Then $prev(P) = a0b20$ and $prev(T) = b0a2b20cc$. Also, $psuffix(T, 3) = a0b20cc$ and P p-matches T at location 3.

We will assume that the patterns in D and text are given as strings on $\Sigma \cup \mathsf{N}$ that represent *prev* applied to the original patterns and text. Thus in Example 1, we would assume that P is presented as $a0b20$ and T is presented as $b0a2b20cc$. As explained in [6] *prev* can be computed from left to right in linear time, so this is a reasonable assumption. This assumption is used only to simplify the theoretical exposition; in our implementation the strings are on the alphabet $\Sigma \cup \Pi$ and we choose disjoint subsets of ASCII to be Σ and to be Π.

The set of states in our automaton will be the set of prefixes of strings in D. For a string $u = u_1 \ldots u_m$ on $\Sigma \cup \mathsf{N}$, define $shorten(u)$ character by character as follows. Each character in Σ is mapped to itself by *shorten*. If $u_i = r \in \mathsf{N}$ then *shorten* maps u_i to r if $r < i$, and maps u_i to 0 if $r \geq i$. For example, $shorten(a3b27b3c90) = a0b20b3c00$.

Lemma 3. *Let* $W = w_1, \ldots, w_m$ *be a string on* $\Sigma \cup \Pi$ *and let* $U = u_1, \ldots, u_m = prev(W)$. *Let* W_i *be the suffix* $w_i, \ldots, w_m$ *and let* U_i *be the suffix* $u_i, \ldots, u_m$. *Then* $prev(W_i) = shorten(U_i)$

Corollary 4. *Pattern* P *matches a substring* S *of the text if and only if* $P = shorten(S)$.

We extend the definitions of *goto, fail, output* to *pgoto, pfail, poutput*. For $a \in \Sigma$ and $x = \epsilon$ define $pgoto(x, a)$ to be a if that state exists and ϵ otherwise. For $a \in \mathsf{N}$ and $x = \epsilon$ define $pgoto(x, a) = 0$ if state 0 exists, and ϵ if state 0 does not exist. For $a \in \Sigma$ and $x \neq \epsilon$, define $pgoto(x, a) = xa$, if state xa exists; otherwise $pgoto(x, a)$ is undefined. If $a \in \mathsf{N}$ and $x \neq \epsilon$ let $q(x, a) = 0$ if $a > |x|$ and let $q(x, a) = a$ otherwise. Then $pgoto(x, a)$ is $x \cdot q(x, a)$ if that state exists and undefined otherwise. Define $pfail(x)$ to be the longest proper suffix y such that $shorten(y)$ is a state in the automaton.

From the definitions above and the previous Lemma, we get the following characterization of *pfail*.

Corollary 5. *Suppose that* U *is an automaton state,* $U = prev(W)$ *and* W_i *is the longest proper suffix of* W *such that* $prev(W_i)$ *is a state in the automaton. Then* $prev(W_i) = shorten(U_i) = pfail(U)$.

For each state x, $poutput(x)$ is the set of patterns P such that P equals *shorten* applied to some suffix of x.

The search algorithm to solve the PMPM problem is very similar to the the one given in [1]. The correctness proof (omitted in this abstract) is new because *pgoto* and *pfail* are fundamentally different from *goto* and *fail*.

```
SEARCH(T = t₁ ... tₙ)
    state ← ϵ
    for i ← 1 to n do
        While pgoto(state, tᵢ) is undefined
            state ← pfail(state)
        state ← pgoto(state, tᵢ)
        If poutput(state) is nonempty
            print location i and poutput(state)
```

Lemma 6. *The search loop maintains the invariant that after* t_i *is scanned, state is the longest state such that there is a suffix* x *of* $t_1 \ldots t_i$ *with* $state = shorten(x)$.

Lemma 7. *The time to scan the text is* $O(t \log \sigma + tocc)$, *where* σ *is now the number of regular and parameterized characters occurring in the original patterns.*

The following algorithm constructs the automaton states and *pgoto* by adding one pattern at a time. Recall that each pattern is presented in *prev* notation.

```
BUILD-PGOTO(D = {P₁, ..., Pₖ})
    Create the empty state ϵ
    for i ← 1 to k do
```

```
/*Insert pattern Pᵢ = Pᵢ[1...lᵢ]*/
    state ← ε
    j ← 1
    newstate ← false
    do /* go through existing states */
        if pgoto(x, Pᵢ[j]) is defined then
            state ← pgoto(x, Pᵢ[j])
            j ← j + 1
        else
            newstate ← true
    while newstate = false and j ≤ lᵢ
    for r ← j until lᵢ do
        create state state · Pᵢ[r] storing its label length
        pgoto(state, Pᵢ[r]) ← state · Pᵢ[r]
        state ← state · Pᵢ[r]
    add Pᵢ to output(state).
for all c ∈ Σ ∪ {0} such that pgoto(ε, c) is undefined
    pgoto(ε, c) ← ε
```

As in [1], we do not actually store the state labels but instead, number the states consecutively. Unlike [1] we do store the length of the label, so that we can compute *shorten* quickly.

Lemma 8. *The algorithm to construct pgoto is correct and runs in $O(d \log \sigma)$ time.*

Each state has one incoming *pgoto* edge whose symbol is the last character of the state name, so the *pgoto* edges form an outgoing tree rooted at ϵ.

To compute *pfail* and fill in *poutput* we visit the states in the *pgoto* tree in breadth-first order. To get started $pfail(c) \leftarrow \epsilon$ for all states c with one character names. If we have computed $pfail(z)$ for all states z such that $|z| < |x|$ we can compute $pfail(x)$ using:

```
BUILD-PFAIL(x)
    Suppose x = yc
    found ← false
    do
        temp ← pfail(y)
        if pgoto(temp, c) is defined
            pfail(x) ← pgoto(temp, c)
            found ← true
        else if temp = ε
            pfail(x) ← ε
            found ← true
        y ← temp
    while found = false
    poutput(x) ← poutput(x) ∪ poutput(pfail(x))
```

Lemma 9. *If build-pfail(x) is applied in increasing order of length of x, then all the values of pfail are computed correctly. The time to compute all values is $O(d \log \sigma)$.*

Lemma 10. *The initialization of poutput at the end of* build-pgoto *and the increment at the end of* build-pfail *ensure that for each state* x, *poutput(x) contains precisely those patterns* P *such that* P *equals shorten applied to a suffix of* x.

Theorem 11. *It is possible to solve the PMPM problem using preprocessing time* $O(d \log \sigma)$ *and scanning time* $O(t \log \sigma + tocc)$.

Corollary 12. *If there is only one pattern (as in the PFPM problem), the preprocessing and search times can be improved to* $O(d)$ *and* $O(t)$ *respectively.*

3 Dynamic Dictionary Matching of Parameterized Patterns

To modify the previous algorithms for the dynamic case we must be able to compute *pgoto*, *poutput*, and *pfail* when the set of patterns is changing. Managing *pgoto* is straightforward and will be discussed briefly. In the dynamic setting we will not be able to store the *poutput* function; instead we reuse a technique for the DDM algorithm in [11] to do the output on the fly. To compute the *pfail* function we would like to use the reduction from failure function maintenance to balanced parenthesis maintenance in [11, 4], but the rules for the numeric characters in the state labels invalidate the original reduction.

As in the AC algorithm and the algorithm in the previous section, we store the *pgoto* function as a directed tree rooted at ϵ, where the nodes correspond to automaton states, whose implicit labels are *prev* applied to a pattern prefix. Every edge is labeled with a character from $\Sigma \cup \mathsf{N}$, although we will be later augmenting Σ to do the output and compute *pfail*. The fact that the dictionary is dynamic has minimal impact on the way the *pgoto* function is computed, and we update *pgoto* when patterns are inserted as in the previous section. We store with each state a pointer to its parent state in the *pgoto* tree. For each state x we keep a count of how many patterns P have x as a prefix. When a pattern is deleted, we visit the states for prefixes of the pattern from longest to shortest. For each prefix of the pattern, we decrement the pattern counter of that state and delete the state if the counter decreases to 0 (meaning that no other patterns use that state).

To recognize patterns and do output, we make two modifications to the pattern inputs. First, we assume that the empty string ϵ is always a pattern in the dictionary, although we never output ϵ as a matched pattern. Second we add a distinct symbol \$ to Σ and we assume that \$ is the largest symbol in the lexicographic order on Σ. We append to each pattern the special symbol \$ before inserting the pattern. We use the symbol $D\$$ to represent the dictionary with \$ appended to every pattern. For each pattern P, there will be a state corresponding to the string $P\$$ at which we recognize P; we keep a pointer to the pattern at that state to be able to print the pattern. We extend the definition of *pgoto*, so that $pgoto(P, \$) = P\$$, for each pattern P. We use the following dictionary as an example to explain various definitions and concepts of our automaton.

Example 2. Suppose $\Sigma = \{a, b, \$\}$. Let $\hat{D} = \{\$, b\$, b0\$, a0b2\$\}$ be a sample dictionary where every pattern is appended with the special symbol \$.

We call a state label a *normal prefix* if it is a prefix of some pattern in $D\$$. For each proper prefix x of a pattern we also define an *extended prefix*, $x\$$, by appending the character $\$$. The corresponding states are called normal or extended. In Example 2, the set of normal prefixes is $\{\epsilon, a, a0, a0b, a0b2, b, b0, \$, b\$, b0\$, a0b2\$, \}$ and the set of extended prefixes is $\{\$, b\$, b0\$, a\$, a0\$, a0b\$, a0b2\$\}$. Notice that some prefixes are both normal and extended. In this example, $pgoto(b0, \$) = b0\$$ because $b0$ is a pattern, but $pgoto(a0b, \$)$ is not defined because $a0b$ is not a full pattern.

We modify the definition of $pfail$ to accommodate the extended prefixes as follows: Let w be a state label (normal or extended). Define $pfail(w) = shorten(x)$ such that $|x| < |w|$ and x is the longest suffix of w such that $shorten(x)$ is a *normal* prefix. In Example 2, $pfail(a0b2\$) = b0\$$ because $b0 = shorten(b2)$.

We recognize patterns as follows. When we reach a position of a text, and compute $pgoto$ for that position, we pretend that the next symbol is a $\$$. If we can make another $pgoto$ transition to some *normal* prefix ending with a $\$$, then we know that a pattern has been matched at that position, since any normal prefix ending with a $\$$ must be a pattern in the dictionary. By applying $pfail$ repeatedly, we can report all the matching patterns in the order from the longest pattern to the shortest. When we reach the pattern $\epsilon\$$, we stop without reporting that empty pattern.

The basic difficulty in extending the automata approach to the DDM problem is how to maintain the $fail$ function when doing insertions and deletions. We summarize the key ideas used to maintain $fail$ for the DDM problem, so that we can significantly modify them to handle $pfail$ in the parameterized version.

Let $* \notin \Sigma$ be a new symbol such that $* > a$ for any $a \in \Sigma$ in lexicographic comparison; in particular $* > \$$. For every prefix $w \in \Sigma^*$ we define $*w$ as the *complement* of w. We call w a *regular prefix*, and $*w$ a *complementary prefix*. Let S be the set of regular and complementary prefixes of patterns. We define an ordering $<$ on S, such that regular prefix x is a suffix of regular prefix y if and only if $x < y < *x$. An ordering that satisfies this property has the virtues that

Lemma 13 [11]. *Let $w, x \in S$ be arbitrary regular prefixes.*
1. $w < x < *w$ if and only if $w < *x < *w$.
2. If we replace a regular prefix with a '(' and its complement with a ')' then the prefixes of S in the $<_{inv}$ order yield a list of well balanced parentheses.

If we store the parenthesis list of part 2 of Lemma 13, then $fail(y)$ can be computed by finding the left parenthesis $(_y$ that y is mapped to, finding nearest enclosing parentheses of $(_y$, and then finding the string x that has been mapped to the nearest enclosing left parentheses [11, 4].

Unfortunately, the alphanumeric *prev* prefixes we used as the state labels in the previous section do not have the property that $pfail(x)$ is always a suffix of x. To solve the suffix problem we define a new labeling function called $next : (\Sigma \cup \Pi)^* \rightarrow (\Sigma \cup \mathbb{N})^*$ which can be viewed as a "dual" of *prev*. The function $next$ maps each character in Σ to itself. The function $next$ maps all but the last occurrence of $c \in \Pi$ to the number of characters until the next occurrence of c. The last occurrence of $c \in \Pi$ is mapped to 0 by $next$. For example, $next(axbxaybbyxy) = a2b6a3bb200$. In the full paper we describe a simple linear-time function PREV-TO-NEXT that converts labels from *prev* representation to *next* representation.

Since $next(x)$ is a suffix of $next(ax)$ we can now restate the characterization of $pfail$ in Corollary 5.

Corollary 14. *For any automaton state x, $pfail(x)$ is the longest automaton state y such that* PREV-TO-NEXT*(y) is a proper suffix of* PREV-TO-NEXT*(x).*

The preceding corollary suggests that we should use the $next$ labeling of states to compute $pfail$ as follows. We define a total ordering on the $next$ labeling of states and their complements and call it the *inverted order* denoted by $<_{inv}$.

For technical reasons needed later, we assume that any $c \in \Sigma$ is $>_{inv}$ any $m \in$ N in the ordering of characters. We assume that the numbers are ordered in the opposite of their usual order, so $0 >_{inv} 1 >_{inv} 2, \ldots$. For two distinct strings w and x, $w <_{inv} x$ if $(\text{PREV-TO-NEXT}(w))^R$ comes before $(\text{PREV-TO-NEXT}(x))^R$ in the lexicographic ordering on $(\Sigma \cup N)^*$, where x^R is the reverse of the string x.

Since PREV-TO-NEXT(x) is a suffix of PREV-TO-NEXT(ax), it follows that regular prefix x is a suffix of regular prefix y if and only if $x <_{inv} y <_{inv} *x$. and Lemma 13 applies to our $<_{inv}$ ordering.

Example 3. Consider the list of prefixes of $\hat{D}$ in Example 2. Adding in the complementary prefixes, the set S using $prev$ labels is as follows in $<_{inv}$ order: ϵ, $a0$, $*a0$, $b0$, $a0b2$, $*a0b2$, $*b0$, a, $*a$, b, $a0b$, $*a0b$, $*b$, $\$$, $a0\$$, $*a0\$$, $b0\$$, $a0b2\$$, $*a0b2\$$, $*b0\$$, $a\$$, $*a\$$, $b\$$, $a0b\$$, $*a0b\$$, $*b\$$, $*\$$, $*$.

If we convert each prefix to its $next$ label, the order of corresponding prefixes is ϵ, $a0$, $*a0$, $b0$, $a2b0$, $*a2b0$, $*b0$, a, $*a$, b, $a0b$, $*a0b$, $*b$, $\$$, $a0\$$, $*a0\$$, $b0\$$, $a2b0\$$, $*a2b0\$$, $*b0\$$, $a\$$, $*a\$$, $b\$$, $a0b\$$, $*a0b\$$, $*b\$$, $*\$$, $*$ and we can check the order more easily.

The number of extended prefixes is at most equal to the number of normal prefixes, so the number of regular prefixes is $O(d)$. The number of complementary prefixes is exactly equal to the number of regular prefixes. Therefore, the total number of prefixes in our dictionary structure is only $O(d)$.

To compute $pfail$ on the regular prefixes we use the parenthesis mapping suggested in part 2 of Lemma 13. More specifically, each regular prefix is mapped to a left parenthesis. The complement of each regular prefix is mapped to a right parenthesis. Each left parenthesis that corresponds to prefix x has a bidirectional pointer between it and the state s_x in the automaton that represents prefix x. To compute $pfail(y)$ for a regular prefix y we do the following steps: Let $(_y$ be the left parenthesis that the state for y points to. Find the nearest enclosing parenthesis pair $(_z,)_{*z}$ of $(_y$. Find the state z that $(_z$ points to. Then $pfail(y)$ is z.

Extended prefixes that are not regular are not mapped to anything in the parenthesis list. The improved algorithm for parenthesis queries in [4] is not applicable because in parameterized matching the alphabet of the prev labels includes N and is therefore unbounded.

We use a height balanced tree, such as an a-b tree [14] with a distinct leaf representing each prefix in S to store the list. We assume that the leaves are linked, so that each leaf can find its neighbors in constant time. The a-b tree enables us to insert or delete a new prefix in $O(\log d)$ comparisons and $O(\log d)$ other operations. In our implementation we use a 2-3 tree.

We supplement each node v of the tree with the values of two functions BALANCE(v) and SUFF(v) to be defined shortly. The function BALANCE was first proposed by Güting and Wood in their data structure called a *parenthesis tree* [10].

The parenthesis tree maintains a list of d well-balanced parentheses as its leaves supporting the operations. Each operation takes $O(\log d)$ worst-case time [10].

$P\_Insert(p_1, p_2, p_1', p_2')$: Insert the matching pair (p_1, p_2) with p_1 immediately after p_1' and p_2 immediately before p_2'.

$P\_Delete(p_1, p_2)$: Delete the matching pair (p_1, p_2).

$P\_Nearest(p_1)$: Return the nearest enclosing parentheses of p_1.

In the parenthesis tree of Güting and Wood the parentheses are represented as leaves in a balanced binary tree. Each pair of matching parentheses are connected to each other by a separate link. The parenthesis tree is augmented with BALANCE information in the following way: For every node p in the tree, let the pair BALANCE(p) = $\langle close(p), open(p) \rangle$ denote the number of unmatched closing and unmatched opening parentheses in its subtree.

For each node p of the tree, the BALANCE(p) can be computed recursively as follows:

(a) *p is a leaf.* Then

$$\text{BALANCE}(p) = \begin{cases} \langle 0, 1 \rangle \text{ if } p \text{ is '(',} \\ \langle 1, 0 \rangle \text{ if } p \text{ is ')'.} \end{cases}$$

(b) *v is an internal node with children v_1 and v_2.* We compute BALANCE(v) = BALANCE(v_1) $\oplus$ BALANCE(v_2), where

$$\langle i, j \rangle \oplus \langle k, l \rangle = \begin{cases} \langle i, j - k + l \rangle \text{ if } j \geq k, \\ \langle i - j + k, l \rangle \text{ otherwise.} \end{cases}$$

We can extend the scheme of Güting and Wood to work with trees with multiple arity (> 2), by observing that the $\oplus$ operator is associative. For a node v with children $v_1, v_2, \ldots v_k$, BALANCE(v) = BALANCE(v_1) $\oplus$ BALANCE(v_2) $\oplus \cdots \oplus$ BALANCE(v_k). We allow some leaves (corresponding to the extended prefixes that are not normal) to not represent parentheses; for such a leaf v, we define BALANCE(v) = $\langle 0, 0 \rangle$. It still makes sense to ask for the nearest enclosing parentheses of such a leaf.

Now we define SUFF. Let x_1, x_2 be two state labels in *prev* notation; then SUFF(x_1, x_2) = 1 + the length of the longest common suffix of PREV-TO-NEXT(x_1) and PREV-TO-NEXT(x_2). For example, if PREV-TO-NEXT(x_1) is a suffix of PREV-TO-NEXT(x_2), then SUFF(x_1, x_2) = $1 + |x_1|$.

Let l_1 and l_2 be two leaves with labels x_1 and x_2. We extend the definition of SUFF to say that SUFF(l_1, l_2) is the same as SUFF(x_1, x_2). We further extend SUFF to just one node argument. Define SUFF of the rightmost leaf to be ∞. For any other leaf l_1 with right neighbor l_1' define SUFF(l_1) =SUFF(l_1, l_1'). For an internal node v with leftmost descendant leaf label x^l and rightmost descendant leaf label x^r define SUFF(v) to be 1 + the length of the longest common suffix of PREV-TO-NEXT(v^l) and PREV-TO-NEXT(v^r). For node v let its leftmost and rightmost descendant leaves be v^l and v^r.

Lemma 15. *If v is an internal node with children $v_1, v_2, \ldots v_k$ then $\text{SUFF}(v) =$*
$\min(\min(\text{SUFF}(v_1), \text{SUFF}(v_1^r)), \min(\text{SUFF}(v_2), \text{SUFF}(v_2^r)), \ldots, \min(\text{SUFF}(v_k), \text{SUFF}(v_k^r)))$.

For each node v we can either store $\text{SUFF}(v^r)$ directly at v or we can store a pointer to v^r and compute $\text{SUFF}(v^r)$ with one level of indirection. Given this information, both $\text{BALANCE}(v)$ and $\text{SUFF}(v)$ can be computed by doing an associative "sum" on the values at the children of v using $\oplus$ for BALANCE and min for SUFF. When the a-b tree is rebalanced all the nodes that change children can have their BALANCE and SUFF values updated using standard techniques as in [10]. Insertions and deletions in the a-b tree, including function value updates, take time proportional to the height of the tree.

The $<_{inv}$ ordering on the strings ensures that any parenthesis pair we insert or delete preserves balance in the parenthesis string. The outermost parenthesis pair corresponds to the prefixes $(\epsilon, *\epsilon)$ on which we never do a *P_Nearest* query; thus every *P_Nearest* query will find an enclosing pair.

One novel feature of our PDDM algorithm is how we insert a node for a new prefix xa into the tree. We use a search down the a-b tree to figure out where x should go. Our aim is to find the largest prefix smaller than xa in the $<_{inv}$ order on leaves, so that we know where to insert xa. We start the search at the root of the a-b tree and proceed towards the leaves. Suppose yb is the label associated with an internal node of the tree. At the next level we take the right child of the node if $yb <_{inv} xa$ and the left child otherwise. With this scheme we need $O(\log d)$ tree comparisons to find where to insert xa.

We cannot afford to compare $\text{PREV-TO-NEXT}(xa)$ against $\text{PREV-TO-NEXT}(yb)$ directly for two reasons. First, we do not explicitly compute or store the *next* labels. Even if we did store these labels, they would be arbitrarily long strings and could not be compared in $O(1)$ time. Therefore we do all the comparisons of x against existing nodes in a roundabout way using the $<_{inv}$ ordering and the rules specified in the following Lemma. In constant time we can find the last character a of xa, and the parent x of xa in the *pgoto* tree. This is implemented by making the links in the *pgoto* tree bidirectional.

Lemma 16. *To compare xa and yb in the $<_{inv}$ order we use the following rules.*

1. *If $a \neq b \in \Sigma$, then $xa <_{inv} yb$ if and only if $a < b$.*
2. *If $a = b \in \Sigma$, then $xa <_{inv} yb$ if and only if $x <_{inv} y$.*
3. *If $a \in \Sigma, b \in \mathbb{N}$, then $yb <_{inv} xa$.*
4. *If $a \in \mathbb{N}, b \in \Sigma$, then $xa <_{inv} yb$.*
5. *If $a = b \in \mathbb{N}$, then $xa <_{inv} yb$ if and only if $x <_{inv} y$.*
6. *If $a < b \in \mathbb{N}$ and $\text{SUFF}(x, y) < a$, then $xa <_{inv} yb$ if and only if $x <_{inv} y$.*
7. *If $0 \neq a < b \in \mathbb{N}$ and $\text{SUFF}(x, y) \geq a$, then $xa <_{inv} yb$.*
8. *If $0 = a < b \in \mathbb{N}$, $\text{SUFF}(x, y) \geq b$, and $\text{SUFF}(x, y) > |xa|$, then $yb <_{inv} xa$.*
9. *If $0 = a < b \in \mathbb{N}$, $\text{SUFF}(x, y) \geq b$, and $\text{SUFF}(x, y) = |xa|$, then $xa <_{inv} yb$.*
10. *If $0 = a < b \in \mathbb{N}$, $\text{SUFF}(x, y) < b$, then $xa <_{inv} yb$ if and only if $x <_{inv} y$.*

The last five rules have symmetric counterparts for $b < a \in \mathbb{N}$.

Some of the comparison rules require only comparisons between characters and the relative order of other leaves, while others require that we compute SUFF. For the comparisons that do not require SUFF we can gain efficiency by storing the list of leaves in the ordered list data structure of Dietz and Sleator [9]. Using this data structure we can do the comparisons that do not involve SUFF in $O(1)$ time. Our implementation does them in $O(1)$ amortized time. For those comparisons where we need to compute SUFF, we used standard techniques for balanced dynamic trees [8] to compute SUFF and do the comparison in time $O(\log d)$, proportional to the height of the tree.

When we insert a leaf l and its corresponding parenthesis updating BALANCE is straightforward going all the way up the tree using $O(1)$ time per node. We use the following lemma to compute SUFF for l and its neighbor to the left. Once we compute these two values it is straightforward to update the values of SUFF on the path up the tree in $O(1)$ time per node.

Lemma 17. *Suppose xa and yb are the prev labels of two adjacent leaves. Then* SUFF(xa, yb) *is given by:*

1. *1 if $a, b \in \Sigma$ and $a \neq b$.*
2. *1 + SUFF(x, y), if $a = b$.*
3. *1 if $a \in \Sigma, b \in N$ or if $a \in N, b \in \Sigma$.*
4. *1 + SUFF(x, y) if $a, b \in N$, $a \neq b$ and SUFF$(x, y) < \min(a, b)$.*
5. *1 + $\min(a, b)$ if $a, b \in N$, $a \neq b$ and SUFF$(x, y) \geq \min(a, b) > 0$.*
6. *1 + $\min(\max(a, b), $SUFF$(x, y))$ if $a, b \in N$, $a \neq b$ and $\min(a, b) = 0$.*

Pseudocode for SEARCH, INSERT, and DELETE, which are used for searching a text, inserting a the prefixes of a pattern in the a-b tree, and deleting the prefixes of a pattern from the a-b tree respectively is given in an appendix. We use prefixes instead of states for clarity.

Theorem 18. *Let D be a dictionary of parameterized patterns over an alphabet $\Sigma \cup \Pi$. We can search a parameterized text T for patterns of D in time $O((t + tocc) \cdot (\log \sigma + \log d))$, where tocc is the total number of patterns reported. We can insert a pattern P in time $O(p(\log^2 d + \log \sigma))$. We can delete a pattern P in time $O(p(\log d + \log \sigma))$. Moreover, we require only $O(d)$ space to store the automaton.*

Proof. (Sketch) The search algorithm is essentially the same as in the non-dynamic case, but we have a different representation of $pfail$. As long as we can maintain the $<_{inv}$ order on the prefixes, the equivalence of $pfail$ computation and nearest enclosing parentheses queries holds. Earlier Lemmas proved the correctness of the algorithms for comparisons and insertions.

The running time analysis is similar to the static case. The extra log factor in searching comes from the cost of computing $pfail$. One extra log factor in the insertion and deletion cost comes from the cost of updating the tree which is proportional to the height of the tree. The second extra log factor in the insertion cost comes from the cost of doing a comparison.

In our implementation, we preprocessed the initial dictionary by insertion of the individual patterns.

References

1. A. V. Aho and M. J. Corasick. Efficient string matching: An aid to bibliographic search. *Comm. ACM*, 18:333–340, 1975.
2. A. Amir and M. Farach. Adaptive dictionary matching. *Proc. of the 32nd IEEE Annual Symp. on Foundation of Computer Science*, pages 760–766, 1991.
3. A. Amir, M. Farach, R. Giancarlo, Z. Galil, and K. Park. Dynamic dictionary matching. Manuscript, 1991.
4. A. Amir, M. Farach, R. M. Idury, J. A. La Poutré, and A. A. Schäffer. Improved dynamic dictionary matching. *Proc. of the Fourth Ann. ACM-SIAM Symp. on Discrete Algorithms*, pages 392–401, 1993. Full paper to appear in *Information and Computation*.
5. A. Amir, M. Farach, and S. Muthukrishnan. Alphabet dependence in parameterized matching. To appear in *Information Processing Letters*, 1994.
6. B. Baker. A theory of parameterized pattern matching: algorithms and applications. *Proc. of the 25th Ann. ACM Symp. on Theory of Computing*, pages 71–80, 1993.
7. M. T. Chen and J. Seiferas. Efficient and elegant subword tree construction. In A. Apostolico and Z. Galil, editors, *Combinatorial Algorithms on Words*, chapter 12, pages 97–107. NATO ASI Series F: Computer and System Sciences, 1985.
8. T. H. Cormen, C. E. Leiserson, and R. L. Rivest. *Introduction to Algorithms*. MIT Press, 1990.
9. P. Dietz and D. D. Sleator. Two algorithms for maintaining order in a list. In *Proc. of the 19th Ann. ACM Symp. on Theory of Computing*, pages 365–372, 1987. To appear in *J. Comp. Syst. Sci.*
10. R. H. Güting and D. Wood. The parenthesis tree. *Information Sciences*, 27:151–162, 1982.
11. R. M. Idury and A. A. Schäffer. Dynamic dictionary matching with failure functions. In *Proc. of the Third Symp. on Combinatorial Pattern Matching, Lecture Notes Comp. Sci. 644*, pages 276–287, 1992. Full paper to appear in *Theoretical Computer Science*.
12. D. E. Knuth, J. H. Morris, and V. R. Pratt. Fast pattern matching in strings. *SIAM J. Comp.*, 6:323–350, 1977.
13. E. M. McCreight. A space-economical suffix tree construction algorithm. *J. ACM*, 23:262–272, 1976.
14. K. Mehlhorn. *Data Structures and Algorithms 1: Sorting and Searching*. Springer-Verlag, 1984.
15. B. Meyer. Incremental string matching. *Information Processing Letters*, 21:219–227, 1985.
16. P. Weiner. Linear pattern matching algorithm. *Proc. of the 14th IEEE Annual Symp. on Switching and Automata Theory*, pages 1–11, 1973.

A   Pseudocode for dynamic Search, Insert, and Delete

Algorithm 1 *Pseudocode for searching a text.*

> SEARCH($T = t_1 \ldots t_n$)
> $state \leftarrow \epsilon$
> for $i \leftarrow 1$ to n do
>   While $pgoto(state, t_i)$ is undefined
> $state \leftarrow pfail(state)$
> $state \leftarrow pgoto(state, t_i)$
>   /* Pretend a $ is read to check if any patterns match */
> $temp \leftarrow pgoto(state, \$)$
>   If $temp$ is not normal then $temp \leftarrow pfail(temp)$
>   While $temp \neq \$$ do /* Report all non-empty patterns */
>   /* Since $temp$ ends in $ we have matched a pattern */
>   Print the pattern that $temp$ points to
> $temp \leftarrow pfail(temp)$ /* See if any smaller patterns match */

Algorithm 2 *Pseudocode for inserting a pattern into the tree for a dictionary.*

> INSERT($P = p_1 \ldots p_m$) /* $p_m = \$$ */
> Suppose $p_1 \ldots p_j$ is the longest prefix of P shared by some other pattern.
> Increment the reference count for the prefixes of $p_1 \ldots p_j$.
> For $i \leftarrow j + 1$ to m do
>   Let $x = p_1 \ldots p_{i-1}$. Let $a = p_i$. /* xa is being inserted. x is already in S */
>   Insert a leaf and left parenthesis for xa updating BALANCE and SUFF values as needed
>   Insert a leaf and right parenthesis $*xa$ updating BALANCE and SUFF values as needed
>   Insert a leaf for $xa\$$ and a left parenthesis if a is the last character updating BALANCE and SUFF values as needed
>   Insert a leaf for $*xa\$$ and a right parenthesis if a is the last character updating BALANCE and SUFF values as needed

Algorithm 3 *Pseudocode for deleting a pattern from the tree for a dictionary.*

> DELETE($P = p_1 \ldots p_m$) /* $p_m = \$$ */
> Suppose $p_1 \ldots p_j$ is the longest prefix of P shared by some other pattern.
> Decrement the reference count for the prefixes of $p_1 \ldots p_j$.
> For $i \leftarrow m$ downto $j + 1$ do
>   Let $x = p_1 \ldots p_i$ /* x is a normal prefix */
>   If $x\$$ is still in S Then
>   *delete* $x\$$ and $*x\$$ updating BALANCE and SUFF values as needed
>   *delete* x and $*x$ updating BALANCE and SUFF values as needed

Approximate String Matching with Don't Care Characters

Tatsuya Akutsu *

Mechanical Engineering Laboratory,
1-2 Namiki, Tsukuba, Ibaraki, 305 Japan.

Abstract. This paper presents parallel and serial approximate matching algorithms for strings with don't care characters. They are based on Landau and Vishkin's approximate string matching algorithm and Fisher and Paterson's exact string matching algorithm with don't care characters. The serial algorithm works in $O(\sqrt{km}\, n \log |\Sigma| \log^2 \frac{m}{k} \log\log \frac{m}{k})$ time, and the parallel algorithm works in $O(k \log m)$ time using $O(\sqrt{\frac{m}{k}}\, n \log |\Sigma| \log \frac{m}{k} \log\log \frac{m}{k})$ processors on a CRCW-PRAM, where n denotes the length of a text string, m denotes the length of a pattern string, k denotes the maximum number of differences, and Σ denotes the alphabet (i.e. the set of characters). Several extensions are also described.

Keywords: approximate string matching, don't care characters, sequence analysis

1 Introduction

Approximate string matching is important not only from a theoretical viewpoint but also from a practical viewpoint. In particular, it is important in molecular biology since exact matching is not sufficient [4, 8]. While several variants are considered, the *string matching with k differences* is the most important one [7, 10]. Let $T = t_1 \cdots t_n$ be a text string and $P = p_1 \cdots p_m$ be a pattern string over an alphabet Σ, where $|\Sigma| \leq m$ can be assumed without loss of generality. A *difference* is one of the following:

(A) A character of the pattern corresponds to a different character of the text.
(B) A character of the pattern corresponds to no character in the text.
(C) A character of the text corresponds to no character in the pattern.

---
* Partially supported by the Grant-in-Aid for Scientific Research on Priority Areas, "Genome Informatics", of the Ministry of Education, Science and Culture of Japan.

If the minimum number of differences between the pattern string P and any substring of the text string T ending at t_j, is less than or equal to k, we say that P occurs at position j of T with at most k differences. Then, the problem is defined as follows: given a text string T, a pattern string P and an positive integer k $(1 \leq k \leq m)$, find all positions of T where P occurs with at most k differences.

[**Example 1**] P ="bcdefgh" occurs at position 8 of T ="abxdyeghij" with differences 3 by the following correspondence:

```
T    a    b    x    d    y    e         g    h    i    j
P         b    c    d         e    f    g    h
          (A)       (C)       (B)
```

Landau and Vishkin have developed an $O((k + \log m)n)$ time algorithm for string matching with k differences [10], while Galil and Park have made theoretical and practical improvements [7].

Although efficient algorithms have been developed, string matching with k differences seems insufficient for such applications as motif search in molecular biology. In motif search, *don't care* characters are frequently used [4, 8], where the don't care character is a character which matches any character. In this paper, '*' denotes the don't care character. Exact string matching with don't cares was studied about 20 years ago, Fisher and Paterson developed an $O(n \log |\Sigma| \log^2 m \log \log m)$ time algorithm based on the fast computation method of convolutions [5]. Their technique has been applied to various pattern matching problems [1, 2]. Abrahamson generalized their algorithm and developed algorithms for the *generalized string matching* [1]. Manber and Baeza-Yates considered an exact matching problem with a sequence of don't cares, in which the length of the sequence of don't cares is not fixed [11].

Combining Landau and Vishkin's algorithm with Fisher and Paterson's algorithm, we have developed serial and parallel algorithms for string matching with k differences, in which don't care characters may appear both in a text string and in a pattern string. The serial algorithm works in $O(\sqrt{km}\, n \log |\Sigma| \log^2 \frac{m}{k} \log \log \frac{m}{k})$ time, and the parallel algorithm works in $O(k \log m)$ time using $O(\sqrt{\frac{m}{k}}\, n \log |\Sigma| \log \frac{m}{k} \log \log \frac{m}{k})$ processors, where a CRCW-PRAM is assumed as a model of a parallel computer. While suffix trees are used in Landau and Vishkin's algorithm, tables constructed using Fisher and Paterson's algorithm are used in our algorithms. This paper describes these algorithms and extensions for more general patterns of strings.

[**Example 2**] P ="bc*eghi" occurs at position 8 of T ="a*cdefgij" with differences 2 by the following correspondence:

```
T    a    *    c    d    e    f    g         i    j
P         b    c    *    e         g    h    i
                    (C)       (B)
```

2 String matching with k differences

In this section, we overview Landau and Vishkin's approximate string matching algorithm [10].

2.1 A simple dynamic programming algorithm

In this subsection, we describe an $O(mn)$ time algorithm for string matching with k differences. It was developed by many persons independently, and is based on the dynamic programming technique.

Let $D[i,j]$ $(0 \leq i \leq m$ and $0 \leq j \leq n)$ be the minimum number of differences between $p_1 \cdots p_i$ and any substring of T ending at t_j. Then, it is easy to see that $D[i,j]$ is determined by

$$D[i,j] = \min(D[i-1,j]+1, \ D[i,j-1]+1, \ D[i-1,j-1]+h),$$

where $h = 0$ if $t_j = p_i$ and $h = 1$ otherwise. Thus, the following procedure solves string matching with k differences in $O(mn)$ time. Note that all occurrences can be enumerated by outputting all j's such that $D[m,j] \leq k$.

Procedure $SimpleDynamic(P,T)$
begin
    for all j such that $0 \leq j \leq n$ **do** $D[0,j] \leftarrow 0$;
    for all i such that $0 \leq i \leq m$ **do** $D[i,0] \leftarrow i$;
    for $i = 1$ **to** m **do**
        for $j = 1$ **to** n **do**
            begin
                if $p_i = t_j$ **then** $h \leftarrow 0$ **else** $h \leftarrow 1$;
                $D[i,j] \leftarrow \min(D[i-1,j]+1, \ D[i,j-1]+1, \ D[i-1,j-1]+h)$
            end
end

[**Example 3**] Let $P =$"caab" and $T =$"bccabad". Then, the following table shows the values of $D[i,j]$'s.

	b	c	c	a	b	a	d
	0	0	0	0	0	0	0	0
c	1	1	0	0	1	1	1	1
a	2	2	1	1	0	1	1	2
a	3	3	2	2	1	1	1	2
b	4	3	3	3	2	1	2	2

2.2 Landau and Vishkin's algorithm

While the simple dynamic programming algorithm takes $O(mn)$ time, Landau and Vishkin's algorithm takes an $O((k + \log m)n)$ time [10]. Their algorithm computes the same information as in the matrix $D[i, j]$ of the simple dynamic programming algorithm, using the *diagonals* of the matrix. A diagonal d of the matrix consists of all $D[i, j]$'s such that $j - i = d$.

For a number of differences e and a diagonal d, $L[d, e]$ denotes the largest row i such that $D[i, j] = e$ and $j - i = d$. In the case of Example 3, $L[3, 0] = 0$, $L[3, 1] = 3$ and $L[3, 2] = 4$. Note that the value of $D[i, j]$ such that $j - i = d$ grows monotonically as i grows. Thus, for string matching with k differences, we need only compute the values of $L[d, e]$'s such that $e \leq k$. Since the number of diagonals is $O(n)$, the number of $L[d, e]$'s required to be computed is $O(kn)$. Moreover, Landau and Vishkin showed that $L[d, e]$'s could be computed by the following procedure.

Procedure $LandauVishkin(P, T)$
begin
    for all d such that $0 \leq d \leq n$ **do** $L[d, -1] \leftarrow -1$;
    for all d such that $-(k + 1) \leq d \leq -1$ **do**
    begin
        $L[d, |d| - 1] \leftarrow |d| - 1$; $L[d, |d| - 2] \leftarrow |d| - 2$
    end;
    for all e such that $-1 \leq e \leq k$ **do** $L[n + 1, e] \leftarrow -1$;
    for $e = 0$ **to** k **do**
        for all d such that $-e \leq d \leq n$ **do**
        begin
            $row \leftarrow \max(L[d, e - 1] + 1, L[d - 1, e - 1], L[d + 1, e - 1] + 1)$;
            $row \leftarrow \min(row, m)$;
            while $row < m$ **and** $row + d < n$ **and** $p_{row+1} = t_{row+1+d}$ **do** $-(\$)$
                $row \leftarrow row + 1$;
            $L[d, e] \leftarrow row$;
            if $L[d, e] = m$ **then** output "There is an occurrence ending at t_{d+m}"
        end
end

If procedure $LandauVishkin(P, T)$ is implemeted as it is, it takes $O(mn)$ time. However, Landau and Vishkin showed that the part $(\$)$ can be computed in $O(1)$ time, if the suffix tree associated with $T \cdot P \cdot "\$"$ is already constructed; where $s_1 \cdot s_2$ denotes the concatenation of s_1 and s_2, and '$' is a character which does not appear in the pattern string or in the text string. Note that a suffix tree associated with a string of length n is constructed in $O(n)$ time for a fixed size alphabet and in $O(n \log n)$ time for a general alphabet [3, 12]. Using this technique, $LandauVishkin(P, T)$ works in $O(kn)$ time for a fixed size alphabet and in $O((k + \log m)n)$ time for a general alphabet [10], where the time for the construction of a suffix tree is included.

3 Approximate matching with don't cares

In this section, we describe serial and parallel algorithms for the k differences problem with don't cares. For two characters p and q, we write $p \sim q$ if $p = q$, $p =$'*', or $q =$'*' holds. For two strings $s = s_1 \cdots s_j$ and $u = u_1 \cdots u_k$, we write $s \sim u$ if $(\forall i)(s_i \sim u_i)$ and $j = k$ hold.

First, note that procedure $LandauVishkin(P, T)$ works correctly for strings with don't cares if the part '$p_{row+1} = t_{row+1+d}$' in ($\$$) is replaced by '$p_{row+1} \sim t_{row+1+d}$'. However, suffix trees can not be used to compute this modified part efficiently. Thus, we use a table $W[r, j]$ instead of a suffix tree.

3.1 Utilization of a table

Let M be an integer where the value of M is to be determined later. We assume without loss of generality that M divides m. Let P^r denote the substring $p_{(r-1)M+1} p_{(r-1)M+2} \cdots p_{rM}$ of P. Let T^j denotes the substring $t_j t_{j+1} \cdots t_{j+M-1}$ of T. If $j + M - 1 > n$, T^j denotes the empty string. Then, the value of $W[r, j]$ is the maximum number h such that $P^r \cdot P^{r+1} \ldots P^{r+h-1} \sim T^j \cdot T^{j+M} \ldots T^{j+(h-1)M}$ holds. If such h does not exist, $W[r, j] = 0$.

If $W[r, j]$'s are already computed for all $1 \le r \le \frac{m}{M}$ and $1 \le j \le n$, the part corresponding to ($\$$) can be computed efficiently by the following procedure ($\#$):

> **while** (row mod M) $\ne 0$ and $row < m$ and $row + d < n$ and $-(\#1)$
> $p_{row+1} \sim t_{row+1+d}$ **do** $row \leftarrow row + 1$;
> **if** (row mod M) $= 0$ and $0 < row < m$ and $row + d < n$ **then** $-(\#2)$
> $row \leftarrow row + M \times W[\frac{row}{M} + 1, row + 1 + d]$; $-(\#2')$
> **while** $row < m$ and $row + d < n$ and $p_{row+1} \sim t_{row+1+d}$ **do** $-(\#3)$
> $row \leftarrow row + 1$;

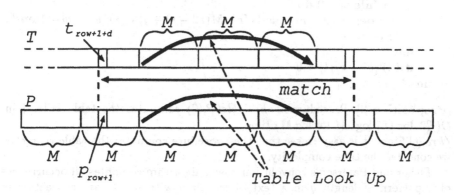

Fig. 1. Utilization of the table $W[r, j]$.

Let $ModifiedLV(P,T)$ denote the modified procedure of $LandauVishkin(P,T)$ where ($\$$) is replaced by (#). Since (#) can be done in $O(M)$ time per execution, the following proposition holds (see Fig. 1).

[Proposition 1] Assume that a table $W[r,j]$ is already constructed. Then, $ModifiedLV(P,T)$ solves the k differences problem with don't care characters in $O(knM)$ time.

3.2 Construction of the table

The table $W[r,j]$ can be constructed using the convolution based algorithm by Fisher and Paterson [5]. The following procedure constructs the table. Note that the value of $W[r,j]$ is defined as -1 if $r > \frac{m}{M}$ or $j > n$ holds.

Procedure $MakeTable(P,T)$
begin
  for all r such that $1 \le r \le \frac{m}{M}$ **do**   –(a)
    for all j such that $1 \le j \le n$ **do**   –(a1)
      if $P^r \sim T^j$ **then** $W[r,j] \leftarrow 1$ **else** $W[r,j] \leftarrow 0$;
  for all j such that $1 \le j \le n$ **do** $MakeTableSub(1,j)$;   –(b)
  for all r such that $2 \le r \le \frac{m}{M}$ **do**   –(c)
    for all j such that $1 \le j \le M$ **do** $MakeTableSub(r,j)$
end

Procedure $MakeTableSub(r,j)$
begin
  $w \leftarrow 0$; $r1 \leftarrow r$;
  repeat
    begin
      $j1 \leftarrow M(r1 - r) + j$;
      if $W[r1,j1] = 1$ **then** $w \leftarrow w + 1$
      else if $W[r1,j1] \le 0$ **then**
        while $w > 0$ **do**
          begin $r2 = r1 - w$; $W[r2, M(r2 - r) + j] \leftarrow w$; $w \leftarrow w - 1$ **end**;
      $r1 \leftarrow r1 + 1$
    end
  until $W[r1 - r, j1] < 0$
end

[Proposition 2] Procedure $MakeTable(P,T)$ computes the table $W[r,j]$ in $O(\frac{mn}{M} \log |\Sigma| \log^2 M \log\log M)$ time.
(Proof) Since it is easy to see that the procedure computes the table correctly, we consider the time complexity.

Fisher and Paterson's convolution based algorithm computes all occurrences of a pattern of length q in a text of length p with don't care characters in $O(p \log |\Sigma| \log^2 q \log\log q)$ time [1, 5]. Thus, using Fisher and Paterson's algorithm, part (a1) can be done in $O(n \log |\Sigma| \log^2 M \log\log M)$ time per execution. Since part (a1) is repeated $\frac{m}{M}$ times, the total time required for part (a) is $O(\frac{mn}{M} \log |\Sigma| \log^2 M \log \log M)$.

Since $MakeTableSub(r,j)$ takes $O(\frac{m}{M})$ time per execution, part (b) takes $O(n \times \frac{m}{M}) = O(\frac{mn}{M})$ time and part (c) takes $O(\frac{m}{M} \times M \times \frac{m}{M}) = O(\frac{m^2}{M})$ time. Thus, the total time required for $MakeTable(P,T)$ is $O(\frac{mn}{M} \log |\Sigma| \log^2 M \log \log M)$. $\square$

Letting $M = \sqrt{\frac{m}{k}}$ and combining propositions 1 and 2, we get the following theorem.

[**Theorem 1**] The k differences problem with don't care characters can be solved in $O(\sqrt{km}\, n \log |\Sigma| \log^2 \frac{m}{k} \log \log \frac{m}{k})$ time.

3.3 Parallel algorithm

In [10], a parallel version of procedure $LandauVishkin(P,T)$ is described. It works in $O(k)$ time using $O(n)$ processors, except for the construction of the suffix tree. Here, we consider a parallel version of our algorithm.

First, we consider procedure $ModifiedLV(P,T)$. Since the other parts are not modified from $LandauVishkin(P,T)$, we may consider part (#) only. It is easy to see that part (#) can be done in $O(\log M)$ time using $O(M)$ processors per execution. Thus, $ModifiedLV(P,T)$ works in $O(k \log M)$ time using $O(nM)$ processors.

Next, we consider the construction of the table. For a text string of length p and a pattern string of length q, exact matching with don't cares can be done in $O(\log q)$ time using $O(p \log |\Sigma| \log q \log \log q)$ processors [6]. Thus, part (a) can be done in $O(\log M)$ time using $O(\frac{mn}{M} \log |\Sigma| \log M \log \log M)$ processors. $MakeTableSub(r,j)$ can be done using the list ranking technique. Thus, $MakeTableSub(r,j)$ can be done in $O(\log \frac{m}{M})$ time using $O(\frac{m}{M})$ processors per execution even if a simple parallel list ranking algorithm [9] is used. It follows that parts (b) and (c) can be done in $O(\log \frac{m}{M})$ time using $O(\frac{mn}{M})$ processors. Therefore, the table $W[r,j]$ can be constructed in $O(\log m)$ time using $O(\frac{mn}{M} \log |\Sigma| \log M \log \log M)$ processors.

Letting $M = \sqrt{\frac{m}{k}}$, $ModifiedLV(P,T)$ can be done in $O(k \log m)$ time using $O(\sqrt{\frac{m}{k}}\, n)$ processors and the construction of the table can be done in $O(k \log m)$ time using $O(\sqrt{\frac{m}{k}}\, n \log |\Sigma| \log \frac{m}{k} \log \log \frac{m}{k})$ processors. Therefore, the following theorem holds.

[**Theorem 2**] The k differences problem with don't care characters can be solved in $O(k \log m)$ time using $O(\sqrt{\frac{m}{k}}\, n \log |\Sigma| \log \frac{m}{k} \log \log \frac{m}{k})$ processors on a CRCW-PRAM.

4 Extensions

In motif search, more complex patterns are used [4, 8] than those discussed previously. For example, the number of consecutive don't care characters is sometimes specified instead of simply repeating don't care characters. For another example, expressions such as $\langle a\,c\rangle$ are sometimes used, where $\langle a\,c\rangle$ denotes a character which matches 'a' or 'c'. In this section, we show that the k differences problem can be solved in $o(mn)$ time for small k even if such extended patterns are used.

4.1 Approximate matching with integer characters

In this subsection, we consider the k *differences problem with integer characters*, in which any positive integer number (representing the number of don't care characters) may appear as a character in a pattern string. For example, "ab1db3c" denotes "ab*db* * *c". We assume without loss of generality that two consecutive integer characters do not appear. For a string s, s' denotes the string in which each integer character is replaced by corresponding don't care characters. Of course, the k differences problem with integer characters can be treated by replacing P with P'. However, P' may become much longer than P, so we have developed an algorithm for the k differences problem with integer characters, by modifying the serial algorithm described in Section 3.

First, we consider the modification for the table $W[r,j]$. In this case, the table is defined as follows. Let $L(r,h)$ denote the length of $(P^r)'\cdot(P^{r+1})'\cdots(P^{r+h-1})'$. Then, the value of $W[r,j]$ is $L(r,h)$ where h is the maximum number such that $(P^r)'\cdot(P^{r+1})'\cdots(P^{r+h-1})' \sim t_j t_{j+1}\cdots t_{j+L(r,h)-1}$ holds. Then, the modification for $MakeTable(P,T)$ and $MakeTableSub(r,j)$ is straightforward, and the time complexity for constructing the table is $O(\frac{mn}{M}\log|\Sigma|\operatorname{polylog}(n))$ since the length of $(P^r)'$ may be $O(n)$.

Next, we consider how to modify the main routine $(ModifiedLV(P,T))$, where we assume without loss of generality that the length of P' is at most n. In this case, each occurrence of m in $ModifiedLV(P,T)$ is replaced by m' which denotes the length of P', and each occurrence of p_{row} is replaced by p'_{row} which denotes the row'th character of P'. Note that the size of the table $L[d,e]$ remains $O(kn)$.

The other modifications are done for part (#) only. Parts (#1) and (#3) are modified so that, if p'_{row+1} is '*', row is updated to points to the next position of P' such that $p'_{row+1}\neq$'*'; instead of executing '$row \leftarrow row+1$'. The condition '$(row \bmod M)=0$' in (#2) is replaced by the condition that there exists r such that '$row+1$' points to the first character of $(P^r)'$. The condition in (#1) is replaced in a similar way. Part (#2') is replaced by '$row \leftarrow row+W[r,row+1+d]$'. Note that the modified part of (#) can be done in $O(M)$ time per execution, if appropriate preprocessing is done for updating row in (#1) and (#3), where this preprocessing can be done in $O(n)$ time. Thus, the modified procedure takes $O(knM)$, not including the time for the table construction.

Therefore, letting $M=\sqrt{\frac{m}{k}}$, the k differences problem with integer characters can be solved in $O(\sqrt{km}\,n\log|\Sigma|\operatorname{polylog}(n))$ time.

4.2 Approximate matching with generalized string patterns

Abrahamson considered the *generalized string matching* problem [1], which is a generalization of string matching with don't cares. In generalized string matching, such expressions as $\langle x_1 x_2 \cdots \rangle$ and $[x_1 x_2 \cdots]$ may appear in a pattern string; where $\langle x_1 x_2 \cdots \rangle$ denotes a character which matches any character of $x_1, x_2, \cdots$; and $[x_1 x_2 \cdots]$ denotes a character which matches any character except $x_1, x_2, \cdots$. We call such characters *generalized characters*. Similar expressions sometimes appear in motif searches. While Abrahamson considered the exact matching problem, we consider the *generalized string matching problem with k differences*.

In this problem, let $P = p_1 p_2 \cdots p_h$ be a pattern string, where p_i may be a generalized character; let $T = t_1 t_2 \cdots t_n$ be a text string, where don't care characters or generalized characters must not appear; and let m be the length of the expression of the pattern string. In this paper, we only consider the case where $m \leq n$ holds.

First, we consider the case for a fixed size alphabet. Note that $m = O(h)$ holds in this case. The algorithm is almost the same as the serial algorithm described in Section 3. While Fisher and Paterson's algorithm is used in Section 3, Abrahamson's generalized string matching algorithm for a fixed size alphabet [1] is used here. The modification for the other parts is straightforward. Abrahamson's algorithm for a fixed size alphabet works in $O(p \, \text{polylog}(q))$ time where p is the length of the text string and q is the length of the expression of the pattern string. Using a similar discussion as in Section 3, we can obtain an $O(\sqrt{km} \, n \, \text{polylog}(m))$ time algorithm for generalized string matching with k differences for a fixed size alphabet.

Next, we consider the case for a general alphabet. The algorithm is similar to the one for a fixed size alphabet, although in this case, Abrahamson's algorithm for a general alphabet [1] is used. Since Abrahamson's algorithm for a general alphabet works in $O(\sqrt{q} \, p \, \text{polylog}(q))$ time, the construction of the table takes $O(\frac{\sqrt{m} hn}{M} \, \text{polylog}(m))$ time. The main procedure takes $O(knM \log m)$ time, since testing whether or not a character matches a generalized character takes $O(\log m)$ time using a binary search technique. Using a similar discussion as in Section 3 and letting $M = k^{-\frac{1}{2}} m^{\frac{1}{4}} h^{\frac{1}{2}}$, we can obtain an $O(k^{\frac{1}{2}} m^{\frac{1}{4}} h^{\frac{1}{2}} n \, \text{polylog}(m))$ time algorithm for a general alphabet. Since h is $O(m)$ in the worst case, it is an $O(k^{\frac{1}{2}} m^{\frac{3}{4}} n \, \text{polylog}(m))$ time algorithm. Note that it is worse than the simple dynamic programming algorithm when k is large.

5 Concluding remarks

Algorithms for generalized approximate string matching problems have been presented in this paper. Although the algorithms work in $o(nm)$ time for small k, they are not practical since the convolution based exact matching algorithms are not practical. However, for small size patterns, Abrahamson developed a practical algorithm for generalized string matching [1], which can also be used for string matching with don't cares. Using this, practical algorithms for generalized approximate string matching might be developed.

Other problems remain for the presented algorithms, and the most important one is that their space complexities are not small. Thus, more space economical algorithms should be developed. Faster algorithms should also be developed since the presented algorithms are not necessarily optimal.

References

1. K. Abrahamson. "Genaralized string matching". *SIAM Journal on Computing*, Vol. 16, pp. 1039–1051, 1987.
2. A. Amir and G. Landau. "Fast parallel and serial multidimensional approximate array matching". *Theoretical Computer Science*, Vol. 81, pp. 97–115, 1991.
3. A. Apostolico, C. Iliopoulos, G. M. Landau, B. Schieber, and U. Vishkin. "Parallel construction of a suffix tree with applications". *Algorithmica*, Vol. 3, pp. 347–365, 1988.
4. C. Branden and J. Tooze. *Introduction to Protein Structure*. Garland Publishing Inc., New York, 1991.
5. M. Fisher and M. Paterson. "String matching and other products". In *Complexity of Computation (SIAM-AMS Proceedings)*, volume 7, pp. 113–125, 1974.
6. Z. Galil and R. Giancarlo. "Data structures and algorithms for approximate string matching". *Journal of Complexity*, Vol. 4, pp. 33–72, 1988.
7. Z. Galil and K. Park. "An improved algorithm for approximate string matching". *SIAM Journal on Computing*, Vol. 19, pp. 989–999, 1990.
8. G. Heijne. *Sequence Analysis in Molecular Biology - Treasure Trove or Trivial Pursuit*. Academic Press, Inc., San Diego, 1987.
9. J. JáJá. *An Introduction to Parallel Algorithms*. Addison-Wesley, Massachusetts, 1992.
10. G. M. Landau and U. Vishkin. "Fast parallel and serial approximate string matching". *Journal of Algorithms*, Vol. 10, pp. 157–169, 1989.
11. U. Manber and R. Baeza-Yates. "An algorithm for string matching with a sequence of don't cares". *Information Processing Letters*, Vol. 37, pp. 133–136, 1991.
12. P. Weiner. "Linear pattern matching algorithms". In *Proceedings of IEEE Symposium on Switching and Automata Theory*, pp. 1–11, 1973.

Matching with Matrix Norm Minimization *

Shouwen Tang[1], Kaizhong Zhang[2], Xiaolin Wu[2]

[1] Department of Computer Science, Beijing Computer Institute, Beijing, China
[2] Department of Computer Science, University of Western Ontario, London, Ontario,
N6A 5B7 Canada

Abstract. Given $(r_1, r_2, ... r_n) \in R^n$, for any $I = (I_1, I_2, ... I_n) \in Z^n$,
let $E_I = (e_{ij})$, where $e_{ij} = (r_i - r_j) - (I_i - I_j)$, find $I \in Z^n$ such
that $\|E_I\|$ is minimized, where $\|.\|$ is a matrix norm. This is a matching
problem where, given a real-valued pattern, the goal is to find the best
discrete pattern that matches the real-valued pattern. The criterion of
the matching is based on the matrix norm minimization instead of simple
pairwise distance minimization. This matching problem arises in optimal
curve rasterization in computer graphics and in vector quantization of
data compression. Until now, there has been no polynomial-time solution
to this problem. We present a very simple $O(nlgn)$ time algorithm to
solve this problem under various matrix norms.

1 Introduction

In this paper we consider the following problem. For $r = (r_1, r_2, ... r_n) \in R^n$, and
$I = (I_1, I_2, ... I_n) \in Z^n$, let $E_I = (e_{ij})$, where $e_{ij} = (r_i - r_j) - (I_i - I_j)$. Given
$r \in R^n$, find $I \in Z^n$ such that $\|E_I\|$ is minimized, where $\|.\|$ is a matrix norm. We
call this problem matrix norm minimization problem. In this paper, we present
a very simple $O(nlgn)$ time algorithm to solve the matrix norm minimization
problem under various matrix norms.

Given a real-valued (or high-resolution) pattern $r = (r_1, r_2, ... r_n)$, we want
to find the integer-valued (or low-resolution) pattern $I = (I_1, I_2, ... I_n)$ which
matches the real-valued pattern the best. Intuitively the best solution would be
just choosing I_i to be the nearest integer to r_i. However, in some applications it
is desirable to consider a criterion that takes into account not only the pairwise
distance $r_i - I_i$ but also the relative distance $(r_i - r_j) - (I_i - I_j)$.

One application is for curve rasterization (digitization) in computer graphics,
where a two-dimensional continuous curve is converted into a sequence of pixels
on raster displays. The simplest curve rasterization scheme is to choose individual
pixels that are closest to the original curve [1]. However, in terms of human
perception of geometric figures, it is desirable to rasterize the curve in such a
way that the distortion in curve dynamics is minimized. Let $(r_1, r_2, ... r_n)$ be the
sampling of a continuous curve and $(I_1, I_2, ... I_n)$ be a digitized curve, the norm of

---

* Research supported by the Natural Sciences and Engineering Research Council of
Canada grant OGP0046373.

the error matrix E_I defined as above was proposed in [4] as a global and dynamic error measure for curve rasterization. Empirical evidence was shown that $\|E_I\|$ is a subjectively better rasterization criterion than the simple nearest-neighbor mapping. The problem was addressed again in [5] in the context of antialiasing. However, no polynomial time algorithm was given for minimizing $\|E_I\|$. Another application is for vector quantization which is a very important method for data compression. In essence, vector quantization can be considered to be a problem of selecting the best low-resolution pattern from a set of candidates to match a high-resolution input pattern. Therefore the results presented in this paper can also be used for vector quantization.

The paper is organized as follows. In the next section we introduce definitions. Section 3 contains the simplified formulas for matrix norms. Then in section 4, we give the main results concerning the optimal solutions. Finally we present a simple algorithm and give a simple example of curve rasterization in section 5.

2 Preliminaries

2.1 Definitions

In this paper a lower case letter, x, y, ..., denotes an $n \times 1$ vector and an upper case letter, A, B, ..., denotes an $n \times n$ matrix for some unspecified $n > 1$. We use x_i to denote the i-component of vector x. We use (a_{ij}) to denote a matrix whose (i, j)-component is a_{ij}.

The most often used vector norms are 1-norm, $\|x\|_1 = \sum_{i=1}^{n} |x_i|$, 2-norm, $\|x\|_2 = \sqrt{\sum_{i=1}^{n} x_i^2}$, and ∞-norm, $\|x\|_\infty = \max_i |x_i|$.

Matrix norms can be defined by vector norms. In the following we will use vector 1-norm, 2-norm and ∞-norm to define matrix norms. We assume that λ_{max} is the largest eigenvalue of $A^t A$, where $A = (a_{ij})$.

The most often used matrix norms are the following: $\|A\|_1 = \max_{\|x\|_1 = 1} \|Ax\|_1$, $\|A\|_2 = \max_{\|x\|_2 = 1} \|Ax\|_2$, $\|A\|_\infty = \max_{\|x\|_\infty = 1} \|Ax\|_\infty$. It is well known that $\|A\|_1 = \max_j \sum_{i=1}^{n} |a_{ij}|$, $\|A\|_2 = \sqrt{\lambda_{max}}$, and $\|A\|_\infty = \max_i \sum_{j=1}^{n} |a_{ij}|$.

Another popular matrix norm known as a *Frobenius* norm or an *Euclidean* norm is defined as follows:

$$\|A\|_F = \sqrt{\sum_{i=1}^{n} \sum_{j=1}^{n} a_{ij}^2}$$

According to these definitions if we exchange two rows or two columns of the matrix, the matrix norms remain the same.

For n real numbers $r_1, r_2, ..., r_n$, the *center* refers to $c(r_1, r_2, ..., r_n) = \bar{r} = \frac{\sum_{i=1}^{n} r_i}{n}$, the *radius* refers to $r(r_1, r_2, ..., r_n) = \max\{\bar{r} - \min_i r_i , \max_i r_i - \bar{r}\}$, the *moment* refers to $m(r_1, r_2, ..., r_n) = \sum_{i=1}^{n} (r_i - \bar{r})^2$.

2.2 The formulation

Given $r_1, r_2, ...r_n \in R^n$, we would like to choose $I_1, I_2, ...I_n$ such that $(r_i - r_j) - (I_i - I_j)$ is "minimized". If we consider a matrix $E = ((r_i - r_j) - (I_i - I_j))$, then what we want to do is to minimize the norm of E.

Formally, given $(r_1, r_2, \ldots, r_n) \in R^n$, for any $(I_1, I_2, \ldots, I_n) \in Z^n$, let $b_i = r_i - I_i$, $e_{ij} = b_i - b_j$ and $E = (e_{ij})$. Our goal is to find $(K_1, K_2, \ldots, K_n) \in Z^n$ such that the norm of E, $\|E\|$, is minimized.

Given $(r_1, r_2, \ldots, r_n) \in R^n$, we say $(K_1, K_2, \ldots K_n) \in Z^n$ minimizes some matrix norm, $\| \cdot \|$, if $\|((r_i - r_j) - (K_i - K_j))\| \leq \|((r_i - r_j) - (I_i - I_j))\|$, for any $(I_1, I_2, \ldots, I_n) \in Z^n$. When $(K_1, K_2, \ldots K_n)$ minimize matrix norm $\| \cdot \|$, we also say $(K_1, K_2, \ldots K_n)$ is an optimal solution for $\| \cdot \|$.

We say $(b_1, b_2, \ldots, b_n)$ is an optimal remainder for a given $(r_1, r_2, \ldots, r_n)$ and a matrix norm $\|\cdot\|$ if for $1 \leq i \leq n$ $b_i = r_i - K_i$ such that $(K_1, K_2, \ldots, K_n) \in Z^n$ is an optimal solution for $(r_1, r_2, \ldots, r_n)$ and the matrix norm $\| \cdot \|$.

By definition the order of the b_i's is immaterial because exchanging b_i and b_j means exchanging row i and row j and then exchanging column i and column j of matrix E. This does not change the value of the norm of E.

Since $e_{ij} = -e_{ji}$, $\|E\|_1 = \|E\|_\infty$ for $E = (e_{ij})$. Interestingly we can show, in the next section, that $\sqrt{2} \cdot \|E\|_2 = \|E\|_F$. Therefore we have only two different matrix norms for $E = (e_{ij} = b_i - b_j)$, namely, $\|E\|_1$ and $\|E\|_2$.

The immediate consequences of the definition are the following.

Theorem 1. *Given* $(r_1, r_2, \ldots, r_n) \in R^n$, *if* $(K_1, K_2, \ldots K_n) \in Z^n$ *is an optimal solution, then so is* $K_1 + I, K_2 + I, \ldots K_n + I$ *for any* $I \in Z$.

Theorem 2. $(K_1, K_2, \ldots K_n) \in Z^n$ *is an optimal solution for* $(r_1, r_2, \ldots, r_n) \in R^n$ *if and only if* $(K_1, K_2, \ldots K_n)$ *is an optimal solution for* $(r_1 + r, r_2 + r, \ldots, r_n + r)$ *for any* $r \in R$.

3 Formulas for matrix norm

In this section, we give the formulas for various norms of the matrix $E = (e_{ij} = b_i - b_j)$.

Proposition 3. *Let* $b_1 \leq b_2 \leq \cdots \leq b_n$ *and* $s_k = \sum_{i=1}^n |b_i - b_k|$ *for* $1 \leq k \leq n$, *then*

$$s_1 \geq s_2 \geq \cdots \geq s_{\lfloor (n+1)/2 \rfloor} \leq \cdots \leq s_{n-1} \leq s_n.$$

Proof: Since $s_k = \sum_{i=1}^k (b_k - b_i) + \sum_{i=k+1}^n (b_i - b_k)$ and $s_{k+1} = \sum_{i=1}^k (b_{k+1} - b_i) + \sum_{i=k+1}^n (b_i - b_{k+1})$, therefore for $1 \leq k \leq n-1$, $s_{k+1} = s_k + (2k - n)(b_{k+1} - b_k)$. Hence $s_1 \geq s_2 \geq \cdots \geq s_{\lfloor (n+1)/2 \rfloor} \leq \cdots \leq s_{n-1} \leq s_n$. $\square$

Lemma 4. $\|E\|_1 = n \cdot r(b_1, b_2, \ldots, b_n)$.

Proof: Let $s_k = \sum_{i=1}^{n} |b_i - b_k|$ for $1 \leq k \leq n$. Let m and M be integers such that $1 \leq m, M \leq n$, $b_m = \min_i b_i$, and $b_M = \max_i b_i$. By definition $\|E\|_1 = max_j\{s_j\}$. By Proposition 3 $\max_j\{s_j\} = \max\{s_m, s_M\}$. Therefore

$$
\begin{aligned}
\|E\|_1 &= \max\{s_m, s_M\} \\
&= \max\{\sum_{i=1}^{n}(b_i - b_m), \sum_{i=1}^{n}(b_M - b_i)\} \\
&= \max\{n \cdot (\bar{b} - b_m), n \cdot (b_M - \bar{b})\} \\
&= n \cdot \max\{\bar{b} - b_m, b_M - \bar{b}\} \\
&= n \cdot r(b_1, b_2, \ldots, b_n).
\end{aligned}
$$

$\square$

Lemma 5. $\|E\|_2 = \sqrt{n \cdot \sum_{i=1}^{n}(b_i - \bar{b})^2}$.

Proof: Notice that

$$
E = (b_i - b_j) = ((b_i - \bar{b}) - (b_j - \bar{b})) = \begin{pmatrix} b_1 - \bar{b} \\ b_2 - \bar{b} \\ \vdots \\ b_n - \bar{b} \end{pmatrix} (1, 1, \ldots, 1) - \begin{pmatrix} 1 \\ 1 \\ \vdots \\ 1 \end{pmatrix} (b_1 -
$$

$\bar{b}, b_2 - \bar{b}, \ldots, b_n - \bar{b})$.

Let $f = \begin{pmatrix} 1 \\ 1 \\ \vdots \\ 1 \end{pmatrix}$ and $g = \begin{pmatrix} b_1 - \bar{b} \\ b_2 - \bar{b} \\ \vdots \\ b_n - \bar{b} \end{pmatrix}$.

Since $f^t \cdot g = g^t \cdot f = 0$, we have that $Eg = -\sum_{i=1}^{n}(b_i - \bar{b})^2 \cdot f$ and $Ef = n \cdot g$.
So, $E^t Ef = n \sum_{i=1}^{n}(b_i - \bar{b})^2 \cdot f$ and $E^t Eg = n \sum_{i=1}^{n}(b_i - \bar{b})^2 \cdot g$.
On the other hand, $E x = 0$ if vector x satisfies $x \perp f$ and $x \perp g$.
Hence the largest eigenvalue of $E^t E$ is $n \cdot \sum_{i=1}^{n}(b_i - \bar{b})^2$. $\square$

Lemma 6. $\|E\|_F = \sqrt{2}\|E\|_2$

Proof: Since $\sum_{i=1}^{n} \sum_{j=1}^{n}(b_i - b_j)^2 = 2n \cdot \sum_{i=1}^{n}(b_i - \bar{b})^2$, by Lemma 5 $\|E\|_F = \sqrt{2}\|E\|_2$. $\square$

4 Properties of Optimal Solution

In this section, without loss of generality, we assume that $b_1 \leq b_2 \leq \ldots \leq b_n$. First, we will prove the following proposition which is needed in this section.

Proposition 7. If $b_n - \bar{b} \geq 1/2$, then $m(b_1, b_2, \ldots, b_n - 1) < m(b_1, b_2, \ldots, b_n)$. If $\bar{b} - b_1 \geq 1/2$, then $m(b_1 + 1, b_2, \ldots, b_n) < m(b_1, b_2, \ldots, b_n)$.

Proof: We give the proof of the first assertion only.
The center of sequence $b_1, b_2, \ldots, b_n$ is $\bar{b} = (1/n) \cdot \sum_{i=1}^{n} b_i$. The centre of sequence $b_n - 1, b_1, \ldots, b_{n-1}$ is $(1/n) \cdot (b_n - 1 + \sum_{i=1}^{n-1} b_i). = (1/n) \cdot (-1 + \sum_{i=1}^{n} b_i). = \bar{b} - 1/n$.

The moment of the sequence $b_1, b_2, \ldots, b_n$ is $\sum_{i=1}^{n}(b_i - \bar{b})^2$. The moment of the sequence $b_n - 1, b_1, \ldots, b_{n-1}$ is

$$
\begin{aligned}
&(b_n - 1 - (\bar{b} - 1/n))^2 + \sum_{i=1}^{n-1}(b_i - (\bar{b} - 1/n))^2 \\
&= 1 - 2 \cdot (b_n - \bar{b}) - 2/n + \sum_{i=1}^{n}(b_i - (\bar{b} - 1/n))^2 \\
&= 1 - 2 \cdot (b_n - \bar{b}) - 2/n + \sum_{i=1}^{n}((b_i - \bar{b})^2 + (1/n)^2) \\
&= 1 - 2 \cdot (b_n - \bar{b}) - 1/n + \sum_{i=1}^{n}(b_i - \bar{b})^2 \\
&\leq \sum_{i=1}^{n}(b_i - \bar{b})^2 - 1/n \\
&< \sum_{i=1}^{n}(b_i - \bar{b})^2.
\end{aligned}
$$

$\square$

Lemma 8. If $b_1, b_2, \ldots, b_n$ is the optimal remainder for $\|E\|_2$, then $b_n - b_1 < 1$.

Proof: If $b_n - b_1 \geq 1$, then either $\bar{b} - b_1 \geq 1/2$ or $b_n - \bar{b} \geq 1/2$. If $\bar{b} - b_1 \geq 1/2$ then $b_1 + 1, b_2, \ldots, b_n$ is a better remainder by the above proposition and lemma 5, which is a contradiction. The same argument shows that $b_n - \bar{b} < 1/2$. Hence $b_n - b_1 < 1$. $\square$

The following definitions and proposition are needed for the next lemma.

Assume that $b_1 \leq b_2 \leq \cdots \leq b_n$. Denote sequence $b_{i+1}, \ldots, b_n, b_1 + 1, b_2 + 1, \ldots, b_i + 1$ by S_i where $0 \leq i \leq n$. Note that sequence S_0 is $b_1, b_2, \ldots, b_n$ and sequence S_n is $b_1 + 1, b_2 + 1, \ldots, b_n + 1$. Also if $b_n - b_1 \leq 1$ then each sequence S_i, where $0 \leq i \leq n$, is a sorted sequence.

Define $m_i = \min_b\{b \in S_i\}$, $M_i = \max_b\{b \in S_i\}$ and $\bar{c}_i = center(S_i)$ where $0 \leq i \leq n$. From the definition $m_0 = b_1$, $M_0 = b_n$, $m_n = b_1 + 1$ and $M_n = b_n + 1$. Since $b_1 \leq b_2 \leq \cdots \leq b_n$, we know that, for $1 \leq i \leq n-1$, $m_i = \min\{b_{i+1}, b_1 + 1\}$ and $M_i = \min\{b_n, b_i + 1\}$. Also it is clear that, for $1 \leq i \leq n$, $\bar{c}_i = \bar{c}_{i-1} + 1/n = \bar{c}_0 + i/n$.

Proposition 9. If $b_n - b_1 \geq 1$ then $M_{i+1} - m_i \leq b_n - b_1$ for $0 \leq i \leq n$.

Proof: If $i = 0$, then $m_0 = b_1$ and $M_1 = \max\{b_n, b_1 + 1\}$, therefore $M_1 - m_0 \leq b_n - b_1$. If $i > 0$, then $m_i = \min\{b_{i+1}, b_1 + 1\}$ and $M_{i+1} = \max\{b_n, b_{i+1} + 1\}$. Therefore $M_{i+1} - m_i \leq \max\{b_n - b_{i+1}, b_n - (b_1 + 1), b_{i+1} + 1 - b_{i+1}, b_{i+1} + 1 - (b_1 + 1)\} = \max\{b_n - b_{i+1}, b_n - b_1 - 1, 1, b_{i+1} - b_1\} \leq b_n - b_1$. $\square$

Lemma 10. If $b_1 \leq b_2 \leq \ldots \leq b_n$ is the optimal remainder for $\|E\|_1$, then $b_n - b_1 < 1$.

Proof: Suppose that $b_n - b_1 = d \geq 1$ and $\bar{c}_0 - m_0 \geq M_0 - \bar{c}_0$.

First we can show that for sequence S_{n-1}, $\bar{c}_{n-1} - m_{n-1} \leq M_{n-1} - \bar{c}_{n-1}$. It is easy to see that $\bar{c}_{n-1} = \bar{c}_0 + (n-1)/n$ and $m_{n-1} = \min\{b_1 + 1, b_n\} = b_1 + 1$. Therefore $\bar{c}_{n-1} - m_{n-1} = \bar{c}_0 + (n-1)/n - (b_1 + 1) = \bar{c}_0 - b_1 - 1/n < \bar{c}_0 - m_0$. If $\bar{c}_{n-1} - m_{n-1} > M_{n-1} - \bar{c}_{n-1}$ then $r(b_n, b_1 + 1, b_2 + 1, \ldots, b_{n-1} + 1) < r(b_1, b_2, \ldots, b_n)$. This means that $b_1, b_2, \ldots, b_n$ is not the optimal remainder. Hence $\bar{c}_{n-1} - m_{n-1} \leq M_{n-1} - \bar{c}_{n-1}$.

Since $\bar{c}_0 - m_0 \geq M_0 - \bar{c}_0$ and $\bar{c}_{n-1} - m_{n-1} \leq M_{n-1} - \bar{c}_{n-1}$, there exist an i, $0 \leq i \leq n-2$, such that $\bar{c}_i - m_i \geq M_i - \bar{c}_i$ and $\bar{c}_{i+1} - m_{i+1} \leq M_{i+1} - \bar{c}_{i+1}$. By Proposition 9, $d \geq M_{i+1} - m_i = M_{i+1} - \bar{c}_{i+1} + \bar{c}_{i+1} - m_i = M_{i+1} - \bar{c}_{i+1} + 1/n + \bar{c}_i - m_i$. This means that either $M_{i+1} - \bar{c}_{i+1} < d/2$ or $\bar{c}_i - m_i < d/2$. Therefore either $r(b_{i+1}, \ldots, b_n, b_1 + 1, \ldots, b_i + 1) < r(b_1, b_2, \ldots, b_n)$ or $r(b_{i+2}, \ldots, b_n, b_1 + 1, \ldots, b_{i+1} + 1) < r(b_1, b_2, \ldots, b_n)$. This contradicts to the fact that $b_1, b_2, \ldots, b_n$ is the optimal remainder.

Similarly if $b_n - b_1 = d \geq 1$ and $\bar{c}_0 - m_0 \leq M_0 - \bar{c}_0$, we can show that $b_1, b_2, \ldots, b_n$ is not the optimal remainder. $\square$

Combining the results of Lemma 8 and Lemma 10, we have the following theorem.

Theorem 11. *If $b_1, b_2, \ldots, b_n$ is the optimal remainder for $\|E\|_x$, where $x \in \{1, 2, \infty, F\}$, then $\max_i b_i - \min_i b_i < 1$.*

Lemma 12. *Let $(J_1, J_2, \ldots, J_n) \in Z^n$ be an optimal solution for $(r_1, r_2, \ldots, r_n) \in R^n$. Then there is an optimal solution $(I_1, I_2, \ldots, I_n) \in Z^n$ such that $-1 < r_i - I_i < 1$ and $I_i = J_i + H$, for some $H \in Z$, for $1 \leq i \leq n$. Furthermore if we let $b_i = r_i - I_i$ and $b_m = \max_{i=1}^n b_i$, then $0 \leq b_m < 1$.*

Proof: Let $J_1, J_2, \ldots, J_n$ be an optimal solution. Let $e_i = r_i - J_i$ for $1 \leq i \leq n$, $e_m = \max_{i=1}^n e_i$ and $H = \lfloor e_m \rfloor$. By Theorem 1, $I_1 = J_1 + H, I_2 = J_2 + H, \ldots, I_n = J_n + H$ is also an optimal solution. Now $b_i = r_i - I_i = r_i - (J_i + H) = e_i - H, 1 \leq i \leq n$. Since $b_m = \max_{i=1}^n b_i = e_m - \lfloor e_m \rfloor$, therefore $0 \leq b_m < 1$. By Theorem 11 $b_m - b_i < 1$, therefore $-1 \leq b_m - 1 < b_i \leq b_m < 1$. $\square$

Theorem 13. *Let $(J_1, J_2, \ldots, J_n) \in Z^n$ be an optimal solution for $(r_1, r_2, \ldots, r_n) \in R^n$. There is an optimal solution $(I_1 = J_1 + H, I_2 = J_2 + H, \ldots, I_n = J_n + H)$ such that if $r_i - \lfloor r_i \rfloor \leq r$, then $I_i = \lfloor r_i \rfloor$ and if $r_i - \lfloor r_i \rfloor > r$, then $I_i = \lceil r_i \rceil$, where $H \in Z$ and $0 \leq r < 1$.*

Proof: Consider the optimal solution $I_1, I_2, \ldots, I_n$ in Lemma 12, for which either $I_i = \lfloor r_i \rfloor$ or $I_i = \lceil r_i \rceil$ for $1 \leq i \leq n$.

Moreover let $b_i = r_i - I_i$ for $1 \leq i \leq n$ and $r = \max_{i=1}^n b_i$. By Theorem 11, $r - 1 < r_i - I_i <= r$ for $1 \leq i \leq n$. If $r_i - \lfloor r_i \rfloor \leq r$ then $r_i - \lceil r_i \rceil \leq r - 1$ and therefore $I_i = \lfloor r_i \rfloor$. If $r_i - \lfloor r_i \rfloor > r$ then $r - 1 < r_i - \lceil r_i \rceil < r$ and therefore $I_i = \lceil r_i \rceil$. $\square$

Corollary 14. *Let $(I_1, I_2, \ldots, I_n) \in Z^n$ be an optimal solution for $(r_1, r_2, \ldots, r_n) \in R^n$. If $r_i - r_j = K$, where $K \in Z$, then $I_i - I_j = K$.*

5 Algorithm and Discussion

In this section, based on Theorem 13, we will present a simple algorithm for minimum matrix norms. We will then briefly discuss the properties of the optimal solution and the application of minimum matrix norm to curve rasterization.

5.1 Algorithms

We are now ready to give a very simple algorithm based on Theorem 13. Given $(r_1, r_2, ..., r_n) \in R^n$, let $b_i = r_i - \lfloor r_i \rfloor$ for $1 \leq i \leq n$. Without loss of generality we assume that $b_1, b_2, ..., b_n$ are sorted in nondecreasing order. Let $(I_1, I_2, ..., I_n) \in Z^n$ be an optimal solution for $(r_1, r_2, ..., r_n) \in R^n$. By Lemma 12, we can assume that $-1 < r_i - I_i < 1$ for $1 \leq i \leq n$. By Theorem 13, there is a real number r, $0 \leq r < 1$, such that $I_i = \lfloor r_i \rfloor$ if $b_i \leq r$ and $I_i = \lceil r_i \rceil$ if $b_i > r$. In order to find the optimal solution, what we need to do is to check all the possible values for r. We can decompose interval $[0, 1)$ into the following $n + 1$ intervals: $[0, b_1), [b_1, b_2), [b_2, b_3), ..., [b_{n_1}, b_n), [b_n, 1)$. Since b_i, $1 \leq i \leq n$, are sorted, for any interval, we only need to check one value for r. Therefore we need only to choose r from at most $n + 1$ value, $0, b_1, b_2, ..., b_n$. This leads directly to the following simple algorithm for minimum matrix norms.

Algorithm framework

Input: $r_1, r_2, ...r_n$
Output: $I_1, I_2, ...I_n$
1. compute $b_i = r_i - \lfloor r_i \rfloor$.
2. sort b_i resulting $c_1, c_2, ...c_n$ in nondecreasing order.
3. for each i, $0 \leq i \leq n$, compute matrix norm, according to formulas in Lemma 4 and Lemma 5, for $c_{i+1} - 1...c_{n-1} - 1, c_n - 1, c_1, ...c_i$.
4. if $c_{m+1} - 1...c_{n-1} - 1, c_n - 1, c_1, ...c_m$ is the minimum from step 3), then let $h = c_m$.
5. if $r_i - \lfloor r_i \rfloor \leq h$ $\quad I_i = \lfloor r_i \rfloor$
   else $\quad I_i = \lceil r_i \rceil$

Let us consider the time complexity of this algorithm framework. The time complexity of step 1), 4) and 5) is $O(n)$. The time complexity of step 2) is $O(n lg n)$. The time complexity of step 3) in its current form can be bounded by $O(n^2)$. However, we can reduce the complexity from $O(n^2)$ to $O(n)$ with careful arrangement of the computation.

In the following we will only give the algorithm for $\|E\|_1$ due to space limitation. The algorithm for $\|E\|_2$ is similar. The algorithm has time complexity of $O(n lg n)$. The space complexity is clearly $O(n)$.

```
Algorithm for ||E||₁:
Input: r₁, r₂, ..., rₙ in a[1..n] of real.
Output: I₁, I₂, ..., Iₙ in I[1..n] of integer.
    for i = 1 to n    b[i] = a[i] − ⌊a[i]⌋
    sort(b)
    max = b[n]; min = b[1];
    avg = Σⁿᵢ₌₁ b[i]/n;
    m = max(avg − min, max − avg);
    h = b[n]; b[0] = b[n] − 1;
    for i = n to 1 step -1
        min = b[i] − 1; max = b[i − 1];
        avg = avg − 1/n;
        tm = max(avg − min, max − avg);
        if tm < m    m = tm; h = b[i − 1];
    for i=1 to n
        if a[i] − ⌊a[i]⌋ ≤ h    I[i] = ⌊a[i]⌋;
        else    I[i] = ⌈a[i]⌉;
```

5.2 Discussion

Let us briefly discuss the properties of the optimal solution and then consider an example.

By the Corollary 14, if the difference between r_i and r_j is in fact an integer, then the optimal solution retains the same difference (or relationship). If all $r_i - r_j$ are integers, then there is a real number r such that $r_i = \lfloor r_i \rfloor + r$ for $1 \le i \le n$. Therefore by Theorem 13, $r_1 - r, r_2 - r, ..., r_n - r$ is the optimal solution. This means that if the input is already in the digital form the optimal solution will be the input itself.

By Theorem 2 the optimal solution is translation invariant. This is a desirable property. The reason is that since $r_1 + r, r_2 + r, ..., r_n + r$ represents the same pattern after a translation, the digitized pattern should also remain the same. This is a property that nearest integer scheme dose not have.

Finally let us consider a concrete example. We now compare the conventional nearest-neighbor method and our matrix norm minimization method for rasterization. The curve $f(x) = \frac{10}{\pi} \cdot sin(\frac{\pi}{11}x)$ is to be rasterized. In Figure 1, the first ten curves are the discrete images of $f(x) + r$ by the conventional method, where $r \in \{0.0, 0.1, 0.2, ..., 0.9\}$. There are two problems with the conventional method. Firstly, the solution is not shift invariant. Same function has different discrete rasterizations. Secondly, since the conventional method dose not consider the dynamic context of the curve, it rasterizes function $\frac{10}{\pi} \cdot sin(\frac{\pi}{11}x) + 0.5$ into a discrete triangular waveform that cannot be visually recognized as a *sin* function. The last two curves in Figure 1 are the discrete images computed under the two matrix norms by our method. Note that in this example two matrix norms produce the same optimal solution. It is apparent from the figure that matrix norm method gives the visually more pleasant rasterization of the original curve $\frac{10}{\pi} \cdot sin(\frac{\pi}{11}x) + r$.

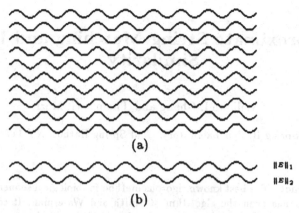

(a)

$\|\varepsilon\|_1$

$\|\varepsilon\|_2$

(b)

Fig. 1. Comparison of the rasterizations by conventional method (a) and our new method (b) for $f(x) = \frac{10}{\pi} \cdot sin(\frac{\pi}{11}x) + r$, where $r \in \{0.0, 0.1, 0.2, ..., 0.9\}$. The new algorithm produces the same discrete image for two matrix norms discussed.

By matrix norm method, optimal curve rasterization is done in two steps. The first step arranges the pixels by considering the global property of the curve, and then the second step puts the chosen pixel pattern in the raster plane to minimize the distance between the pixels and the curve.

We would like to mention that for this application usually we do not need high precision for $r_i - \lfloor r_i \rfloor$ due to the finite raster device resolution. In this situation, we can use bucket sort for step 2 of our algorithm, reducing the running time of the algorithm to linear.

References

1. J. E. Bresenham, "Algorithm for computer control of digital plotting", *IBM Syst. J.*, vol. 4, pp. 25-30, 1965.

2. C. E. Kim, 'On the cellular convexity of complexes', *IEEE Trans. Pattern Analysis and Machine Intelligence*, vol. 3, pp.617-625, 1981

3. A. Rosenfeld 'Arcs and curves in digital pictures', *J. of ACM* Vol. 20, No. 1, pp.81-87, 1973

4. X. Wu and J. Roken, 'Dynamic error measure for curve scan-conversion', *Proc. Graphics/Interface '89*, London, Ontario, pp.183-190, 1989

5. X. Wu 'An efficient antialiasing technique', *Computer Graphics* vol. 25, no. 4, pp.143-152, 1991

Approximate String Matching and Local Similarity *

William I. Chang and Thomas G. Marr

Cold Spring Harbor Laboratory, Cold Spring Harbor, NY 11724 USA

Abstract. The best known rigorous method for biological sequence comparison has been the algorithm of Smith and Waterman. It computes in quadratic time the highest scoring local alignment of two sequences given a (nonmetric) similarity measure and gap penalty. In this paper, we describe how the distance-based sublinear expected time algorithm of Chang and Lawler can be extended to solve efficiently the local similarity problem. We present both a new theoretical result, *polynomial-space, constant-fraction-error matching* that is provably optimal, and a practical adaptation of it that produces nearly identical results as Smith-Waterman, at speedups of 2X (PAM 120, roughly corresponding to 33% identity) to 10X (PAM 90, 50% identity) or better. Further improvements are anticipated. What makes this possible is the addition of a new constraint on *unit score* (average score per residue), which filters out both very short alignments and very long alignments with unacceptably low average. This program is part of a package called *Genome Analyst* that is being developed at CSHL.

1 Introduction and Summary of Results

1.1 Sequence comparison based on similarity

The fundamental problem of protein sequence comparison is to judge whether two or more sequences share structural and/or functional features, based on similarities observed in their amino acid sequences. Current understanding of the underlying processes of structure and function is not sufficient for a completely rigorous solution to this problem. Nevertheless, two developments in particular have combined to produce a method that is reasonably rigorous and successful. The first of these attacks the central problem of assigning a score to the matching of a single pair of residues, according to chemical properties or statistical analysis of allowed mutations in known homologous sequences. The second is the rigorous treatment of optimal alignment of regions by dynamic programming.

The widely used *point-accepted mutation* (PAM) matrices of Dayhoff et al. (1979; originally 1968) were calculated on the basis of 1600 accepted point mutations in 71 groups of closely related proteins (<15% different). This model of amino acid substitution assumes that the nonlethal mutations follow the same

* Supported by Department of Energy grant DE-FG02-91ER61190 and National Institutes of Health grant 1R01 HG0020301A1 to T.G. Marr.

rate and distribution as in the original data, *extrapolated* to evolutionarily distant sequences and beyond. Generally, 50% identity over a putative domain (not the entire sequences) is considered extremely significant; in the PAM scale this corresponds to about 90 PAMs. Occasionally, a particular 33% identity (120 PAMs) alignment is claimed to be significant. But the standard or "default" scoring matrix is actually PAM250, which in this model corresponds to the statistics of very distant (<20% identity) relationships. The use of PAM250 is partly historical, and partly due to its empirical success at uncovering distant homologs. However, alignments produced by PAM250 with low gap penalty are usually indistinguishable from noise. Although some known homologs are indeed 250 PAMs apart, in practice such relationships are nearly impossible to detect and to analyze, requiring much more sensitive methods (such as Argos and Vingron, 1990; Vingron and Waterman, 1993), and often with user-input.

Recently, Altschul (1991) has put forth a unifying theory of amino acid substitution matrices as devices for distinguishing a "target" frequency of aligned pairs from a "background" frequency of random amino acids. Specifically, the score for aligning a pair (x, y) has the form $\log q_{xy}/p_x p_y$, where q_{xy} is the characteristic frequency of pairing x with y among alignments representing homologs, and $p_x p_y$ is the frequency of pairing x with y by chance. The likelihood or "odds" ratio $q_{xy}/p_x p_y$ compares the probability of the event occurring under two alternative hypotheses; taking logarithm replaces multiplication with addition, so the total additive score of an alignment remains a "log-odds" score. A given scoring matrix is optimal for identifying homologs at its target frequency, against a background of similarities due to chance (Karlin and Altschul, 1990; also observed empirically by others). It is extremely important to note the following distinction: PAM x is a scoring matrix; x PAMs is an evolutionary distance, defined by the target frequency of PAM x. In short, PAM x measures the likelihood that the aligned portion of two sequences are x PAMS apart evolutionarily. The best matrix for scoring an x PAMs alignment is PAM x, if x were known (Altschul, 1991).

Needleman and Wunsch (1970) formulated an explicit criterion for the optimality of a global alignment, scoring matches, mismatches, insertions, and deletions, and provided a quadratic time algorithm. Here "global" can mean matching the entirety of one sequence (actually, all prefixes) against substrings of another. Smith and Waterman (1981) solved the local alignment problem by introducing a "zero trick": if an entry of the dynamic programming table is negative, then the optimal local alignment cannot go through this entry because the first part would lower the score; one may therefore replace it with zero, in effect cutting off the prefixes. (This simple trick is known in computer science folklore as the *maximum subvector* algorithm.) Gotoh (1982) then showed that affine gap penalty (separate costs for number and lengths of gaps) is about as efficiently solved as is linear gap penalty, by using three states per cell (instead of the naive four). The identification of multiple, similar segments was achieved by Waterman and Eggert (1987). Despite a slightly imprecise definition (i.e. procedural and nondeterministic under ties) of k best nonoverlapping alignments,

261

this is "probably the most useful dynamic programming algorithm for current problems in biology" (Waterman, 1989). Huang and Miller (1991) introduced linear space optimizations that allow very large DNA sequences to be compared.

The most obvious problem with using Smith-Waterman for database searching is speed—a single search used to take several hours. For this reason, faster heuristic methods (BLAST, FASTA) are far more popular than rigorous dynamic programming. The penalty of using a heuristic method is not only the potential loss of accuracy (which is incompletely understood), but also the loss of precision in describing the findings and nonfindings of a highly tweaked heuristic program. Although Pearson has tuned FASTA over many years to be a powerful tool, and in fact claims (1991) that it is "as sensitive as" Smith-Waterman, we feel many of the techniques used to improve FASTA (such as combining several alignments) can be more profitably applied to a rigorous algorithmic foundation—provided it is fast enough. Currently our own implementation of Gotoh-Smith-Waterman runs at about one-third the speed of FASTA (most sensitive setting, -o and ktup = 1). It takes just ten minutes on a 160 Mhz DEC Alpha computer to compare a 400-residue sequence against the SwissProt database of about 12 million residues. The Email server blitz@EMBL-Heidelberg.de (Sturrock and Collins, 1993) is about 13 times faster, running on a 4,000-processor MasPar computer. The BLAST program is also very fast, but only finds gapless alignments.

There is a less obvious "fault" with the Smith-Waterman algorithm: the optimal local alignment can have arbitrarily low (positive) *unit score* (mean score per aligned pair). If an alignment is significant, one would expect its unit score to be fairly high. Indeed, one can easily compute from the given PAM x matrix the expected unit score (called "relative entropy") of its target frequency x PAMs (Altschul, 1991). If the observed unit score of the alignment in question is much lower than the relative entropy, then the alignment must be much more distant than x PAMs. It is generally futile to look for alignments much more distant than x PAMs using PAM x as the scoring matrix. Instead, one should look for alignments with the right unit score (characteristic of x PAMs) as well as total score (at least 30 "bits" for database search)—but this is not possible using the Smith-Waterman algorithm. The desired alignment can be "shadowed" by one that is higher scoring but also much longer. In a database search, Smith-Waterman will discover every sequence that contains an alignment of the desired total score. One may attempt to use the *nonoverlapping* suboptimal alignments algorithm to generate all such alignments and apply the unit score test, but this does not work if the desired alignment is shadowed by one that does overlap it. Indeed, shadowing takes place even in the normal course of computing a Smith-Waterman dynamic programming matrix—the highest scoring local alignment ending at a given cell may not be one with the highest average.

To solve this problem, one can apply the Needleman-Wunsch algorithm separately to every suffix of the query sequence, so alignments ending at a particular cell have a fixed length (defined to be the length of the query substring). The drawback is of course that this is cubic, though significant optimization is possible in practice. This is related to previous attempts at defining a *distance-based* optimal local alignment, to be discussed in the next section. An alternative approach is to subtract the relative entropy from each entry of the scoring matrix. This also has the effect of greatly enhancing the power of database filtering, at the risk of chopping alignments into short pieces, which must be combined in some way. It is noteworthy that much of the power of *heuristic* methods such as FASTA are derived from careful fine-tuning of such a step.

Useful surveys on this subject include Altschul (1991), Argos et al. (1991), George et al. (1990), and Myers (1991b).

1.2 Sequence comparison based on distance

Pattern matching is one of the classical problems of computer science. Fast algorithms and tight lower bounds are known (Knuth, Morris, Pratt, 1975; Yao, 1979; etc.) for the *exact* matching of a pattern (of size m) inside a text string (of size n), over some alphabet (of size b). The situation with nonexact, or *approximate* matching is less satisfactory. Until Chang and Lawler's (1990) (sub-) linear expected time filtering result for (sub-) logarithmic-fraction error, approximate string matching was thought to be hard. The $O(kn)$ algorithm of Landau and Vishkin (1988) for k differences was a breakthrough, but no better than classical dynamic programming when the error rate is a constant fraction of the pattern size. By relaxing the error rate requirement slightly, to $k < m/(\log_b m + O(1))$, Chang and Lawler found an algorithm that is much faster *on average* for random text. (Subsequently several $O(m/\log m)$-error filters have appeared, but one should make explicit the constant factor in the *big-O* notation; for example $m/3 \log_b m$ is quite easy by hashing, but also rather weak.) This is achieved by filtering the text with a device called *matching statistics* (locally the longest exact match), and calling a dynamic programming subroutine only when the filter cannot *prove* that there is no match. In this paper, we improve that result even further, to *constant-fraction* $k < \epsilon_b m$ error, although the constants ϵ_b for which we can *provably* apply the algorithm are relatively small $(1 - e/\sqrt{b}$ as $b \to \infty)$. Nevertheless, this is the first algorithm for constant-fraction error matching that is significantly sub-quadratic (in fact $O(n)$ on average) and does not use exponential space (Ukkonen, 1985) or an inverted index of the text (Myers 1991a, $O(\epsilon m n^{f(\epsilon,b)} \log n)$ time). It will also be noted that this algorithm is in fact *optimal*, i.e. meets a lower bound to within a constant factor.

Because it is easier to be rigorous with a distance (metric) than with a similarity (nonmetric), computer science has traditionally focused on distance measures, such as (Hamming) mismatch distance or (Levenshtein) edit distance. For these special measures, very efficient optimizations have been invented (several are surveyed in Chang and Lampe, 1992). Currently the fastest algorithm for edit distance (arbitrary k), kn.clp (Chang and Lampe, 1992) partitions columns of the dynamic programming matrix (or indeed the "edit distance" matrix; see Figure 1c for this new result) into segments containing increasing arithmetic progressions. By doing only one unit of work per segment, kn.clp achieves an observed speed of $O(kn/(\sqrt{b} - 1))$, that is, much faster with larger alphabets (see Figure 1a). In fact, we have found edit-distance based methods (kn.clp in particular) very useful in our work at Cold Spring Harbor Laboratory. We have *extremely* fast programs for *motif finding* (clp) where each position of the pattern can match several (chemically similar) amino acids or a variable length gap; *redundancy detection* (differ) where an error threshold is given; *overlap detection* (overlap) allowing some fraction error in the overlap (see Figure 1). The contig assembly program (CAP) of Huang (1991) for the task of piecing together overlapping DNA fragments (with errors) is based on Chang and Lawler (1990). Wu, Manber, Myers (1993) describes a program similar to clp (its "shell" is used by clp) that is even a bit faster when the machine has a large (512K) data cache. The agrep program of Wu and Manber (1992) is also extremely fast at extracting records from unindexed databases.

Immediately after presenting the Chang and Lampe (1992) results, we realized that the combinatorial lemma used to prove the $O(kn)$ average-case running time can be applied to a "bootstrapping" algorithm described already in Chang (1991), to produce the constant-fraction-error optimality result mentioned above. Briefly, this algorithm treats short ($\alpha \cdot \log_b m$) segments of the text as patterns, and the original pattern as text (see Figure 3). The best match between each segment and the entire original pattern is computed, and used to bound the local alignment. This generalizes and strengthens the *matching statistics* filtering method used by Chang and Lawler (1990) to discard entire sections of the text. Only polynomial space m^α is required, where α is independent of m and n, but dependent on b and the error rate ϵ_b. Interestingly, this new method is equally applicable to both distance and similarity. (We have recently learned that the adaptation to similarity measure was independently discovered by Myers at about the same time; however no analytical results are given in Myers 1994.) Although it bears superficial resemblance to previous tuple-based methods, it is much more powerful. For the purpose of protein matching at high PAMs, however, this filter is still not strong enough by itself (a fact also noted by Myers). Indeed, the main contribution of this paper is the introduction of the unit score constraint to this general method, resulting in 75% (PAM 120) to 99.9% (PAM 40) filtering of the database.

While it is easy to define optimal local alignment for a similarity measure, a suitable definition for distance is nonobvious. Sellers in his 1981 paper that defined the pattern matching problem (by finessing a technical difficulty with the size of output being potentially quadratic) also defined "intervals which most resembled eachother locally." Goad and Kanehisa (1982), Sellers (1984) transformed the distance problem back to similarity, using a scoring function of the form $\epsilon \cdot \text{length} - \text{distance}$ (positive alignments therefore satisfy distance/length $< \epsilon$). However, because there are too many *locally* optimal (as opposed to optimal) alignments, definitions were modified (sometimes by complex procedural constraints) in order to reduce the number of such alignments. Note that an alignment with positive total score can in fact contain several negative and positive sub-alignments found this way (see Figure 5). This happens because the Smith-Waterman algorithm finds the highest scoring alignment, not the longest positive alignment. If length is to be taken into consideration of significance, then local alignments must be concatenated. In other words, we have the same problem as described in the previous section: how to combine several alignments, found by using a scoring matrix minus its relative entropy, into one whose average as well as total score is high once the relative entropy is added back. It is not known how to do this efficiently. However, as indicated earlier, the Needleman-Wunsch type algorithm (with optimization) can be applied to each suffix of the pattern.

We will again try to finess this difficulty. Chang and Lawler (1990) defined the (k, l) *substring* matching problem as matching any length l substring of the pattern, allowing up to k errors. Chang and Lawler sketched both a linear expected time filter and a dynamic programming subroutine that first attempts to match the pattern *as text* against the local regions left by the filter (instead of considering separately all length l substrings of the pattern). We now generalize this problem to allow substrings of length l *or greater* with at most ϵ-fraction error. By applying the Smith-Waterman zero-trick to the matching statistics minus ϵ per letter, and then scanning backwards to find maximal positive regions, we can extend the algorithm for (k, l) matching to (ϵ, l) matching. The key is that filtering is one-dimensional, so scanning backwards does not present the same multiple choices as trying to extend a Smith-Waterman local alignment (one-dimensional diagonal embedded inside a two-dimensional matrix). The time needed for dynamic programming, though potentially very large, will be averaged out in many applications where the matching statistics filter is highly selective. Furthermore, by modifying and strengthening the filter as described above, this method can be applied to local similarity search with unit score as well as total score constraints.

The mathematics is sketched in the following sections: *optimal approximate matching* and *substring matching*. Finally, a local similarity application is discussed. The cited papers (especially Myers 1991b, Chang et al.) have extensive bibliographies.

2 Optimal Approximate Matching

The k-differences string matching problem. Given text $T[1, \ldots, n]$, pattern $P[1, \ldots, m]$, the number k of differences (substitutions, insertions, or deletions) allowed in a match, find all locations in the text were a match occurs.

Dynamic programming algorithm (Sellers). Compute an $m + 1$ by $n + 1$ table whose entry $D(i, j)$ is the minimum number of edit operations needed to transform the length i prefix of P into *some* substring of T ending at position j, using the following recurrence: $D(i, 0) = i$; $D(0, j) = 0$; and

$$D(i, j) = \min\{1 + D(i - 1, j), 1 + D(i, j - 1), I_{ij} + D(i - 1, j - 1)\}$$

where $I_{ij} = 0$ if $P[i] = T[j]$; 1 otherwise. There is a match ending at position j if and only if $D(m, j) \leq k$.

Linear expected time algorithm (LET). Set $S_1 = 1$ and for $j \geq 1$ compute $S_{j+1} = S_j + 1 +$length of longest exact match $T[S_j, \ldots] = P[\ldots]$. For $j = 1, 2, \ldots$, if $S_{j+k+2} - S_j \geq m - k$ apply the dynamic programming algorithm to $T[S_j, \ldots, S_{j+k+2} - 1]$. (Figure 2a.)

The lengths of longest exact matches required by LET (matching statistics) is efficiently computed using a data structure called the suffix tree (see Figure 2b).

Theorem (Chang and Lawler). *Assume T is uniformly random over a b letter alphabet. There exist constants c_1, c_2 such that if $k < k^* = m/(\log_b m + c_1) - c_2$, then the expected running time of* LET *is $O(n)$.*

The proof involves detailed probabilistic analysis using a technique called Chernoff bound, which shows it is highly unlikely for the sum of i.i.d. random variables to be high above the expectation. A little bit more work gives the following sublinear result when $k < k^*/2 - 3$ (k^* is given above).

Sublinear expected time algorithm (SET). Partition T into regions of length $(m-k)/2$. Any substring of T that matches P must contain the whole of at least one region. Starting from the left end of each region R, compute $k + 1$ longest exact matches, say ending at some position p. If position p is beyond R, apply the dynamic programming algorithm to a stretch of text beginning $(m + 3k)/2$ letters to the left of region R and ending at position p. (Figure 3a.)

Theorem (Chang and Lawler). *The average case complexity of k error (edit distance, Hamming distance, or longest common subsequence metric) approximate string matching is $O((n/m)k \log_b m)$, when k is bounded by the threshold $m/(\log_b m + O(1))$. In particular, when $k = o(m/ \log_b m)$ the complexity is $o(n)$.*

The l-tuple version given below is both easier to implement and more powerful, provided $k \leq \epsilon m$ (ϵ to be specified later) and there is enough memory to store the m^α precomputed lower bounds on matching the l-tuples against the pattern ($l = \alpha \log_b m$). It improves upon previous tuple based methods, being both more rigorous and provably more powerful. We also note that it is quite different from the "Four Russians" trick of precomputing dynamic programming "state transitions" (cf. Ukkonen, 1985; Wu, Manber, Myers, 1993), in that we tabulate lower bounds to be used later (and prove they are sufficiently strong). In practice, we have found the space requirement to be reasonable.

Optimal sublinear expected time algorithm (OPT). Define asm(S, P) as the best-match distance between S and any substring of P. Precompute and store in a table asm(S, P) for every l-tuple S. Then modify SET as follows: instead of longest exact matches, look up asm(S, P), S ranging over consecutive, nonoverlapping l-tuples, until the sum is greater than k. If the area covered by the tuples reach beyond R, apply dynamic programming. (Figure 3b.)

Correctness is clear. In order to prove the algorithm is fast, we will show $E[\text{asm}(S, P)] \geq \epsilon l$ for some constant ϵ (i.e. constant fraction error is detected). In fact we show $\Pr[\text{asm}(S, P)] < \epsilon l$ can be made sufficiently small. It then follows easily that only $O(k)$ work is done on average, to detect $k + 1$ errors. We use the following combinatorial lemma to bound the probability that a random string S matches any substring of P closely.

Lemma 1. *There exist constants $0 < c, d < 1$ and a such that $\Pr[$two strings of length $l = \alpha \log_b m$, one of which being random, have a common subsequence of length $cl] < (a/l) \cdot m^{-d\alpha}$.*

Proof. For convenience assume cl is an integer. By Stirling's formula $l!/(cl)!(l-cl)! = ((1+o(1))/(\sqrt{2\pi c(1-c)}l) \cdot c^{-cl}(1-c)^{-(1-c)l}$. The desired probability $p \leq \sum b^{-cl}$ where the summation is over all size cl bipartite matchings of positions. Hence

$$p \leq \binom{l}{cl}^2 \cdot b^{-cl} = \frac{1+o(1)}{2\pi c(1-c)l} \cdot (c^c(1-c)^{1-c})^{-2l} \cdot b^{-cl}.$$

This last expression decreases exponentially in l if $(c^c(1-c)^{1-c})^{-2} \cdot b^{-c} < 1$. Since this goes to b^{-1} as c goes to 1, it is b^{-d} for some choice of $c, d \in (0, 1)$. As $b \to \infty$, it suffices to choose $c > e/\sqrt{b}$, where e is the base of the natural logarithm. Choose $a > (2\pi c(1-c))^{-1}$, sufficiently large to overcome the error term in Stirling's formula. *Q.E.D.*

Lemma 2. *If S is an l-tuple such that $asm(S, P) < \epsilon l$, then S has a common subsequence of size $l - \epsilon l$ with some l-tuple of P.*

Proof. Assume S is edit distance $< \epsilon l$ from some substring Q of P. If the length of $Q \leq l$, then clearly S has a common subsequence of size $l - \epsilon l$ with an l-tuple extension of Q. If the length of Q equals $l' > l$, then S has a common subsequence with Q of size $l' - \epsilon l$, which implies S has a common subsequence of size $l' - \epsilon l - (l' - l)$ with the l-tuple prefix of Q. *Q.E.D.*

Let c, d be as in Lemma 1. Let $\epsilon = 1 - c$ and choose $\alpha > 4/d$ sufficiently large so that $\Pr[$random l-tuple has common subsequence of size cl in common with any given l-tuple of $P] < m^{-4}$. Then, $\Pr[asm(S, P) < \epsilon l] < \sum_{i=1}^{m} \cdot \Pr[S$ has common subsequence of size cl in common with the i-th l-tuple of $P] < m^{-3}$. Then, the chance that there is even one $asm(S, P) < \epsilon l < m^{-2}$, so the dynamic programming work is negligible. The expected running time is $O(n)$ if $k = \epsilon m$ (constant fraction error); $O((n/m) \cdot (k/\epsilon)) = O(kn/m)$ if $\log_b m < k < \epsilon m$; and $O((n/m) \log_b m)$ if $k < \log_b m$. The cases can be combined, and in fact this algorithm is optimal.

Weak Approximate String Matching Theorem. *There exist constants $\epsilon_b > 0$ such that for all $k \leq \epsilon_b m$, k error approximate string matching has average case complexity $Theta((n/m)(k + \log_b m))$. The $\Omega(kn/m)$ lower bound applies when a match does not exist and $k > \log_b m$; the $\Omega((n/m) \log_b m)$ lower bound applies to almost all patterns (in the sense of Yao, 1979) when $k < \log_b m$. Furthermore, an algorithm exists that uses $m^{O(1)}$ space, where the exponent depends on b and ϵ_b but is independent of m and n.*

Proof. Only the lower bound remains to be shown. When there is no match, any algorithm must examine at least k letters out of every block of size m, before it can declare there is no match. (One can construct examples where fewer letters need to be examined in order to pronounce there *is* a match.) Yao's classical $O((n/m)\log_b m)$ lower bound proof using "negative certificates" for exact matching also applies to approximate matching. The precise statement is technical, and we refer the reader to Yao (1979). *Q.E.D.*

3 Substring Matching

The (k, l) substring matching problem (Chang and Lawler). Find all locations in the text where a length l substring of P occurs, with at most k differences.

The (ϵ, l) substring matching problem. Find all locations where a substring of P of length at least l occurs, with at most ϵ fraction differences.

The difficulty is that an (ϵ, l) match that is actually longer than l can quite possibly have no substring that satisfies the (k, l) problem with $k = \epsilon l$. Indeed a match of length $l' > l$ can have no proper substring that also matches (for example, the string *abbbbba* when $P = a^m, \epsilon = 3/4, l = 5$). First we sketch the classical *maximum subvector* algorithm, and then modify it to find all *unit score maximal subvectors* (see Figure 4).

Maximum subvector sum problem. Given a vector of numbers $a[1, \ldots, n]$, find a subvector $a[i, \ldots, j]$ whose sum is maximum.

Linear time algorithm (zero trick). Let $s = m = M = 0$. For $i = 1, \ldots, n$, if $s + a[i] \geq 0$ then add $a[i]$ to s, and replace m, M by $\max(m, s), \max(M, s)$ respectively; if $s + a[i] < 0$ then set $s = m = 0$. M is the maximum subvector sum.

Proof of correctness. In a left-to-right scan, the vector is partitioned into alternating nonnegative and entirely-negative regions such that within each nonnegative region all prefix sums are nonnegative, but adding the next element to the right makes the sum negative. Now suppose $a[i, \ldots, j]$ is the subvector with smallest i whose sum equals the maximum subvector sum. Clearly $a[i] \geq 0$. We claim $a[i]$ is the beginning of a nonnegative region. If not, the rest of the region to the left of $a[i]$ has nonnegative (prefix) sum, so can be added to the subvector (contradiction). Next we claim $a[i, \ldots, j]$ cannot extend beyond this region. If it did, deleting the region and one more element (this sum must be negative) would increase the subvector sum (contradiction). So $a[i, \ldots, j]$ is a prefix of some nonnegative region, and the above algorithm will find it. *Q.E.D.*

Unit score maximal subvectors problem. Find all subvectors whose mean is at least a given unit score and which cannot be extended.

Linear time algorithm. By subtracting the given unit score from every element of the vector we can assume the unit score is zero. We need two scans, the first is the maximum subvector sum algorithm augmented with a *stack* that holds, for each nonnegative region, its *rightmost decreasing prefix sums* and where they occur (i.e. the maximum prefix sum for the region, then the maximum to its right, etc., each taken to be furthest to the right in case to ties). The second pass is right-to-left, to determine the leftmost extension possible for each rightmost maximum that is saved on the stack (unless no subvector with the given right endpoint can be maximal). Move the left endpoint to the beginning of the next nonnegative region unless the subvector sum would turn negative. Then, move the left endpoint one place at a time as long as the sum stays nonnegative. The total work is linear because the search for left endpoints never backs up to the right.

Proof of correctness. We claim the right endpoint of a nonnegative maximal subvector must be one of the rightmost maximum prefix sum locations on the stack; otherwise the subvector can be extended to the next such location (the last element of a nonnegative region is always on the stack). Its left endpoint must be either the beginning of a nonnegative region (same proof as before), or inside a negative region. The key observation is that any subvector whose right endpoint is inside a negative region must have negative sum. So, once a sum turns negative (and the left endpoint is not just to the right of a nonnegative region), any further extension is futile. Thus, all the right endpoints of nonnegative maximal subvectors will be properly considered, and the corresponding left endpoints discovered by the algorithm. *Q.E.D.*

To solve the (ϵ, l) substring matching problem (when ϵ is sufficiently small), we will apply the maximal positive subvectors algorithm to a suitably transformed filtering method (subtract each distance lower bound from ϵ·length to create a similarity upper bound). There is however a complication caused by the "granularity" of the tuple filter (the matching statistic filter does not have this complication). Namely, the match we are looking for may start or end in the middle of tuples, so the bounds at the ends are not applicable. Recall that a match can possibly have no proper substring that also matches—so we cannot just look for a slightly shorter match. The solution is to relax ϵ slightly as well—an (ϵ, l) match can be guaranteed to contain an (ϵ', l') match that ends at tuple boundaries, if $\epsilon' > \epsilon$ and $l' < l$ are appropriately chosen. (Take the most that one could lose at the ends, relax ϵ to cover it.)

In conclusion, we have an algorithm whose filtering method is guaranteed not to miss any match. The dynamic programming step (applied separately to every suffix of the pattern) is expensive, but potentially infrequent, especially with the addition of more sophisticated secondary filters (work in progress).

4   Local Similarity

We now sketch how to apply the above methods to protein database search, using PAM 90 and PAM 120 scoring matrices as examples. Recall that the target frequency of PAM 120 is about 33% identity, about the lowest level where one could still claim homology without additional information. The relative entropy of PAM 120 is one *bit* per residue (actually 0.98; see Altschul, 1991), where the unit *bit* comes from taking logarithm to the base 2 of the target over background odds ratio. The best random local alignments of a query sequence against the protein databases have scores about 25 bits; alignments with scores lower than 20–25 bits are not distinguishable from random noise, even if they are biologically important. That is, there are hundreds if not thousands of random alignments with those scores, so any signal below 20–25 bits is essentially undetectable. This is a basic limitation of the method. What it means, is that we can ignore low-scoring alignments, or alignments that are too short to score 20 bits. A 120 PAMs alignment has to have length $L = 20$, in order to score 20 bits (one bit per residue on average). We are therefore looking for alignments with both length (at least 20) and unit score (one bit per residue) constraints. For PAM 90, the relative entropy is about 1.4 bits, so $L = 14$ residues are needed to yield 20 bits.

A tuple size of five (pentamers) is appropriate for database search, given that protein databases currently contain about 20 million residues (there are 3.2 million pentamers but 64 million hexamers). Both time and space are reasonable for pentamers. There turns out to be a complication: if the query sequence is too long, the best match between a random pentamer (using amino acid distribution) and it will score higher than the relative entropy! For PAM 120, we divide the query sequence into blocks of size $S = 40$ that overlap by $H = 20$, so every alignment of size $L = 20$ is entirely contained within some block. For PAM 90, the corresponding parameters are $S = 50$ and $H = 15$.

For each block, we compute a table of pentamer scores (3.2 million in number), i.e. every pentamer's best match against the given block. Only one table is kept at a time. As the database is scanned (once per block), the table of pentamer scores for the current block gives upper bounds on alignment scores. For each database sequence and each possible offset (0–4), we use the maximal subvectors algorithm to find regions of length at least L (rounded down to a multiple of 5) that satisfy the unit score constraint, then form the union of all such regions. Smith-Waterman is performed on the union against the block. If a local alignment is found that scores 20 bits (extremely rarely), Smith-Waterman is called for the entire query and database sequences. In preliminary experiments, this program produced very nearly identical results as Smith-Waterman, but is faster. Because the expected score of a random pentamer against a single block is much less than the relative entropy, and in order to trigger Smith-Waterman several consecutive pentamers must score high, the filtering efficiency ranges from fairly to extremely high: 75% for PAM 140, 95% for PAM 90, >99% for PAM 40. Taking into account the overlap of blocks and table construction, the overall speedups are 2X for PAM 140 (33% identity) and 10X for PAM 90 (50% identity). Further improvements are expected, and a variation for PAM 250 (<20% identity) using dyanmic programming on pentamer scores is being developed.

Preferably, database searches are made using several scoring matrices: at least PAM 40, PAM 90, and PAM 120, in that order. Database entries containing 40 PAMs local alignments with the query sequence are not searched again for 90 PAMs or 120 PAMs alignments, etc. This way, a search using the PAM x matrix has the specific purpose of uncovering x PAMs local alignments. It is clear that distant alignments are more difficult to discover. For example, treating the query sequence as a single block usually suffices for PAM 40, because the relative entropy (2.3 bits per residue) is much higher than the noise of random pentamer scores. The total time required to do multiple PAM searches is therefore not much worse than a single search using the highest PAM, because most of the work is done at the highest PAM. Finally, we note that although (affine) gap penalty is not discussed, it is included in the scoring and in the pentamer upper bounds (i.e. best match allowing gaps).

This program is now part of a package called *Genome Analyst* that is being developed at CSHL. Other components include an interactive display, with scrolling windows and multiple views of the data, that uses character-cell graphics. It works effectively through both high-speed Ethernet and dial-up phone connections. The user-interface is kept simple, even though fairly powerful facilities are available for selectively hiding portions of the data (such as low-scoring alignments). Redundant data (such as identical alignments against several database entries) are combined but not discarded. Such filtering can pare down a BLAST or FASTA report by a factor of ten, without losing any information. Because searching a large database can produce a massive amount of data that must be scrutinized for useful information, the importance of concise and effective visual display of the data cannot be overstated. Access to databases and powerful search facilities are available on-line as well. For example, several alignments can be merged to produce a PROSITE-type pattern, which is then compared against the entire database in just seconds.

Acknowledgment
We would like to thank David Cuddihy and E. Corprew Reed for programming support.

References

1. S.F. Altschul, Amino Acid Substitution Matrices from an Information Theoretic Perspective, *J. Molecular Biology*, 219(1991), pp. 555-565.
2. S.F. Altschul, W. Gish, W. Miller, E.W. Myers, and D.J. Lipman, A Basic Local Alignment Search Tool, *J. Molecular Biology*, 215(1990), pp. 403-410.
3. P. Argos and M. Vingron, Sensitive Comparison of Protein Amino Acid Sequences, in R.F. Doolittle, ed. *Methods in Enzymology Volume 183*, Academic Press (1990), pp. 352-365.
4. P. Argos, M. Vingron, and G. Vogt, Protein sequence comparison: methods and significance, *Protein Engineering* 4(1991), pp. 375-383.
5. W.I. Chang, *Approximate Pattern Matching and Biological Applications*, Ph.D. thesis, U.C. Berkeley, August 1991. Also available as Computer Science Division Reports UCB/CSD 91/653-654.
6. W.I. Chang and J. Lampe, Theoretical and Empirical Comparisons of Approximate String Matching Algorithms, *Proc. Combinatorial Pattern Matching '92*, Tucson, AZ, April 29-May 1, 1992, Lecture Notes in Computer Science 644, Springer-Verlag, pp. 172-181.
7. W.I. Chang and E.L. Lawler, Approximate String Matching in Sublinear Expected Time, *Proc. 31st Annual IEEE Symposium on Foundations of Computer Science*, St. Louis, MO, Oct. 22-24, 1990, pp. 116-124.

8. W.I. Chang and W.L. Lawler, Sublinear Expected Time Approximate String Matching and Biological Applications, *Algorithmica*, in press.

9. V. Chvátal and D. Sankoff, Longest Common Subsequences of Two Random Sequences, *Technical Report STAN-CS-75-477*, Stanford University, Computer Science Department, 1975.

10. M.O. Dayhoff, R.M. Schwartz, and B.C. Orcutt, A Model of Evolutionary Change in Proteins, in M.O. Dayhoff, ed., *Atlas of Protein Sequence and Structure* vol. 5. suppl. 3., Nat. Biomed. Res. Found., Washington, D.C., pp. 345–352, 1979.

11. R.F. Doolittle, ed. *Molecular Evolution: Computer Analysis of Protein and Nucleic Acid Sequences, Methods in Enzymology Volume 183*, Academic Press (1990).

12. D.G. George, W.C. Barker, and L.T. Hunt, Mutation Data Matrix and Its Uses, in R.F. Doolittle, ed. *Methods in Enzymology Volume 183*, Academic Press (1990), pp. 333–351.

13. W.B. Goad and M.I. Kanehisa, Pattern Recognition in Nucleic Acid Sequences I, A General Method for Finding Local Homologies and Symmetries, *Nucl. Acids Res.* 10(1982), pp. 247–263.

14. O. Gotoh, An improved algorithm for matching biological sequences, *J. Mol. Biol.* 162(1982), pp. 705–708.

15. X. Huang, A Contig Assembly Program Based on Sensitive Detection of Fragment Overlaps, *Genomics*, 1992.

16. X. Huang and W. Miller, A time-efficient, linear-space local similarity algorithm. *Advances in Applied Mathematics* 12(1991), pp. 337–357.

17. S. Karlin and S.F. Altschul, Methods for assessing the statistical significance of molecular sequence features by using general scoring schemes, *Proc. Nat. Acad. Sci., USA*, 87(1990), 2264–2268.

18. D.E. Knuth, J.H. Morris, and V.R. Pratt, Fast Pattern Matching in Strings, *SIAM J. Comput.* 6:2 (1977), pp. 323–350.

19. G.M. Landau and U. Vishkin, Fast String Matching with k Differences, *J. Comp. Sys. Sci.* 37(1988), pp. 63–78.

20. E.W. Myers (1991a), A Sublinear Algorithm for Approximate Keyword Matching, Technical Report TR90-25, Computer Science Dept., University of Arizona, Tucson, September 1991.

21. E.W. Myers (1991b), An Overview of Sequence Comparison Algorithms in Molecular Biology, Technical Report TR91-29, Computer Science Dept., University of Arizona, Tucson, December 1991.

22. E.W. Myers, Algorithmic Advances for Searching Biosequence Databases, to appear in S. Suhai, ed., *Computational Methods in Genome Research*, Plenum Press (1994).

23. S.B. Needleman and C.E. Wunsch, A General Method Applicable to the Search for Similarities in the Amino Acid Sequence of Two Proteins, *J. Mol. Biol.* 48(1970), pp. 443–453.

24. W.R. Pearson, Searching protein sequence libraries: comparison of the sensitivity and selectivity of the Smith-Waterman and FASTA algorithms, *Genomics* 11(1991), pp. 635–650.

25. W.R. Pearson and D.J. Lipman, Improved tools for biological sequence comparison, *Proc. Natl. Acad. Sci. USA* 85(1988), pp. 2444–2448.

26. D. Sankoff and J.B. Kruskal, eds., *Time Warps, String Edits, and Macromolecules: The Theory and Practice of Sequence Comparison*, Addison-Wesley (1983).

27. P.H. Sellers, The Theory and Computation of Evolutionary Distances: Pattern Recognition, *J. Algorithms* 1(1980), pp. 359–373.

28. P.H. Sellers, Pattern Recognition in Genetic Sequences by Mismatch Density, *Bull. Math. Biol.* 46(1984), pp. 501-514.
29. T.F. Smith and M.S. Waterman, Identification of Common Molecular Subsequences, *J. Mol. Biol.* 147(1981), pp. 195-197.
30. S.S. Sturrock and J.F. Collins (1993), MPsrch version 1.3, Biocomputing Research Unit, University of Edinburgh, UK.
31. E. Ukkonen, Finding Approximate Patterns in Strings, *J. Algorithms* 6(1985), pp. 132-137.
32. M. Vingron and M.S. Waterman, Parametric Sequence Alignments and Penalty Choice: Case Studies, manuscript, 1993.
33. M.S. Waterman, Sequence Alignments, in M.S. Waterman, ed., *Mathematical Methods for DNA Sequences*, CRC Press (1989), pp. 53-92.
34. M.S. Waterman and M. Eggert, A new algorithm for best subsequence alignments with applicaiton to tRNA-rRNA comparison, *J. Mol. Biol.* 197(1987), pp. 723-728.
35. S. Wu and U. Manber, Fast Text Searching Allowing Errors, *Comm. ACM* 35(1992), pp. 83-91.
36. S. Wu, U. Manber, and E.W. Myers, A Sub-quadratic Algorithm for Approximate Limited Expression Matching, Technical Report TR92-36, Computer Science Dept., University of Arizona, Tucson, December 1992.
37. A.C. Yao, The Complexity of Pattern Matching for a Random String, *SIAM J. Comput.* 8(1979), pp. 368-387.

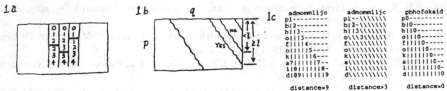

1a

1b   q

1c

```
        admommlljc    admommlljc    pbhofokaid
p1---------   p1--\\\\\\\   p0---------
b|2--------   b|2-\\\\\\\   b|0--------
h||3-------   h||3\\\\\\\   h||0-------
o|||3------   o\\|3\\\\\\   o|||0------
f|||4-----   f\\\\\\\\\   f|||0-----
o||||5----   o\\\\\\\\\   o||||0----
k|||||6---   k\\\\\\\\\   k|||||0---
a7|||||7--   a\\\\\\\\\   a||||||0--
i|8||||||8-   i\\\\\\\\\   i|||||||0-
d|89|||||9   d\\\\\\\\\   d||||||||0

distance=9    distance>3    distance=0
```

Figure 1a. clp

Pattern matching. Computed column by column, the amount of work is proportional to the number of segments (run of consecutive integers). The empirically observed (and conjectured) average run length is $\sqrt{b}$.

Figure 1b. overlap(p,q,k,l)

Overlap detection. A special case of pattern matching is overlap detection: using p as pattern, determine whether q contains p, or a suffix of q matches a prefix of p (of length at least l), allowing a constant *fraction* k/l errors. Then, try q as pattern.

Figure 1c. differ(p,q,10) differ(p,q,3) differ(p,p,10)

Redundancy detection. A new algorithm for edit distance: computed in alternating columns and rows, the amount of work is proportional to the *number* of digits shown; each digit indicates a new segment ($|$ or $-$). Cells that are deduced to be greater than the threshold are shown as $\backslash$.

2a

2b   *abaab*

Figure 2a. $O(n)$ expected time algorithm of Chang and Lawler, based on the matching statistics filter

Figure 2b. Suffix tree of the pattern; summarizes all substrings (and repeats). The set of leaves underneath any given node stores all the positions where the substring represented by the node occurs in the pattern. Matching statistics (longest exact matches) is computed using the suffix tree.

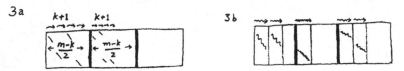

3a

$k+1$ $k+1$

$\leftarrow \frac{m-k}{2} \rightarrow$ $\leftarrow \frac{m-k}{2} \rightarrow$

3b

Figure 3a. $O((n/m)(k \log_b m))$ expected time algorithm of Chang and Lawler

Figure 3b. New $O((n/m)(k + \log_b m))$ expected time algorithm (optimal)

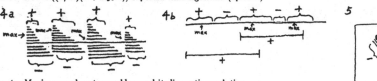

4a   +   +   +   +

max→   max   max   max

4b   +   +   +

max   max

+

5

Figure 4a. Maximum subvector problem and its linear time solution

Figure 4b. Maximal positive subvectors problem and its linear time solution

Figure 5. A long positive alignment can span several positive and negative regions

Polynomial-Time Algorithms for Computing Characteristic Strings

Minoru Ito[1], Kuniyasu Shimizu[2], Michio Nakanishi[3] and Akihiro Hashimoto[3] *

[1] Graduate School of Information Science, Nara Institute of Science and Technology,
Ikoma, Nara 630-01, Japan
[2] Information Systems Engineering Laboratory, TOSHIBA Corporation,
Ome, Tokyo 198, Japan
[3] Department of Information and Computer Sciences, Faculty of Engineering Science,
Osaka University, Toyonaka, Osaka 560, Japan

E-mail : naka@ics.es.osaka-u.ac.jp

Abstract. The difference between two strings is the minimum number of editing steps (insertions, deletions, changes) that convert one string into the other. Let S be a finite set of strings, let T be a subset of S, and let δ be a positive integer. A δ-characteristic string of T under S is a string that is a common substring of T and that has at least δ-differences from any substring of any string in $S - T$. In this paper, the following result is presented. It can be decided in $O(\|T\| + l^2 \cdot |S - T| + l \cdot \delta \cdot \|S - T\|)$ time whether or not there exists a δ-characteristic string of T under S, where l denotes the length of a shortest string in T, $|S - T|$ the cardinality of $S - T$, and $\|T\|$ the size of T. If such a string exits, then all the shortest δ-characteristic strings of T under S can also be obtained in that time.

Keywords: characteristic string, approximate pattern matching, DNA probe

1 Introduction

Let Σ be an alphabet consisting of two or more characters. A *string* is a sequence of characters over Σ. A *substring* of a string $a_1 a_2 \ldots a_n$ is of the form $a_{i+1} a_{i+2} \ldots a_j$, where $0 \le i \le j \le n$. In particular, if $i = j$, then $a_{i+1} a_{i+2} \ldots a_j$ is a string of zero length, which is called the *empty string*, written ε. A *common substring* of a set of strings is a sting that is a substring of any string in the set. To define the difference between two strings, let us introduce the following three operations for a string $a_1 a_2 \ldots a_n$.

1. *Insert* a character $b \in \Sigma$ at a position i to yield $a_1 a_2 \ldots a_i b a_{i+1} \ldots a_n$.

---

* We would like to thank Professor Masao Nasu, Faculty of Pharmaceutical Sciences, Osaka University, for helping us to formalize the problem of designing DNA probes.

2. *Delete* a character a_i to yield $a_1 a_2 \ldots a_{i-1} a_{i+1} \ldots a_n$.

3. *Exchange* a character a_i for another character $b \in \Sigma$ to yield $a_1 a_2 \ldots a_{i-1} b a_{i+1} \ldots a_n$.

Then the *difference* between two strings is the minimum number of operations above to transform one string into the other. If the difference of two strings p and q is δ, then p is said to have δ-*differences* from q. For example, a string BACC has 3-differences from a string ABABBC, as illustrated below.

$$\text{BACC} \xquad{\xRightarrow{\ insert\ }} \text{ABAĊC} \xRightarrow{\ exchange\ } \text{ABAḂC} \xRightarrow{\ insert\ } \text{ABABBC}$$

Let S be a finite set of strings. Let T be a subset of S, called a *target set*. A string p is called a δ-*characteristic string of* T *under* S if p is a common substring of T and has at least δ-differences from any substring of any string in $S - T$. Such a string p does not always exist for any δ, S and T. A δ-characteristic string of T under S is simply called a characteristic string if δ, S and T are understood from context.

Example 1. Let $T = \{\text{ARCHITECT}, \text{MONARCH}\}$, $S - T = \{\text{CHEMIST}, \text{RHETORICIAN}\}$, and $\delta = 2$. Here, ARCH is a characteristic string, since it is a common substring of T and has at least 2-differences from any substring of any string in $S - T$. Similarly, ARC is also a characteristic string. It is, however, noted that neither AR nor RC is a characteristic string, since each string has 1-difference from the substring R in RHETORICIAN. Actually, ARC is a shortest characteristic string. □

In this paper, the following result is presented. Given a positive integer δ, a finite set S of strings, and a target set T in S, it can be decided in $O(\|T\| + l^2 \cdot |S - T| + l \cdot \delta \cdot \|S - T\|)$ time whether or not there exists a characteristic string, where l denotes the length of a shortest string in T, $|S - T|$ the cardinality of $S - T$, and $\|T\|$ the size of description of T (that is, the sum of lengths of the strings in T). Furthermore, if such a characteristic string exists, then all the shortest characteristic strings can be obtained in that time.

The notion of characteristic string is motivated by DNA probes in molecular biology. In the rest of this section, we will explain, as an example, how our result can be used for finding a specific sequence used as a DNA probe for the diagnosis of bacterial infections [2, 8]. DNA is composed of four kinds of bases, adenine, thymine, guanine and cytosine, each of which is represented by a letter, typically A, T, G and C, respectively. These bases have the property that A and T make a complementary pair as well as C and G do. DNA diagnosis for the bacterial infections is a technique using this property, which checks if the DNA probe hybridizes with the nucleotide sequence of the target bacteria in the specimen. To be more precise, the DNA probe assay proceeds in the following way.

(1) A short nucleotide sequence is synthesized to be complementary to a specific substring of chromosomal DNA or ribosomal RNA (rRNA) of the target bacteria species which can be the cause of the bacterial infection. Then the sequence is labeled by a chemical substance, e.g., photobiotin. This sequence is called a DNA probe. For example, if CCAGT is specific to the target bacteria species, its complementary nucleotide sequence GGTCA is a candidate probe for the bacteria species.

(2) Nucleotide sequences extracted from the specimen are mixed with the probe. If a target bacterium exists in the specimen, the probe hybridizes with its complementary part of the bacterium's nucleotide sequence. By detecting the labeling substance, it can be decided what kind of bacterium causes the infection. This method is a rapid sensitive and specific technique compared to the usual method which depends on the cultivation.

It is crucial for the reliable antibiotics to find a specific substring of the nucleotide sequences of the target bacteria species. In practice, even if the complementary sequence of a probe has a few mismatch with the specific substring of the target sequences, the probe forms duplexes with some of the target sequences. Thus, the more the complementary sequence of a probe has differences with the nucleotide sequences of bacteria other than the target species, the detection can be done more accurately. In other words, it is necessary to find such a probe sequence that its complement has more difference with nucleotide sequences other than the target sequences. Usually, a specific part of the target nucleotide sequences has been selected by a skilled researcher by intuition, then examined whether or not that part is useful as a probe on an experimental basis. Recently, a new method is experimented; (1) First, find a specific part from the computer output of a multiple alignment program, e.g.[4], for the target sequences. (2) Then compare it with all or subset of sequences in a database, such as GenBank [16], using a homology search program [10] to see if it is specific to the target; If it does not match with any substring of those sequences other than the target sequences, it is a candidate probe for the target bacteria [9].

The result of this paper can be applied to finding a probe. Let S be a set of nucleotide sequences of the bacteria that will appear during the diagnosis of a bacterial infection, T be a set of nucleotide sequences of the target bacteria species, and δ be the number of mismatches that is allowed for hybridization process. If a characteristic string exists, its complementary string is a candidate probe for the target bacteria. Since a probe is chemically synthesized and synthesizing cost depends on its length, it is important to find a shortest possible characteristic string. Example 2 shows a small model of finding DNA probes.

Example 2. Let $S = \{s_1, s_2, s_3, s_4, s_5\}$ and $T = \{s_1, s_2\}$ as shown below.

s_1: CTACCGTAGTCT
s_2: AACTGACCGTAGAC
s_3: AGTGGGGGACAACAAATT
s_4: GACACGGCCCAGACT
s_5: AGGCGGCTTTTTA

For the case of $\delta = 2$, two shortest characteristic strings, ACCGTA and CCGTAG, are obtained using our algorithm. Thus, their complementary strings, CGGCAT and GGCATC, become candidate probes. For the case of $\delta = 1$, one shortest characteristic string, ACC is obtained. □

2 Preparation

Let $p = a_1 a_2 \ldots a_n$ be a string. Let us denote by $p[i, j]$ a substring $a_{i+1} a_{i+2} \ldots a_j$ of p, where $0 \le i \le j \le n$. (Note that $p[i, i] = \varepsilon$ by definition.) In particular, $p[0, j]$ and $p[i, n]$ are called a *prefix* and a *suffix* of p, respectively. The length of p is denoted $|p|$. (In this case, $|p| = n$.) Let q be another string. The *concatenation* of p and q is denoted pq, and the difference between p and q is denoted $diff(p, q)$. In the rest of this section, we consider how to solve the following problem, which will be used for computing a characteristic string in section 3.

Basic Problem: Let δ be a positive integer. Let p and q be strings. Decide whether or not there exists a prefix of p that has at least δ-differences from any substring of q. If such a prefix exists, then obtain the shortest one. □

In the following we say that a string p' satisfies *Condition A* (with respect to q) if p' has at least δ-differences from any substring of q. The negation of Condition A is denoted *Condition ¬A*; that is, p' satisfies Condition ¬A if there is a substring q' of q such that $diff(p', q') \le \delta - 1$. Then we have the following lemma.

Lemma 1. *If a prefix $p[0, i]$ of p satisfies Condition ¬A, then so does every prefix of $p[0, i]$.*

Proof. Assume that $p[0, i]$ satisfies Condition ¬A. Then there is a substring q' of q such that

$$diff(p[0, i], q') \le \delta - 1. \tag{2.1}$$

There are two cases to be considered.

Case 1: where $q' = \varepsilon$. Since in general, $diff(x, \varepsilon) = |x|$ for any string x, it follows from (2.1) that $|p[0, i]| = i \leq \delta - 1$. Thus $diff(p[0, i'], q') \leq \delta - 1$ for every i' with $0 \leq i' \leq i$. Hence every prefix of $p[0, i]$ satisfies Condition $\neg$A, thereby Lemma 1 follows.

Case 2: where $q' \neq \varepsilon$. Let us denote q' by $q[k, j]$. Then $q' \neq \varepsilon$ implies $k < j$. Since in general, $diff(x, y) \leq diff(xa, yb)$ for all strings x, y and all characters a, b, we have the following by applying the inequality repeatedly to (2.1):

$$diff(p[0, i - l], q[k, j - l]) \leq diff(p[0, i], q[k, j]) \leq \delta - 1, \tag{2.2}$$

where $0 \leq l \leq \min\{i, j - k\}$. If $i \leq j - k$, then (2.2) implies that every prefix of $p[0, i]$ satisfies Condition $\neg$A, and thus Lemma 1 holds. Assume $i > j - k$. Since $q[k, j - l] = \varepsilon$ when $l = j - k$ in (2.2), it holds that $diff(p[0, i - j + k], \varepsilon) \leq \delta - 1$. Thus from Case 1 above, it follows that $diff(p[0, i'], \varepsilon) \leq \delta - 1$ for $0 \leq i' \leq i - j + k$. By this fact together with (2.2), every prefix of $p[0, i]$ satisfies Condition $\neg$A. This completes proving Lemma 1. $\qquad\square$

The following corollary is the contraposition of Lemma 1.

Corollary 1. *If a prefix $p[0, i]$ of p satisfies Condition A, then so does every prefix $p[0, i']$ of p such that $i' \geq i$.* $\qquad\square$

Let $p[0, i]$ be the longest prefix of p that satisfies Condition $\neg$A. By Lemma 1, no prefix of $p[0, i]$ satisfies Condition A. If $i = |p|$, then $p[0, i]$ coincides with p itself, thereby no prefix of p satisfies Condition A. If on the other hand $i < |p|$, then $p[0, i+1]$ must satisfy Condition A, since $p[0, i]$ is the *longest* prefix of p that satisfies Condition $\neg$A. Then it turns out that $p[0, i + 1]$ is the *shortest* prefix of p that satisfies Condition A. Hence, in order to solve the basic problem above, it suffices to obtain the largest index i such that $p[0, i]$ satisfies Condition $\neg$A. For that purpose, let us introduce two tables, $D(p, q)$ and $R_\delta(p, q)$. (These tables were originally introduced in [11] and [14], respectively.)

$D(p, q)$ is a matrix with $|p| + 1$ rows and $|q| + 1$ columns. Its first row and first column are referred to as the 0th row and the 0th column, respectively. Its (i, j)-component, written $d_{i,j}$, is defined by:

$$d_{i,j} = \min_{0 \leq k \leq j} \{ diff(p[0, i], q[k, j]) \}. \tag{2.3}$$

That is, $d_{i,j}$ is the minimum difference between the prefix $p[0, i]$ of p and the suffixes $q[0, j], q[1, j], \ldots q[j, j]$ of the prefix $q[0, j]$ of q. For each l with $-|p| \leq l \leq |q|$, the components $d_{i,j}$ with $j - i = l$ are said to be on *diagonal l*.

Table 1. Example of $D(p,q)$

j		0	1	2	3	4	5	6	7	8	9	10	11
i	$p \setminus q$	R	H	E	T	O	R	I	C	I	A	N	
0		0	0	0	0	0	0	0	0	0	0	0	0
1	M	1	1	1	1	1	1	1	1	1	1	1	1
2	O	2	2	2	2	2	1	2	2	2	2	2	2
3	N	3	3	3	3	3	2	2	3	3	3	3	2
4	A	4	4	4	4	4	3	3	3	4	4	3	3
5	R	5	4	5	5	5	4	3	4	4	5	4	4
6	C	6	5	5	6	6	5	4	4	4	5	5	5
7	H	7	6	5	6	7	6	5	5	5	5	6	6

Example 3. Table 1 depicts $D(p,q)$ for the case of $p =$ MONARCH and $q =$ RHETORICIAN. Note that p and q are strings in T and $S - T$ shown in Example 1, respectively. For example, $d_{2,5} = 1$ holds, since

$$d_{2,5} = \min_{0 \le k \le 5} \{ \mathit{diff}\,(p[0,2], q[k,5]) \}$$

and the difference between $p[0,2] =$ MO and $q[k,5] =$ TO is minimized to one for the case of $k = 3$. Elements on diagonal 3 of $D(p,q)$, namely the components $d_{0,3}, d_{1,4}, d_{2,5}, \cdots, d_{7,10}$, are $0, 1, 1, 2, \cdots, 6$. $\square$

Let us next define $R_\delta(p,q)$ using $D(p,q)$. $R_\delta(p,q)$ is a matrix with δ rows and $\delta + |q|$ columns. Its first row and first column are referred to as the $0th$ row and the $(-\delta + 1)th$ column, respectively. Its (k, l)-component, written $r_{k,l}$, is defined by:

$$r_{k,l} = \max_{d_{i,l+i}=k} \{i\}. \tag{2.4}$$

That is, $r_{k,l}$ is the largest row number in the components of $D(p,q)$ that have value k and that are on diagonal l. By definition of $D(p,q)$, if $r_{k,l} = i$, then (1) $p[0,i]$ has at least k-differences from any suffix of $q[0, l + i]$ and (2) there is a suffix of $q[0, l + i]$ that has exactly k-differences from $p[0,i]$. Note that $r_{k,l}$ is not necessarily defined. In fact, though each column of $R_\delta(p,q)$ corresponds to a diagonal of $D(p,q)$ and the number of diagonals of $D(p,q)$ is $|p|+|q|+1$, $R_\delta(p,q)$ contains only $\delta + |q|$ columns, since all the values of $r_{k,l}$ with $-|p| \le l \le -\delta$ are undefined [12][13].

Example 4. For the case of $p =$MONARCH, $q =$RHETORICIAN and $\delta = 2$, $R_2(p,q)$ is constructed as shown in Table 2 using the table $D(p,q)$ in Example 3. For example, $r_{1,3} = 2$, since the largest row number in the components of $D(p,q)$ that have value one and that are on diagonal 3 is two. $\square$

Table 2. Example of $R_2(p,q)$

$k \setminus l$	-1	0	1	2	3	4	5	6	7	8	9	10	11
0	-	0	0	0	0	0	0	0	0	0	0	0	0
1	1	1	1	1	2	1	1	1	1	1	1	1	-

Let us define:

$$I_{max}(\delta, p, q) = \max_{k,l}\{r_{k,l} + \delta - k - 1\}. \qquad (2.5)$$

It is simply denoted I_{max} if the arguments are understood from context. For example, in Example 4, since $r_{k,l} + \delta - k - 1$ becomes largest when $(k, l) = (1, 3)$ and $r_{1,3} = 2$, $I_{max}(p, q, \delta) = 2$. We have the following simple lemma.

Lemma 2. $I_{max} \geq \delta - 1$.

Proof. Since $p[0, 0] = q[0, 0] = \varepsilon$ by definition, $d_{0,0} = 0$ follows from (2.3). That is, value 0 occurs on diagonal 0 in $D(p, q)$. Then $r_{0,0} \geq 0$ follows from (2.4). Thus (2.5) implies $I_{max} \geq r_{0,0} + \delta - 0 - 1 \geq \delta - 1$. That is, Lemma 2 holds. $\square$

In the following it is shown that if $I_{max} < |p|$, then $p[0, I_{max}]$ is the longest prefix satisfying Condition $\neg A$; otherwise, p itself is the longest prefix satisfying Condition $\neg A$.

Lemma 3. *Every prefix $p[0, i]$ of p with $i \leq I_{max}$ satisfies Condition $\neg A$.*

Proof. Since I_{max} is well-defined by Lemma 2, it follows from (2.5) that $R_\delta(p, q)$ must contain a component $r_{k,l}$ such that

$$I_{max} = r_{k,l} + \delta - k - 1. \qquad (2.6)$$

Thus by (2.3) and (2.4), there is a suffix q' of $q[0, l + r_{k,l}]$ such that $diff(p[0, r_{k,l}], q') = k$. Since $0 \leq k \leq \delta - 1$, it holds that $diff(p[0, r_{k,l}], q') \leq \delta - 1$. Furthermore, since q' is a substring of q, $p[0, r_{k,l}]$ satisfies Condition $\neg A$. Hence by Lemma 1, if $i \leq r_{k,l}$, then $p[0, i]$ satisfies Condition $\neg A$, thereby Lemma 3 holds. Assume $i > r_{k,l}$. Since (1) in general, $diff(xy, z) \leq diff(x, z) + |y|$ for all strings x, y, z and (2) $p[0, i]$ is the concatenation of $p[0, r_{k,l}]$ and $p[r_{k,l}, i]$, it holds that

$$diff(p[0, i], q') \leq diff(p[0, r_{k,l}], q') + |p[r_{k,l}, i]| = k + (i - r_{k,l}).$$

Here, $k + (i - r_{k,l}) = \delta - I_{max} + i - 1$ by (2.6). Thus if $i \leq I_{max}$, then $k + (i - r_{k,l}) \leq \delta - 1$; that is, $diff(p[0, i], q') \leq \delta - 1$. Hence $p[0, i]$ satisfies Condition $\neg A$. This completes proving Lemma 3. $\square$

Since $p[0, |p|] = p$, Lemma 3 implies that if $I_{max} \geq |p|$, then p itself satisfies Condition $\neg A$. In this case, clearly, p is the longest prefix satisfying Condition $\neg A$. On the other hand, the following lemma considers the case of $I_{max} < |p|$.

Lemma 4. *If $I_{max} < |p|$, then $p[0, I_{max}]$ is the longest prefix of p that satisfies Condition $\neg A$.*

Proof. Assume $I_{max} < |p|$. By Lemma 3, $p[0, I_{max}]$ satisfies Condition $\neg A$. Let us prove the maximality of $p[0, I_{max}]$. Assume, contrary, that some prefix $p[0, i]$ of p with $i > I_{max}$ satisfies Condition $\neg A$. Then there is a substring $q[m, j]$ of q such that $diff(p[0, i], q[m, j]) \leq \delta - 1$. Assume without loss of generality that the difference between $p[0, i]$ and $q[m, j]$ is minimum among the substrings of q. Then $d_{i,j} = diff(p[0, i], q[m, j])$ by (2.3). For simplicity, let $k = diff(p[0, i], q[m, j])$. Then $d_{i,j} = k$ implies $r_{k,j-i} \geq i$ by (2.4). Thus $r_{k,j-i} + \delta - k - 1 \geq i + \delta - k - 1$. Furthermore, since $k = diff(p[0, i], q[m, j]) \leq \delta - 1$, it holds that $r_{k,j-i} + \delta - k - 1 \geq i + \delta - (\delta - 1) - 1 = i$. Hence $i > I_{max}$ implies $r_{k,j-i} + \delta - k - 1 > I_{max}$. This, however, contradicts the maximality of I_{max} in (2.5). Consequently, $p[0, I_{max}]$ is the longest prefix of p that satisfies Condition $\neg A$. Lemma 4 follows. $\square$

By Lemmas 3 and 4, we have the following lemma as for the basic problem.

Lemma 5. *Let δ be a positive integer. Let p and q be strings. There exists a prefix of p (with respect to q) that satisfies Condition A if and only if $I_{max} < |p|$. Furthermore, if $I_{max} < |p|$, then $p[0, I_{max} + 1]$ is the shortest prefix of p that satisfies Condition A.* $\square$

Originally, $D(p, q)$ and $R_{\delta}(p, q)$ were introduced for the approximate string matching algorithm, which finds all occurrences of a given pattern p in a given text q with at most δ-differences. It is shown in [6] that $R_{\delta}(p, q)$ can be obtained in $O(|p| + \delta \cdot |q|)$ time. Thus $I_{max}(\delta, p, q)$ can also be obtained in that time. Furthermore, it can be done in that time to decide if $I_{max} < |p|$ and to obtain $p[0, I_{max} + 1]$. Note that the algorithm in [6] runs in $O(\delta \cdot |q|)$ time, since the approximate string matching problem should be meaningful if $|p| < |q|$. However, we must consider the case of $|p| > \delta \cdot |q|$ in the basic problem. Thus it takes not $O(\delta \cdot |q|)$ but $O(|p| + \delta \cdot |q|)$ time.

3 The Algorithm for Computing Characteristic Strings

Let δ be a positive integer, let S be a finite set of strings, and let T be a target set in S. In this section, we will present an algorithm for deciding whether or not

there exists a characteristic string and for obtaining all the shortest characteristic strings if exists. We begin with presenting one necessary and one sufficient
conditions for a characteristic string to exist. For a string p, let us define:

$$J_{max}(\delta, p, S - T) = \max_{q \in S - T}\{I_{max}(\delta, p, q)\}. \tag{3.1}$$

It is simply denoted J_{max} if the arguments are understood from context. Note
that J_{max} is well-defined by Lemma 2. The following lemma gives a necessary
condition for a characteristic string to exist.

Lemma 6. *Assume that there is a δ-characteristic string of T under S. Then
for every $p \in T$, there exists an integer i satisfying the following three conditions:*

Condition 1: $i \leq |p| - \delta$,
Condition 2: $J_{max}(\delta, p[i, |p|], S - T) < |p| - i$, and
Condition 3: $p[i, i + J_{max} + 1]$ is a common substring of T.

Proof. Let p' be a characteristic string. Let $p \in T$. Since p' is a common substring of T by definition, it must hold that $p' = p[i, i + j]$ for some i, j. By (3.1),
there is a string $q \in S - T$ such that

$$J_{max}(\delta, p[i, |p|], S - T) = I_{max}(\delta, p[i, |p|], q). \tag{3.2}$$

Let us first prove Condition 2. By definition, $p'(= p[i, i + j])$ has at least
δ-differences from any substring of q; that is, $p[i, i+j]$ satisfies Condition A with
respect to q. Here, since $p[i, i + j]$ is a prefix of $p[i, |p|]$, it follows from Lemma 5
that $I_{max} < |p[i, |p|]| = |p| - i$. Thus $J_{max} < |p| - i$ by (3.2); Condition 2 follows.

Let us next prove Condition 1. Assume, contrary, $i > |p| - \delta$. Then $\delta - 1 \geq$
$|p| - i$. Furthermore, since $I_{max} \geq \delta - 1$ by Lemma 2, (3.2) implies $J_{max} \geq \delta - 1$.
Thus $J_{max} \geq |p| - i$, which contradicts Condition 2. Hence Condition 1 holds.

Let us finally prove Condition 3. By Lemma 5, $p[i, i + I_{max} + 1]$ is the shortest
prefix of $p[i, |p|]$ that satisfies Condition A with respect to q. Note that $p[i, i + j]$
is also a prefix of $p[i, |p|]$ satisfying Condition A with respect to q. Thus $|p[i, i +$
$I_{max} + 1]| \leq |p[i, i + j]|$, that is, $I_{max} + 1 \leq j$. Hence $J_{max} + 1 \leq j$ by (3.2), which
means that $p[i, i + J_{max} + 1]$ is a prefix of $p[i, i + j]$. Since $p[i, i + j]$ is a common
substring of T, so is every prefix of $p[i, i + j]$. In particular, $p[i, i + J_{max} + 1]$ is a
common substring of T; Condition 3 follows. $\square$

Conversely, the following lemma gives a sufficient condition for a characteristic string to exist, which uses the conditions of Lemma 6.

Lemma 7. *Assume that for some $p \in T$, there is an integer i satisfying Conditions 2 and 3 of Lemma 6. Then $p[i, i + J_{max} + 1]$ is a δ-characteristic string of
T under S that is shortest among the prefixes of $p[i, |p|]$.*

Proof. By (3.1) and Condition 2, every $q \in S - T$ satisfies

$$\mathrm{I}_{\max}(\delta, p[i, |p|], q) \leq \mathrm{J}_{\max}(\delta, p[i, |p|], S - T) < |p| - i = |p[i, |p|]|. \tag{3.3}$$

Thus it follows from Lemma 5 that $p[i, i + \mathrm{I}_{\max} + 1]$ satisfies Condition A with respect to q. By Corollary 1, for every j with $\mathrm{I}_{\max} + 1 \leq j \leq |p| - i$, $p[i, i+j]$ also satisfies Condition A with respect to q. In particular, $p[i, i + \mathrm{J}_{\max} + 1]$ satisfies Condition A with respect to q, since (3.3) implies $\mathrm{I}_{\max} + 1 \leq \mathrm{J}_{\max} + 1 \leq |p| - i$. That is, $p[i, i + \mathrm{J}_{\max} + 1]$ has at least δ-differences from any substring of q. Since q is an arbitrary string in $S - T$, it turns out that $p[i, i + \mathrm{J}_{\max} + 1]$ has at least δ-differences from any substring of any string in $S - T$. By this fact together with Condition 3, $p[i, i + \mathrm{J}_{\max} + 1]$ is a δ-characteristic string of T under S.

It remains to prove the minimality of $p[i, i + \mathrm{J}_{\max} + 1]$. By (3.1), there is a string $q' \in S - T$ such that $\mathrm{I}_{\max}(\delta, p[i, |p|], q') = \mathrm{J}_{\max}$. By Lemma 5, $p[i, i + \mathrm{I}_{\max} + 1]$ is the shortest prefix of $p[i, |p|]$ that satisfies Condition A with respect to q'. Thus if $j < \mathrm{I}_{\max} + 1 = \mathrm{J}_{\max} + 1$, then $p[i, i+j]$ does not satisfy Condition A with respect to q'; that is, $p[i, i+j]$ cannot be a characteristic string. Hence $p[i, i + \mathrm{J}_{\max} + 1]$ is a characteristic string that is shortest among the prefixes of $p[i, |p|]$. Lemma 7 follows. □

By Lemmas 6 and 7, in order to obtain a characteristic string, it suffices to choose an arbitrary string p from T and to examine, for each suffix $p[i, |p|]$ of p, whether or not the suffix satisfies Conditions 1 to 3. Clearly, it is advantageous to choose a shortest string as p. Furthermore, by the following lemma, once Condition 2 is not satisfied by a suffix $p[i, |p|]$, then the condition is no longer satisfied by any suffix of $p[i, |p|]$.

Lemma 8. *If* $\mathrm{J}_{\max}(\delta, p[i, |p|], S - T) \geq |p| - i$ *for some* i, *then* $\mathrm{J}_{\max}(\delta, p[j, |p|], S - T) \geq |p| - j$ *for every* j *with* $i \leq j \leq |p|$.

Proof. Assume $\mathrm{J}_{\max}(\delta, p[i, |p|], S - T) \geq |p| - i$ for some i. By (3.1), there is a string $q \in S - T$ such that $\mathrm{I}_{\max}(\delta, p[i, |p|], q) = \mathrm{J}_{\max}$. Since $\mathrm{I}_{\max} \geq |p| - i = |p[i, |p|]|$, it follows from Lemma 5 that $p[i, |p|]$ does not satisfy Condition A, that is, satisfies condition $\neg$A with respect to q. In the same way as proving Lemma 1, we can prove that every suffix of $p[i, |p|]$ also satisfies condition $\neg$A with respect to q. Hence by Lemma 5, it holds that $\mathrm{I}_{\max}(\delta, p[j, |p|], q) \geq |p[j, |p|]| = |p| - j$ for every j with $i \leq j \leq |p|$. By (3.1), this implies $\mathrm{J}_{\max}(\delta, p[j, |p|], S - T) \geq |p| - j$. Lemma 8 follows. □

By the observations above, we have the following algorithm.

Algorithm 1. Computing all the shortest δ-characteristic strings of T under S.
Input: a positive integer δ, a finite set of strings S, and a target set T in S.
Output: if a characteristic string exits, then return all the shortest characteristic strings; otherwise, return "none."
Method:

1   choose a shortest string, say p, from T
2   **for** $i \leftarrow 0$ **to** $|p| - \delta$ **do**
     begin
3   compute $J_{max}(\delta, p[i, |p|], S - T)$
4   **if** $J_{max} < |p| - i$
5   **then if** $p[i, i + J_{max} + 1]$ is a common substring of T
6   **then** let p_i be $p[i, i + J_{max} + 1]$
7   **else** let p_i be undefined
8   **else** let p_j be undefined for every j with $i \leq j \leq |p| - \delta$; **goto** step 9
     end
9   **if** there is a string p_i defined in step 6
10   **then** return all the defined and shortest strings p_i
11   **else** return "none"   □

Example 5. Let us apply Algorithm 1 to Example 1. That is, let us compute all the shortest characteristic strings of T under S, where $T = \{\text{MONARCH}, \text{ARCHITECT}\}$, $S - T = \{\text{RHETORICIAN}, \text{CHEMIST}\}$, and $\delta = 2$.

In step 1, MONARCH is assigned to p since it is shorter than ARCHITECT. In step 2, consider the case of $i = 0$. In step 3, to compute $J_{max}(p[0, 7], S - T, 2)$ we have to calculate $I_{max}(2, p[0, 7], \text{RHETORICIAN})$ and $I_{max}(2, p[0, 7], \text{CHEMIST})$. $I_{max}(2, p[0, 7], \text{RHETORICIAN}) = 2$ is obtained from Table 2. By generating $R_2(p[0, 7], \text{CHEMIST})$, $I_{max}(2, p[0, 7], \text{CHEMIST}) = 2$ is obtained. Thus, $J_{max}(2, p[0, 7], S - T) = 2$. In steps 4, 5 and 6, p_0 becomes undefined since $p[0, 3] = \text{MON}$ is not a common substring of T. Similarly, for the cases of $i = 1, 2$, and 3, it holds that $p_1 = $ undefined, $p_2 = $ undefined, $p_3 = \text{ARC}$, respectively. For the case of $i = 4$, $J_{max} = 3$ is obtained. Thus p_4 and p_5 become undefined in step 8. Consequently, in step 10, we get ARC as the shortest characteristic string among $p_0, \cdots, p_5$.   □

Let us prove the correctness of Algorithm 1. Assume that there is a δ-characteristic string of T under S. Let p' be one of the shortest characteristic strings. By Lemma 6, p' is denoted $p[i_0, j_0]$ and i_0 satisfies Conditions 1 to 3. Since i_0 satisfies Condition 2; that is, $J_{max}(\delta, p[i_0, |p|], S - T) < |p| - i_0$, it follows from Lemma 8 that $J_{max}(\delta, p[j, |p|], S - T) < |p| - j$ for every j with $j \leq i_0$. Thus the *if* condition of step 4 holds for every j with $j \leq i_0$, thereby step 8 is not executed. Furthermore, since $i_0 \leq |p| - \delta$ by Condition 1, there must be an instant such that i coincides with i_0 in step 2. Then since i_0 satisfies Conditions 2 and 3, step 6 must be executed, thereby p_{i_0} is assigned $p[i_0, i_0 + J_{max} + 1]$. Since (1) by Lemma 7, $p[i_0, i_0 + J_{max} + 1]$ is the shortest characteristic string among the prefixes of $p[i_0, |p|]$ and (2) p' is a prefix of $p[i_0, |p|]$, the minimality of p' implies $p' = p[i_0, i_0 + J_{max} + 1] = p_{i_0}$. Note that by Lemma 7, all the strings p_i defined in step 6 are characteristic strings. Hence the minimality of p' also implies that p_{i_0} should be included in the return value of step 10. That is, Algorithm 1 computes all the shortest characteristic strings. Conversely, assume that there is no δ-characteristic string of T under S. Then by Lemma 7, for any i, at least one of Conditions 2 and 3 does not hold, and hence either step 7 or step 8 must be executed. In either case, p_i must be undefined. Consequently, "none" returns at step 11. This completes proving the correctness of Algorithm 1.

We next estimate the running time of Algorithm 1. Step 1 requires $O(\|T\|)$ time. In step 3, in order to compute $J_{max}(\delta, p[i, |p|], S - T)$, we need the values of $I_{max}(\delta, p[i, |p|], q)$ for all $q \in S - T$ by (3.1). Note that $I_{max}(\delta, p[i, |p|], q)$ can be computed in $O(|p| + \delta \cdot |q|)$ time from the discussions in section 2. Thus $J_{max}(\delta, p[i, |p|], S - T)$ can be computed in $O(\sum_{q \in S-T}(|p| + \delta \cdot |q|)) = O(|p| \cdot |S - T| + \delta \cdot \|S - T\|)$ time. In step 5, we must test if $p[i, i + J_{max} + 1]$ is a common substring of T or not. But the test can be executed in $O(\|T\|)$ time using a well-known string matching algorithm as in [1] and [5]. Hence one execution of steps 3 to 8 requires $O(|p| \cdot |S-T| + \delta \cdot \|S-T\| + \|T\|)$ time. Steps 3 to 8 are executed $O(|p|)$ times. Consequently, Algorithm 1 requires $O(|p|^2 \cdot |S-T| + |p| \cdot \delta \cdot \|S-T\| + |p| \cdot \|T\|)$ time as a whole.

4 A Refinement of the Algorithm

In step 5 of Algorithm 1, it is cumbersome to test, for each i, whether or not $p[i, i + J_{max} + 1]$ is a common substring of T. Note that p is the string chosen from T in step 1. If the *for* loop of step 2 is executed with respect to a common substring of T, then we can skip the test. Based on the idea, we will consider how to refine Algorithm 1 in the following.

A common substring of T is said to be *maximal* if it is not a proper substring of any common substring of T. Let $\{q_1, q_2, \ldots, q_n\}$ be the set of maximal common substrings of T. By definition, for each m with $1 \leq m \leq n$, there is a substring $p[i_m, j_m]$ of p such that $p[i_m, j_m] = q_m$. By the maximality of each q_m, it is clear that for any distinct m_1, m_2, the interval $[i_{m_1}, j_{m_1}]$ does not include (and is not included in) the other interval $[i_{m_2}, j_{m_2}]$. Thus it must hold that (1) $i_{m_1} \neq i_{m_2}$, (2) $j_{m_1} \neq j_{m_2}$, and (3) $i_{m_1} < i_{m_2}$ implies $j_{m_1} < j_{m_2}$. Without loss of generality, assume $i_1 < i_2 < \cdots < i_n$ (thereby $j_1 < j_2 < \cdots < j_n$). Then steps 2 to 8 of Algorithm 1 are refined as follows.

2   **for** $m \leftarrow 1$ **to** n **do**
3   **for** $i \leftarrow i_m$ **to** $\min\{j_m - \delta, i_{m+1} - 1\}$ **do**
    begin
4   compute $J_{\max}(\delta, p[i, j_m], S - T)$
5   **if** $J_{\max} < j_m - i$
6   **then** let p_i be $p[i, i + J_{\max} + 1]$
7   **else** let p_j be undefined for every j with $i \leq j \leq \min\{j_m - \delta,$
          $i_{m+1} - 1\}$; **goto** step 2 (that is, break the *for* loop of step 3)
    end

Note that using not $j_m - \delta$ but $\min\{j_m - \delta, i_{m+1} - 1\}$ in step 3, we can reduce the running time of the refined algorithm, which will be given later. If we use $j_m - \delta$ instead of $\min\{j_m - \delta, i_{m+1} - 1\}$ in step 3, then we can prove the correctness of the refined algorithm in the same way as proving Algorithm 1. Let us prove that the refined algorithm is still correct, if $\min\{j_m - \delta, i_{m+1} - 1\}$ is used in step 3. Consider when $m = m'$ in the *for* loop of step 2. If $j_{m'} - \delta < i_{m'+1}$, then there is no problem. Assume $j_{m'} - \delta \geq i_{m'+1}$. Furthermore, assume that there is an integer i' with $i_{m'+1} \leq i' \leq j_{m'} - \delta$ such that $J_{\max}(\delta, p[i', j_{m'}], S - T) < j_{m'} - \delta$. Then $p_{i'}$ should be defined as $p[i', i' + J_{\max} + 1]$ (but is not defined when $m = m'$). Let k be the largest integer such that $i_{m'+k} \leq i' \leq j_{m'} - \delta$. Clearly, it holds that $J_{\max}(\delta, p[i', j_{m'+k}], S - T) < j_{m'+k} - \delta$. By the maximality of k, $p_{i'}$ is defined as $p[i', i' + J_{\max} + 1]$ in step 6 when $m = m' + k$ and $i = i'$. Hence, the refined algorithm is correct.

We estimate the running time of the refined algorithm. Note that steps 4 to 7 are executed at most $|p|$ times as a whole, since (1) $i_1 < i_2 < \cdots < i_n \leq |p|$ and (2) when $m = m'$ in the *for* loop of step 2, the possible values of i in the *for* loop of step 3 must be in the interval $[i_{m'}, i_{m'+1} - 1]$ (except the last interval $[i_n, j_n]$). Since step 4 requires $O(|p| \cdot |S - T| + \delta \cdot \|S - T\|)$ time, steps 2 to 7 can be executed in $O(|p|^2 \cdot |S - T| + |p| \cdot \delta \cdot \|S - T\|)$ time as a whole. Before executing steps 2 to 7, we must compute the intervals $[i_1, j_1], [i_2, j_2], \ldots, [i_n, j_n]$. However, we can do in $O(\|T\|)$ time using a *suffix tree* of T, as explained below. Therefore, the refined algorithm can be executed in $O(\|T\| + |p|^2 \cdot |S - T| + |p| \cdot \delta \cdot \|S - T\|)$ time.

Originally, a suffix tree is defined for a single string [15]. But it is easy to extend the definition to a set of strings. Let us briefly explain a suffix tree of T. For the target set $T = \{p_1, p_2, \ldots, p_k\}$, we denote $T\$ = \{p_1\$_1, p_2\$_2, \ldots, p_k\$_k\}$, where $\$_1, \$_2, \ldots, \$_k$ are distinct symbols not in the alphabet Σ. Intuitively, a suffix tree of T is a rooted tree such that each vertex represents a substring of a string in $T\$$ and that each internal vertex has two or more children. There is a one-to-one correspondence between the set of leaf vertices and the set of suffixes of strings in $T\$$. Furthermore, each internal vertex v represents the longest common prefix of the suffixes represented by the leaf vertices that are descendants of v. Thus a maximal common substring of T should be represented by a *lowest* internal vertex that has a descendant, for every h with $1 \leq h \leq k$, that is a leaf vertex representing a suffix of $p_h \$_h$. We can find all such lowest internal vertices in time proportional to the size of the suffix tree. Using a linear-time algorithm for constructing a suffix tree of a single string [15], we can construct a suffix tree of T in $O(\|T\|)$ time [3]. Hence we can also find the intervals $[i_1, j_1]$, $[i_2, j_2]$, $\ldots, [i_n, j_n]$ in that time. Consequently, we have the following theorem.

Theorem 1. *Let δ be a positive integer, let S be a finite set of strings, and let T be a target set in S. It can be decided in $O(\|T\| + l^2 \cdot |S - T| + l \cdot \delta \cdot \|S - T\|)$ time whether or not there exists a δ-characteristic string of T under S, where l denotes the length of a shortest string in T. Furthermore, if such a string exists, then all the shortest characteristic strings can also be obtained in that time.* □

5 Conclusions

We presented the following result. Given a positive integer δ, a finite set S of strings, and a target set T in S, it can be decided in $O(\|T\| + l^2 \cdot |S - T| + l \cdot \delta \cdot \|S - T\|)$ time whether or not there exists a characteristic string. Furthermore, if such a string exists, then all the shortest characteristic strings can be obtained in that time. Note that when our result applies to finding DNA probes, it may be reasonable to assume that $l = O(|q|)$ for any $q \in S$, since every string in S represents a nucleotide sequence. Under the assumption, we can find DNA probes (if exists) in $O(\|T\| + \delta \cdot l \cdot \|S - T\|)$ time using our algorithm. If $\delta = 1$, then we obtained a linear-time algorithm for solving the above problem [3], which was not discussed here because of space limitation.

In this paper, we assumed that a characteristic string must be a *common substring* of the target set T. However, when designing a DNA probe, we cannot always find a DNA probe as a complementary sequence of a common subsequence of the target bacteria sequences. In such a case, we must find a DNA probe under the condition that the DNA probe may have a *mismatch*, to a certain extent, with the target bacteria sequences. In other words, if δ' denotes the maximum possible number of mismatches between the DNA probe and the target bacteria sequences for hybridization, then the condition that a characteristic string is a common substring of T should be relaxed to the condition that a characteristic string has at most δ'-differences from any substring of any string in T. It is a future problem to compute a characteristic string under the extension. (In this paper, we considered the case of $\delta' = 0$.)

Our algorithm can also be used for designing a single primer for Polymerase Chain Reaction (PCR). We have already implemented our algorithm on a workstation and are now estimating how it is useful in practice.

References

1. R.S.Boyer and J.S.Moore: "A fast string searching algorithm," Comm. ACM, 20, 10, pp.762–772 (Oct. 1977).
2. R.Dular, R.Kajioka and S.Kasatiya: "Comparison of Gene-Probe Commercial Kit and Culture Technique for the Diagnosis of Mycoplasma pneumoniae Infection," Journal of Clinical Microbiology, 26, 5, pp.1068-1069 (May 1988).
3. M.Hasidume, M.Ito, M.Nakanishi and A.Hashimoto: "A linear-time algorithm for computing a shortest characteristic substring of strings"(in Japanese), IEICE Technical Report, COMP93-36, pp.39–46 (July 1993).
4. D.G.Higgins and P.M.Sharp: "Fast and sensitive multiple sequence alignments on a microcomputer," CABIOS, 5, 2, pp.151–153 (Apr. 1989).
5. D.E.Knuth, J.H.Morris and V.R.Pratt: "Fast pattern matching in strings," SIAM Journal on Computing, 6, 2, pp.323–350 (June 1977).
6. G.M.Landau and U.Vishkin: "Introducing efficient parallelism into approximate string matching and a new serial algorithm," Proc. 18th ACM Symp. on Theory of Computing, pp.220–230 (May 1986).
7. G.M.Landau and U.Vishkin: "Fast parallel and serial approximate string matching," Journal of Algorithms, 10, pp.157–169 (June 1989).
8. A.J.L.Macario and E.C.de Macario: "Gene Probes for Bacteria," Academic Press (1990)
9. M.Nasu, K.Shimada, S.Inaoka, K.Tani and M.Kondo: " Natural bacterial populations in river water determined by 16S and 23S rRNA-targeted oligonucleotide probes," (submitted to Biomedical and Environmental Sciences).
10. W.R.Pearson and D.J.Lipman: "Improved tools for biological sequence comparison," Proc. Natl. Acad. Sci. USA, 85, pp.2444–2448 (Apr. 1988).
11. P.H.Sellers: "The theory and computation of evolutionary distances: Pattern recognition," Journal of Algorithms, 1, pp.359–373 (Dec. 1980).
12. E.Ukkonen: "On approximate string matching," Lecture Notes in Computer Science, 158, pp.487–495 (Aug. 1983).
13. E.Ukkonen: "Finding approximate patterns in strings," Journal of Algorithms, 6, 1, pp.132–137 (Mar. 1985).
14. E.Ukkonen: "Algorithms for approximate string matching," Information and Control, 64, pp.100–118 (Mar. 1985).
15. P.Weiner: "Linear pattern matching algorithms," Proc. IEEE 14th Symposium on Switching and Automata Theory, pp.1-11 (1973)
16. "Genome Databases," Science, 254 (Oct. 1991).

Recent Methods for RNA Modeling
Using Stochastic Context-Free Grammars

Yasubumi Sakakibara[1] *, Michael Brown[1], Richard Hughey[2], I. Saira Mian[3],
Kimmen Sjölander[1], Rebecca C. Underwood[1], David Haussler[1]

[1] Computer and Information Sciences
[2] Computer Engineering
[3] Sinsheimer Laboratories
University of California, Santa Cruz, CA 95064, USA
Email: haussler@cse.ucsc.edu

Abstract. Stochastic context-free grammars (SCFGs) can be applied
to the problems of folding, aligning and modeling families of homologous
RNA sequences. SCFGs capture the sequences' common primary and
secondary structure and generalize the hidden Markov models (HMMs)
used in related work on protein and DNA. This paper discusses our new
algorithm, Tree-Grammar EM, for deducing SCFG parameters automatically from unaligned, unfolded training sequences. Tree-Grammar EM,
a generalization of the HMM forward-backward algorithm, is based on
tree grammars and is faster than the previously proposed inside-outside
SCFG training algorithm. Independently, Sean Eddy and Richard Durbin
have introduced a trainable "covariance model" (CM) to perform similar
tasks. We compare and contrast our methods with theirs.

Tools for analyzing RNA will become increasingly important as *in vitro* evolution and selection techniques produce greater numbers of synthesized RNA
families to supplement those related by phylogeny. Recent efforts have applied
stochastic context-free grammars (SCFGs) to the problems of statistical modeling, multiple alignment, discrimination and prediction of the secondary structure of RNA families. Our approach in applying SCFGs to modeling RNA is
highly related to our work on modeling protein families and domains with HMMs
[HKMS93, KBM+94].

In RNA, the nucleotides adenine (A), cytosine (C), guanine (G) and uracil (U)
interact to form characteristic secondary-structure motifs such as helices, loops
and bulges [Sae84, WPT89]. Intramolecular A-U and G-C Watson-Crick pairs as
well as G-U and, more rarely, G-A base pairs constitute the so-called *biological
palindromes* in the genome. When RNA sequences are aligned, both primary

* Y. Sakakibara's current address is ISIS, Fujitsu Labs Ltd., 140, Miyamoto, Numazu,
  Shizuoka 410-03, Japan. We thank Anders Krogh, Harry Noller and Bryn Weiser
  for discussions and assistance, and Michael Waterman and David Searls for discus-
  sions. This work was supported by NSF grants CDA-9115268 and IRI-9123692 and
  NIH grant number GM17129. This material is based upon work supported under a
  National Science Foundation Graduate Research Fellowship.

and secondary structure need to be considered since generation of a multiple alignment and analysis of folding are mutually dependent. (A multiple alignment of RNA sequences is a list of the sequences with the letters representing nucleotides spaced such that nucleotides considered functionally equivalent have their letters appearing in the same column in the list. To enhance the alignment of some sequences with respect to others, spaces may need to be inserted in them.) Elucidation of common folding patterns among multiple sequences may indicate the pertinent regions to be aligned and vice versa [San85].

Two principal methods have been established for predicting RNA secondary structure (i.e., which nucleotides are base-paired). The first technique, phylogenetic analysis of homologous RNA molecules [FW75, WGGN83], ascertains structural features that are conserved during evolution. The second technique employs thermodynamics to compare the free energy changes predicted for formation of possible secondary structure and relies on finding the structure with the lowest free energy [TUL71, TSF88, Gou87]. Though in principle HMMs could also be used to model RNA, the standard HMM approach treats all positions as having independent distributions and is unable to model the interactions between positions. However, if two positions in an alignment are base-paired, then the bases at these positions will be highly correlated. Since base-pairing interactions play such a dominant role in determining RNA structure and function, any statistical method for modeling RNA that does not consider these interactions will encounter insurmountable problems.

In this work, we use formal language theory to describe a means to generalize HMMs to model most RNA interactions. We compare our method to Eddy and Durbin's use of "covariance models" (CMs) to model RNA [ED94]. CMs are equivalent to SCFGs, but Eddy and Durbin employ different algorithms for training and producing multiple alignments.

As in the elegant work of Searls [Sea92], we view the character strings representing DNA, RNA and protein as sentences derived from a formal grammar. In the simplest kind of grammar, a *regular* grammar, strings are derived from productions (rewriting rules) of the forms $S \rightarrow aS$ and $S \rightarrow a$, where S is a *nonterminal symbol*, which does not appear in the final string, and a is a *terminal symbol*, which appears as a letter in the final string. Searls has shown that base pairing in RNA can be described by a *context-free grammar* (CFG). CFGs are often used to define the syntax of programming languages. A CFG is more powerful than a regular grammar in that it permits a greater variety of productions, such as those of the forms $S \rightarrow SS$ and $S \rightarrow aSa$. As described by Searls, these additional types of productions are crucial to model the base-pairing structure in RNA. (CFGs can not describe all RNA structure, but we believe they can account for enough to make useful models.) Productions of the forms $S \rightarrow A\,S\,U$, $S \rightarrow U\,S\,A$, $S \rightarrow G\,S\,C$ and $S \rightarrow C\,S\,G$ describe the structure in RNA due to Watson-Crick base pairing. Using productions of this type, a CFG can specify the language of biological palindromes.

Searls' original work [Sea92] argues the benefits of using CFGs as models for RNA folding, but does not discuss stochastic grammars or methods for creating the grammar from training sequences. Recent work provides an effective method for building a stochastic context-free grammar (SCFG) to model a family of RNA sequences. Some analogs of stochastic grammars and training methods do appear in Searls' most recent work in the form of costs and other trainable parameters used during parsing [Sea93a, Sea93b, SD93], but we believe that our integrated probabilistic framework may prove to be a simpler and more effective approach.

We have designed an algorithm that deduces grammar parameters automatically from a set of unaligned primary sequences. It is a novel generalization of the *forward-backward algorithm* commonly used to train HMMs. Our algorithm, Tree-Grammar EM, is based on tree grammars [TW68] and is more efficient than the *inside-outside algorithm* [LY90], a computationally expensive generalization of the forward-backward algorithm developed to train SCFGs [Bak79]. Full details are described elsewhere [SBH+93]; here we present a summary.

1 Stochastic Context-Free Grammar Methods

Specifying a probability for each production in a grammar yields a *stochastic grammar*. A stochastic grammar assigns a probability to each string it derives. Stochastic regular grammars are equivalent to HMMs and suggest an interesting generalization from HMMs to SCFGs [Bak79]. In this work, we explore stochastic models for tRNA sequences using a stochastic context-free grammar that is similar to our HMMs [KBM+94] but incorporates base-pairing information.

1.1 Context-free grammars for RNA

A grammar is principally a set of productions (rewrite rules) that is used to generate a set of strings, a *language*. The productions are applied iteratively to generate a string in a process called *derivation*. For example, application of the productions in Figure 1 could generate the RNA sequence CAUCAGGGAAGAUCUCUUG by the following derivation:

Beginning with the start symbol S_0, any production with S_0 left of the arrow can be chosen to have its right side replace S_0. If the production $S_0 \rightarrow S_1$ is selected (in this case, this is the only production available), then the symbol S_1 replaces S_0. This derivation step is written $S_0 \Rightarrow S_1$, where the double arrow signifies application of a production. Next, if the production $S_1 \rightarrow$ C S_2 G is selected, the derivation step is $S_1 \Rightarrow$ C S_2 G. Continuing with similar derivation steps, each time choosing a nonterminal symbol and replacing it with the right-hand side of an appropriate production, we obtain the following derivation terminating with the desired sequence:

$$S_0 \Rightarrow S_1 \Rightarrow CS_2G \Rightarrow CAS_3UG \Rightarrow CAS_4S_9UG$$
$$\Rightarrow CAUS_5AS_9UG \Rightarrow CAUCS_6GAS_9UG$$
$$\Rightarrow CAUCAS_7GAS_9UG \Rightarrow CAUCAGS_8GAS_9UG$$
$$\Rightarrow CAUCAGGGAS_9UG \Rightarrow CAUCAGGGAAS_{10}UUG$$
$$\Rightarrow CAUCAGGGAAGS_{11}CUUG$$
$$\Rightarrow CAUCAGGGAAGAS_{12}UCUUG$$
$$\Rightarrow CAUCAGGGAAGAUS_{13}UCUUG$$
$$\Rightarrow CAUCAGGGAAGAUCUCUUG.$$

$$P = \{ \begin{array}{ll} S_0 \rightarrow S_1, & S_7 \rightarrow G\,S_8, \\ S_1 \rightarrow C\,S_2\,G, & S_8 \rightarrow G, \\ S_1 \rightarrow A\,S_2\,U, & S_8 \rightarrow U, \\ S_2 \rightarrow A\,S_3\,U, & S_9 \rightarrow A\,S_{10}\,U, \\ S_3 \rightarrow S_4\,S_9, & S_{10} \rightarrow C\,S_{10}\,G, \\ S_4 \rightarrow U\,S_5\,A, & S_{10} \rightarrow G\,S_{11}\,C, \\ S_5 \rightarrow C\,S_6\,G, & S_{11} \rightarrow A\,S_{12}\,U, \\ S_6 \rightarrow A\,S_7, & S_{12} \rightarrow U\,S_{13}, \\ S_7 \rightarrow U\,S_7, & S_{13} \rightarrow C \end{array} \}$$

Fig. 1. This set of productions P generates RNA sequences with a certain restricted structure. $S_0, S_1, \ldots, S_{13}$ are nonterminals; A, U, G and C are terminals representing the four nucleotides.

A derivation can be arranged in a tree structure called a *parse tree* (Figure 2). A parse tree represents the syntactic structure of a sequence produced by a grammar. For an RNA sequence, this syntactic structure corresponds to the physical secondary structure (Figure 3).

Formally, a context-free grammar G consists of a set of nonterminal symbols N, an alphabet of terminal symbols Σ, a set of productions P, and the start symbol S_0. For a nonempty set of symbols X, let X^* denote the set of all finite strings of symbols in X. Every CFG production has the form $S \rightarrow \alpha$ where $S \in N$ and $\alpha \in (N \cup \Sigma)^*$, thus the left-hand side consists of a single nonterminal while there is no restriction on the number or placement of nonterminals and terminals on the right-hand side. The production $S \rightarrow \alpha$ means that the nonterminal S can be replaced by the string α. If $S \rightarrow \alpha$ is a production in P, then for any strings γ and δ in $(N \cup \Sigma)^*$, we define $\gamma S\delta \Rightarrow \gamma\alpha\delta$ and we say that $\gamma S\delta$ *directly derives* $\gamma\alpha\delta$ in G. We say the string β can be *derived* from α, denoted $\alpha \overset{*}{\Rightarrow} \beta$, if there exists a sequence of direct derivations $\alpha_0 \Rightarrow \alpha_1, \alpha_1 \Rightarrow \alpha_2, \ldots, \alpha_{n-1} \Rightarrow \alpha_n$ such that $\alpha_0 = \alpha$, $\alpha_n = \beta$, $\alpha_i \in (N \cup \Sigma)^*$, and $n \geq 0$. Such a sequence is called a *derivation*. Thus, a derivation is an order of productions applied to generate a string. The grammar generates the language $\{w \in \Sigma^* \mid S_0 \overset{*}{\Rightarrow} w\}$, the set of all terminal strings w that can be derived from the grammar.

Our work in modeling RNA uses productions of the following forms: $S \rightarrow SS$, $S \rightarrow aSa$, $S \rightarrow aS$, $S \rightarrow S$ and $S \rightarrow a$, where S is a nonterminal and a is a terminal. $S \rightarrow aSa$ productions describe the base pairings in RNA; $S \rightarrow aS$ and $S \rightarrow a$ describe unpaired bases; $S \rightarrow SS$ describe branched secondary structures and $S \rightarrow S$ are used in the context of multiple alignments.

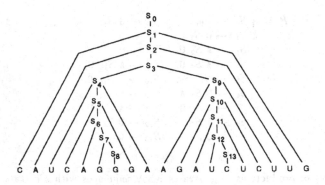

Fig. 2. For the RNA sequence CAUCAGGGAAGAUCUCUUG, the grammar whose productions are given in Figure 1 yields this parse tree which reflects a specific secondary structure (see Figure 3).

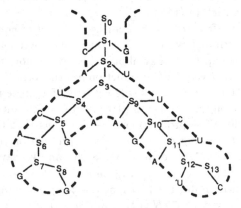

Fig. 3. For the RNA sequence CAUCAGGGAAGAUCUCUUG, the grammar whose productions are given in Figure 1 yields a parse tree (see Figure 2) which reflects this specific secondary structure.

SCFGs are a generalization of HMMs. The three main types of nonterminals in an SCFG correspond to each of the primary states in an HMM: *match,
insert* and *skip* [HKMS93, KBM+94]. The match nonterminals in a grammar
correspond to important structural positions in an RNA molecule. Insert nonterminals also generate nucleotides but with different distributions. These are
used to model loops by inserting nucleotides between important (match) positions. Productions of the form $S \rightarrow S$ are called *skip productions* because they
are used to skip a match nonterminal, so that no nucleotide appears at that position in a multiple alignment. In an SCFG, use of a skip production in parsing
a sequence is equivalent to choice of a delete state in aligning a sequence to an
HMM.

1.2 Stochastic context-free grammars

In an SCFG, every production for a nonterminal S has an associated probability
value such that a probability distribution exists over the set of productions
for S. (Any production with the nonterminal S on the left side is called "a
production for S.") We denote the associated probability for a production $S \rightarrow \alpha$
by $\mathcal{P}(S \rightarrow \alpha)$.

A stochastic context-free grammar G generates sequences and assigns a probability to each generated sequence, thereby defining a probability distribution
on the set of sequences. The probability of a derivation (parse tree) can be calculated as the product of the probabilities of the production instances applied
to produce the derivation. The probability of a sequence s is the sum of probabilities over all possible derivations that G could use to generate s, written as
follows:

$$\text{Prob}(s \mid G) = \sum_{\substack{\text{all derivations} \\ \text{(or parse trees) } d}} \text{Prob}(S_0 \overset{d}{\Rightarrow} s \mid G)$$

$$= \sum_{\alpha_1, \dots, \alpha_n} \Big[\text{Prob}(S_0 \Rightarrow \alpha_1 \mid G) \cdot \text{Prob}(\alpha_1 \Rightarrow \alpha_2 \mid G) \cdot \dots$$

$$\cdot \text{Prob}(\alpha_n \Rightarrow s \mid G) \Big]$$

where terms $\alpha_i \in (N \cup \Sigma)^*$.

Efficiently computing $\text{Prob}(s \mid G)$ presents a problem because the number of possible parse trees for s is exponential in the length of the sequence.
However, a dynamic programming technique analogous to the Cocke-Younger-
Kasami (CYK) or Early parsing methods [AU72] for non-stochastic CFGs can
complete this task in polynomial time (proportional to the cube of the length of
sequence s). We define the negative logarithm of the probability of a sequence
given by the grammar G, $-\log(\text{Prob}(s \mid G))$, as the *negative log likelihood (NLL)*

score of the sequence. The NLL score quantifies how well the sequence *s* fits the grammar—the likelihood that the grammar with its production probabilities would produce *s*.

CFGs have a drawback in that a sequence can sometimes be derived by a CFG in multiple ways. Since alternative parse trees reflect alternative secondary structures (foldings), a grammar may give several possible secondary structures for a single RNA sequence. An SCFG has the advantage that it can provide the most likely parse tree from this set of possibilities. If the productions are chosen carefully and the probabilities are estimated accurately, a parsing algorithm, when given grammar *G* and an RNA sequence *s*, will produce a most likely parse tree for *s* that corresponds to the correct secondary structure for *s*. Indeed, for most of the tRNA sequences we test, the most likely parse trees given by the tRNA-trained grammar match precisely the accepted secondary structures.

We can compute the most likely parse tree efficiently using a variant of the above procedure for calculating $\text{Prob}(s \mid G)$. To obtain the most likely parse tree for the sequence *s*, we calculate

$$\max_{\text{parse trees } d} \text{Prob}(S_0 \overset{d}{\Rightarrow} s \mid G).$$

The dynamic-programming procedure to do this resembles the Viterbi algorithm for HMMs [Rab89]. We also use this procedure to obtain multiple alignments: the parser aligns each sequence by finding the most likely parse tree given by the grammar, yielding an alignment of all nucleotides that correspond to the match nonterminals for each sequence, after which the mutual alignment of the sequences among themselves is determined. (Insertions of varying lengths can exist between match nonterminals, but by inserting enough spaces in each sequence to accommodate the longest insertion, an alignment of all the sequences is obtained.) This is equivalent to multiple alignment in an HMM, where the single most likely path for each sequence is computed.

1.3 Estimating SCFGs from sequences

Both an SCFG's production probabilities and the productions themselves can in principle be chosen using an existing alignment of RNA sequences. Results using this approach were reported in our previous work [SBU+93]. Also, Eddy and Durbin report recent results in which nearly all aspects of the grammar are determined solely from the training sequences [ED94]. In contrast, we make more use of prior information about the structure of tRNA to design an appropriate initial grammar, and then use training sequences only to refine our estimates of the probabilities of the productions used in this grammar.

1.4 The Tree-Grammar EM training algorithm

To estimate the SCFG parameters from unaligned training tRNA sequences, we introduce Tree-Grammar EM (Figure 4), a new method for training SCFGs that uses a generalization of the forward-backward algorithm commonly used to train

HMMs. This generalization, called TG Reestimator, is more efficient than the inside-outside algorithm, which was previously proposed to train SCFGs.

1. Start with an initial grammar G_0.
2. Use grammar G_0 and the CYK-like parsing algorithm to parse the raw input sequences, producing a tree representation for each sequence indicating which nucleotides are base-paired. This set of initial trees is denoted T_0. Set $T_{old} = \emptyset$ and $T_{new} = T_0$.
3. While $T_{new} \neq T_{old}$ do the following: {
   3a. Set $T_{old} = T_{new}$.
   3b. Use T_{old} trees as input to our TG Reestimator algorithm, which iteratively re-estimates the grammar parameters until they stabilize. The grammar with the final stabilized probability values is called new grammar G_{new}.
   3c. Use grammar G_{new} and the CYK-like parsing algorithm to parse the input sequences, producing a new set of trees T_{new}.
}

Fig. 4. Pseudocode for the Tree-Grammar EM training algorithm.

The inside-outside algorithm [LY90, Bak79] is an *expectation maximization* (EM) algorithm that calculates maximum likelihood estimates of an SCFG's parameters based on training data. However, it requires the grammar to be in Chomsky normal form, which is possible but inconvenient for modeling RNA (and requires more nonterminals). Further, it takes time at least proportional to $|s|^3$ per training sequence, whereas the forward-backward procedure for HMMs takes time proportional to $|s|$ per training sequence s, where $|s|$ is the length of s. In addition, the inside-outside algorithm is prone to settling in local minima; this presents a problem when the initial grammar is not highly constrained.

To avoid these problems, we developed an iterative estimator algorithm called *TG Reestimator* (Step 3b in Figure 4). While the running times of both Tree-Grammar EM and the inside-outside algorithm are asymptotically equivalent due to the use of the CYK-like algorithm (Step 3c of Figure 4), in practice Tree-Grammar EM is much more efficient. In Tree-Grammar EM, the inner loop (Step 3b) takes time proportional to $|s|$ per sequence per iteration (the grammar size is constant); in the inside-outside algorithm, the analogous step takes time proportional to $|s|^3$ per sequence per iteration. Since the number of iterations in Step 3b is typically on the order of 100, while the number of iterations of Step 3 is typically two or three, Tree-Grammar EM is more practical for longer RNA sequences.

TG Reestimator requires folded RNA sequences as training examples, rather than unfolded ones. Thus, some tentative "base pairs" in each training sequence have to be identified before TG Reestimator can begin. To do this, we design a rough initial grammar (see Section 1.6) that may represent only a portion of the base-pairing interactions (Step 1) and parse the unfolded RNA training sequences to obtain a set of partially folded RNA sequences (Step 2). Then we estimate a new SCFG using the partially folded sequences and TG Reestimator (Step 3b). Further productions might be added to the grammar at this step, though we have not yet experimented with this. The parameter re-estimation is then repeated. In this way, TG Reestimator can be used even when precise biological knowledge of the base pairing is not available. TG Reestimator constitutes one part of the entire training procedure, Tree-Grammar EM.

The Tree-Grammar EM procedure is based on the theory of stochastic tree grammars [TW68, Fu82]. Tree grammars are used to derive labeled trees instead of strings. Labeled trees can be used to represent the secondary structure of RNA easily [SZ90] (see Figure 2). A tree grammar for RNA denotes both the primary sequence and the secondary structure of each molecule. Since these are given explicitly in each training molecule, the TG Reestimator algorithm does not have to (implicitly) sum over *all* possible interpretations of the secondary structure of the training examples when re-estimating the grammar parameters, as the inside-outside method must do. The TG Reestimator algorithm iteratively finds the best parse for each molecule in the training set and then readjusts the production probabilities to maximize the probability of these parses. The new algorithm also tends to converge faster because each training example is much more informative [SBU+93].

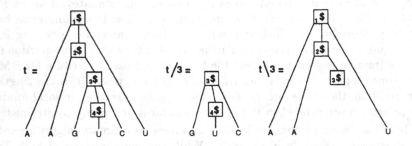

Fig. 5. The folded RNA sequence (AA(GUC)U) can be represented as a tree t (left), which can be broken into two parts such as $t/3$ (middle) and $t \backslash 3$ (right). The root, $_1$\$, and the internal node, $_3$\$, represent A-U and G-C base pairs, respectively.

To avoid unnecessary complexity, we describe this new algorithm in terms of CFGs instead of tree grammars [TW68, Sak92]. A tree is a rooted, directed, connected acyclic finite graph in which the direct successors of any node are linearly ordered from left to right. The predecessor of a node is called the *parent*; the successor, a *child*; and a child of the parent, a *sibling*. A folded RNA sequence can be represented by a labeled tree t as follows. Each leaf node is labeled by one of four nucleotides {A, U, G, C} and all internal nodes are labeled by one special symbol, say \$. The sequence of nucleotides labeled at leaf nodes traced from left to right exactly constitutes the RNA sequence, and the structure of the tree represents its folding structure. See Figure 5 for an example of a tree representation of the folded RNA sequence (AA(GUC)U). We assume all internal nodes in t are numbered from 1 to T (the number of internal nodes) in some order. For an internal node n ($1 \leq n \leq T$), let t/n denote the subtree of t with root n (Figure 5, center) and let $t \backslash n$ denote the tree obtained by removing a subtree t/n from t (Figure 5, right).

The probability of any folded sequence t given by an SCFG $G = (N, \Sigma, P, S_0)$ is calculated efficiently using a dynamic programming technique (as is done with the forward algorithm in HMMs). A labeled tree t representing a folded RNA sequence has the shape of a parse tree, so to parse the folded RNA, the grammar G needs only to assign nonterminals to each internal node according to the productions. Let $in_n(S)$ be the probability of the subtree t/n given that the nonterminal S is assigned to node n and given grammar G, for all nonterminals S and all nodes n such that $1 \leq n \leq T$. We can calculate $in_n(S)$ inductively as follows:

1. Initialization: $in_n(a) = 1$, for all leaf nodes n and all terminals a (all nucleotides). This extension of $in_n(S)$ is for the convenience of the inductive calculation of $in_n(S)$.

2. Induction:

$$in_m(S) = \sum_{\substack{Y_1, \ldots, Y_k \\ \in (N \cup \Sigma)}} in_{n_1}(Y_1) \cdots in_{n_k}(Y_k) \cdot P(S \rightarrow Y_1 \cdots Y_k),$$

for all nonterminals S, all internal nodes m and all m's children nodes $n_1, \ldots, n_k$.

3. Termination: For the root node n and the start symbol S_0,

$$\text{Prob}(t \mid G) = in_n(S_0). \tag{1}$$

We need one more quantity, $out_n(S)$, which defines the probability of $t \backslash n$ given that the nonterminal S is assigned to node n and given grammar G, which we obtain similarly.

1. Initialization: For the root node n,

$$out_n(S) = \begin{cases} 1 \text{ for } S = S_0 \text{ (start symbol)}, \\ \\ 0 \text{ otherwise.} \end{cases}$$

2. Induction:

$$out_m(S) = \sum_{\substack{Y_1, \ldots, Y_k \\ \in (N \cup \Sigma), \\ S' \in N}} in_{n_1}(Y_1) \cdots in_{n_k}(Y_k) \cdot P(S' \rightarrow Y_1 \cdots S \cdots Y_k) \cdot out_l(S'),$$

for all nonterminals S, all internal nodes l and m such that l is m's parent and all nodes $n_1, \ldots, n_k$ are m's siblings. (There is no termination step given in this case because the calculation of $\text{Prob}(t \mid G)$ is given in the termination step for $in_n(S)$.)

Given a set of folded training sequences $t(1), \ldots, t(n)$, we can determine how well a grammar fits the sequences by calculating the probability that the grammar generates them. This probability is simply a product of terms of the form given by (1), i.e.,

$$\text{Prob(sequences} \mid G) = \prod_{j=1}^{n} \text{Prob}(t(j) \mid G), \tag{2}$$

where each term $\text{Prob}(t(j) \mid G)$ is calculated as in Equation (1). The goal is to obtain a high value for this probability, called the *likelihood* of the grammar. The *maximum likelihood* (ML) method of model estimation finds the model that maximizes the likelihood (2). There is no known way to directly and efficiently calculate the best model (the one that maximizes the likelihood) and avoid getting caught in suboptimal solutions during the search. However, the general EM method, given an arbitrary starting point, finds a local maximum by iteratively re-estimating the model such that the likelihood increases in each iteration, and often produces a solution that is acceptable, if perhaps not optimal. This method is often used in statistics.

Thus in Step 1 of Tree-Grammar EM, an initial grammar is created by assigning values to the production probabilities $\mathcal{P}(S \rightarrow Y_1 \cdots Y_k)$ for all S and all $Y_1, \ldots, Y_k$, where S is a nonterminal and Y_i $(1 \leq i \leq k)$ is a nonterminal or terminal. If some constraints or features present in the sequences' actual (trusted) multiple alignment are known, these are encoded in the initial grammar. The current grammar is set to this initial grammar.

In Step 3b of Tree-Grammar EM, using the current grammar, the values $in_n(S)$ and $out_n(S)$ for each nonterminal S and each node n for each folded training sequence are calculated in order to get a new estimate of each production probability, $\hat{\mathcal{P}}(S \rightarrow Y_1 \cdots Y_k) =$

$$\frac{\displaystyle\sum_{\text{all } t} \left(\sum_{\text{all } m} out_m(S) \cdot \mathcal{P}(S \rightarrow Y_1 \cdots Y_k) \cdot in_{n_1}(Y_1) \cdots in_{n_k}(Y_k) / \text{Prob}(t \mid G) \right)}{\text{norm}},$$

which is a double sum over all sequences t and all nodes m, where G is the old grammar and "norm" is the appropriate normalizing constant such that $\sum_{Y_1, \ldots, Y_k} \hat{\mathcal{P}}(S \rightarrow Y_1 \cdots Y_k) = 1$. A new current grammar is created by replacing all $\mathcal{P}(S \rightarrow Y_1 \cdots Y_k)$ with the re-estimated probabilities $\hat{\mathcal{P}}(S \rightarrow Y_1 \cdots Y_k)$.

1.5 Overfitting and regularization

Attempts to estimate a grammar with too many free parameters from a small set of training sequences will encounter the *overfitting* problem—i.e., the grammar fits the training sequences well, but poorly fits other, related (test) sequences. One solution is to use *regularization* to control the effective number of free parameters. We regularize our grammars by taking a Bayesian approach to the parameter estimation problem, similar to our approach with protein HMMs [KBM+94, BHK+93b].

Before training the grammars, we construct a prior probability density for each of their "important" parameter sets. This prior density takes the form of a Dirichlet distribution [SD89]. The "important" productions are of two forms: $S \to aSb$ and $S \to aS$, where terminal symbols $a, b \in \{\text{A}, \text{C}, \text{G}, \text{U}\}$. $S \to aSb$ productions, which generate base pairs, come in groups of 16, corresponding to all possible pairs of terminal symbols. $S \to aS$ productions, which generate nucleotides in loop regions, come in groups of four. For the base-pairing productions, we employ prior information about which productions are most likely. For instance, Watson-Crick pairs are more frequently observed than other base pairs. To calculate precise prior information about base-pair probabilities, we obtain the 16 parameters of a Dirichlet density over possible base-paired position distributions from a large alignment of 16S rRNA sequences [LOM+93]. We similarly use the alignment to calculate a four-parameter Dirichlet prior for nucleotide distributions in loop region positions. (We present further details elsewhere [BHK+93b].) These parameters constitute our regularizer. We add them as "pseudocounts" during each re-estimation step of Tree-Grammar EM (step 3b in Figure 4). Thus, at each iteration, TG Reestimator computes mean posterior estimates of the model parameters rather than maximum likelihood estimates.

In a similar manner, we regularize probability distributions for other production types, including chain rules $S \to S$, branch productions $S \to SS$ and insert productions $S \to aS$. We regularize loops with very large uniform pseudocounts over the four possible nucleotides so that their probability distributions will be fixed at uniform values rather than estimated from the training data. This is equivalent to the regularization we used for the insert states of HMMs [KBM+94]. This further reduces the number of parameters to be estimated, helping to avoid overfitting.

1.6 The initial grammar

In the initial grammar, we model a loop that is typically l nucleotides in length by an HMM model with l match states as described in our previous protein work [KBM+94], except that the four-letter nucleic-acids alphabet replaces the twenty-letter amino-acids alphabet. Nucleotide distributions in such a loop are defined by probabilities of l match-nonterminals' productions. Longer or shorter loops can be derived using special nonterminals and productions that allow position-specific insertions and deletions.

A helix l base pairs in length consists of l nonterminals. Each nonterminal has 16 productions that derive possible base pairs for its position in the helix. Each nonterminal has its own probability distribution over these 16 productions. These distributions, like those for match nonterminals in loops, are initially defined using Dirichlet priors (Section 1.5). Other nonterminals and productions are added to allow deletions of base pairs, enabling helix length variations.

Special treatment of nonterminals involved in branch productions of the form $S \to SS$ can also be included. In particular, we specify that certain branch productions may also, with some probability, omit one of the nonterminals on the right-hand side. This allows the grammar to derive tRNAs that lack certain substructures (such as arms or loops). These probability values are initialized to default prior values and then re-estimated during training on actual sequences, as are all grammar parameters.

1.7 The Eddy and Durbin algorithms

Sean Eddy and Richard Durbin have described an algorithm for obtaining an SCFG's productions themselves, as well as their probabilities (see Figure 6). As discussed in their work [ED94], covariance models (CMs) are describable as SCFGs. To deduce a covariance model's structure (essentially, to choose an SCFG's productions), they use the standard Nussinov/Zuker dynamic-programming algorithm for RNA folding [NPGK78, Zuk89], with the difference that the cost function being optimized is a function of the "mutual information" of two columns in the input multiple alignment (based on Gutell *et al.*'s MIXY procedure [GPH+92]), rather than of the number of base pairs or the thermodynamic stacking energy.

Once a model structure exists, they train the model's parameters (production probabilities) using the Viterbi approximation of EM. This consists of iterating the following two-step procedure until the model parameters stabilize: First, they align each sequence to the model using the same alignment algorithm that we use, a Viterbi approximation to the inside-outside algorithm [LY90, Bak79]. Second, they apply the Viterbi approximation of EM to maximize a Bayesian posterior-probability estimate [ED94]. In this way, training consists of optimizing the alignment scores of the input training sequences. Eddy and Durbin use a variant of their alignment algorithm to perform database searches as well.

---

1. Start with an initial (possibly random) alignment denoted A_0. Set $A_{curr} = A_0$.

2. Use alignment A_{curr} and the Nussinov/Zuker-based RNA folding algorithm to build a covariance model. Estimate its parameters using the Viterbi approximation of EM. Denote this initial model C_0. Set $C_{old} = \emptyset$ and $C_{new} = C_0$.

3. While $C_{new} \neq C_{old}$ do the following: {

    3a. Set $C_{old} = C_{new}$.

    3b. Use alignment A_{curr} and the Nussinov/Zuker-based RNA folding algorithm to restructure the covariance model (choose new productions). Re-estimate the model's parameters using the iterative Viterbi approximation of EM, producing a new model C_{new}.

    3c. Use model C_{new} and the Viterbi inside-outside aligning algorithm to produce a new multiple alignment A_{new}. Set $A_{curr} = A_{new}$.

}

Fig. 6. Pseudocode for the Eddy and Durbin training algorithm.

2 Discussion

In a recent paper [SBH+93], we present our results in detail and compare these with the results of Sean Eddy and Richard Durbin's covariance models. Both methods have been used to produce models that perform three tasks: discriminate tRNA sequences from non-tRNA sequences, produce multiple alignments and ascertain the secondary structure of new sequences. The results show that all our grammars, except one trained on zero sequences, can perfectly discriminate nonmitochondrial tRNA sequences from non-tRNA sequences, and that our multiple alignments are nearly identical to the published tRNA alignments (full details are given elsewhere [SBH+93]). Similarly, Eddy and Durbin's covariance models performed very well in database searches and produced multiple alignments nearly identical to the published. Both methods achieve local optima, rather than global, but the resulting models seem adequate for tRNA. Both methods use an inside-outside-based algorithm as part of the training process: we use one for parsing training trees, while they use one for producing a multiple alignment.

It appears that our basic grammar training algorithm, which is quite different from theirs, may be somewhat faster. Further, our custom-designed grammars and greater emphasis on learned, as opposed to constructed, Bayesian prior probability densities [BHK+93a] may allow us to train with fewer training sequences. However, Eddy and Durbin have developed an exciting new technique to learn the structure of the grammar itself from unaligned training sequences, rather than just learn the probabilities of the productions and rely on prior information to specify the structure of the grammar (as we do). Also, they use a database searching algorithm while we do not.

The SCFG methods discussed in this paper represent a new direction in computational biosequence analysis. SCFGs provide a flexible and highly effective statistical method for solving a number of RNA sequence analysis problems including discrimination, multiple alignment and prediction of secondary structures. They may prove useful in maintaining, updating and revising existing multiple sequence alignments. In addition, a grammar itself may be a valuable tool for representing an RNA family or domain such as group I introns [MW90, MECS90], group II introns [MUO89], RNAse P RNA [BHJ+91, TE93], small nuclear RNAs [GP88] and 7S RNA (signal recognition particle RNA) [Zwi89].

The main difficulties in applying this work to other families of RNA will be the development of appropriate initial grammars and the computational cost of parsing longer sequences. The latter problem can only be solved by the development of fundamentally different parsing methods, perhaps relying more on branch-and-bound methods [LS94] or heuristics. It is currently not clear which approach will be best. The former problem might be solved by the development of effective methods for learning the grammar itself from training sequences. The work of Eddy and Durbin is an important step in this direction [ED94]. Their method relies on correlations between columns in a multiple alignment [GPH+92, Lap92, KB93, Wat89, WOW+90, San85, Wat88] to discover the essential base-pairing structure in an RNA family. Another approach would be to use a method like that proposed by Waterman [Wat89] to find helices in a rough initial multiple alignment, use these helices to design a simple initial grammar in a semi-automated fashion using our high-level RNA grammar specification language, then use the grammar to obtain a better multiple alignment, and iterate this process until a suitable result is obtained. We are currently exploring this approach.

Another important direction for further research is the development of stochastic grammars for tRNA and other RNA families that can be used to search databases for these structures at the DNA level. In order to do this, the grammar must be modified to allow for the possibility of introns in the sequence, and the parsing method must be modified so that it can search efficiently for RNAs that are embedded within larger sequences. Durbin and Eddy have done the latter modifications in their tRNA experiments and report good results in searching the GenBank structural RNA database and 2.2 Mb of *C. elegans* genomic sequence for tRNAs, even without using special intron models. In our earlier work [SBM$^+$94], we reported some very preliminary results on modifying tRNA grammars to accommodate introns. We are currently planning to do further work in this direction. We see no insurmountable obstacles in developing effective stochastic grammar-based search methods, but predict that the main practical problem will be dealing with the long computation time required by the present methods.

Finally, there is the question of what further generalizations of hidden Markov models, beyond SCFGs, might be useful. The key advantage of our method over the HMM method is that it allows us to explicitly deal with the secondary structure of the RNA sequence. By extending stochastic models of strings to stochastic models of trees, we can model the base-pairing interactions of the molecule, which determine its secondary structure. This progression is similar to the path taken by the late King Sun Fu and colleagues in their development of the field of syntactic pattern recognition [Fu82]. Modeling pseudoknots and higher-order structure would require still more general methods. One possibility would be to consider *stochastic graph grammars* (see the introductory survey by Engelfriet and Rozenberg [ER91]) in hopes of obtaining a more general model of the interactions present in the molecule beyond the primary structure. If a stochastic graph grammar framework could be developed that included both an efficient method of finding the most probable folding of the molecule given the grammar and an efficient EM method for estimating the grammar's parameters from folded examples, then extensions of our approach to more challenging problems, including RNA tertiary structure determination and protein folding, would be possible. This is perhaps the most interesting direction for future research suggested by our results.

References

[AU72]   A. V. Aho and J. D. Ullman. *The Theory of Parsing, Translation and Compiling, Vol. I: Parsing.* Prentice Hall, Englewood Cliffs, N.J., 1972.

[Bak79]   J. K. Baker. Trainable grammars for speech recognition. *Speech Communication Papers for the 97th Meeting of the Acoustical Society of America*, pages 547–550, 1979.

[BHJ⁺91]   J. W. Brown, E. S. Haas, B. D. James, D. A. Hunt, J. S. Liu, and N. R. Pace. Phylogenetic analysis and evolution of RNase P RNA in proteobacteria. *Journal of Bacteriology*, 173:3855–3863, 1991.

[BHK⁺93a]   M. P. Brown, R. Hughey, A. Krogh, I. S. Mian, K. Sjölander, and D. Haussler. Dirichlet mixture priors for HMMs. In preparation, 1993.

[BHK⁺93b]   M. P. Brown, R. Hughey, A. Krogh, I. S. Mian, K. Sjölander, and D. Haussler. Using Dirichlet mixture priors to derive hidden Markov models for protein families. In L. Hunter, D. Searls, and J. Shavlik, editors, *Proc. of First Int. Conf. on Intelligent Systems for Molecular Biology*, pages 47–55, Menlo Park, CA, July 1993. AAAI/MIT Press.

[ED94]   S. R. Eddy and R. Durbin. RNA sequence analysis using covariance models. Submitted to *Nucleic Acids Research*, 1994.

[ER91]   J. Engelfriet and G. Rozenberg. Graph grammars based on node rewriting: An introduction to NLC graph grammars. In E. Ehrig, H.J. Kreowski, and G. Rozenberg, editors, *Lecture Notes in Computer Science*, volume 532, pages 12–23. Springer-Verlag, 1991.

[Fu82]   K. S. Fu. *Syntactic pattern recognition and applications*. Prentice-Hall, Englewood Cliffs, NJ, 1982.

[FW75]   G. E. Fox and C. R Woese. 5S RNA secondary structure. *Nature*, 256:505–507, 1975.

[Gou87]   M. Gouy. Secondary structure prediction of RNA. In M. J. Bishop and C. R. Rawlings, editors, *Nucleic acid and protein sequence analysis, a practical approach*, pages 259–284. IRL Press, Oxford, England, 1987.

[GP88]   C. Guthrie and B. Patterson. Spliceosomal snRNAs. *Annual Review of Genetics*, 22:387–419, 1988.

[GPH⁺92]   R. R. Gutell, A. Power, G. Z. Hertz, E. J. Putz, and G. D. Stormo. Identifying constraints on the higher-order structure of RNA: continued development and application of comparative sequence analysis methods. *Nucleic Acids Research*, 20:5785–5795, 1992.

[HKMS93]   D. Haussler, A. Krogh, I. S. Mian, and K. Sjölander. Protein modeling using hidden Markov models: Analysis of globins. In *Proceedings of the Hawaii International Conference on System Sciences*, volume 1, pages 792–802, Los Alamitos, CA, 1993. IEEE Computer Society Press.

[KB93]   T. Klinger and D. Brutlag. Detection of correlations in tRNA sequences with structural implications. In Lawrence Hunter, David Searls, and Jude Shavlik, editors, *First International Conference on Intelligent Systems for Molecular Biology*, Menlo Park, 1993. AAAI Press.

[KBM⁺94]   A. Krogh, M. Brown, I. S. Mian, K. Sjölander, and D. Haussler. Hidden Markov models in computational biology: Applications to protein modeling. *Journal of Molecular Biology*, 235:1501–1531, Feb. 1994.

[Lap92]   Allan Lapedes. Private communication, 1992.

[LOM⁺93]   N. Larsen, G. J. Olsen, B. L. Maidak, M. J. McCaughey, R. Overbeek, T. J. Macke, T. L. Marsh, and C. R. Woese. The ribosomal database project. *Nucleic Acids Research*, 21:3021–3023, 1993.

[LS94]   R. H. Lathrop and T. F. Smith. A branch-and-bound algorithm for optimal protein threading with pairwise (contact potential) amino acid interactions. In *Proceedings of the 27th Hawaii International Conference on System Sciences*, Los Alamitos, CA, 1994. IEEE Computer Society Press.

[LY90] K. Lari and S. J. Young. The estimation of stochastic context-free grammars using the inside-outside algorithm. *Computer Speech and Language*, 4:35–56, 1990.

[MECS90] F. Michel, A. D. Ellington, S. Couture, and J. W. Szostak. Phylogenetic and genetic evidence for base-triples in the catalytic domain of group I introns. *Nature*, 347:578–580, 1990.

[MUO89] F. Michel, K. Umesono, and H. Ozeki. Comparative and functional anatomy of group II catalytic introns–a review. *Gene*, 82:5–30, 1989.

[MW90] F. Michel and E. Westhof. Modelling of the three-dimensional architecture of group I catalytic introns based on comparative sequence analysis. *Journal of Molecular Biology*, 216:585–610, 1990.

[NPGK78] R. Nussinov, G. Pieczenik, J. R. Griggs, and D. J. Kleitman. Algorithms for loop matchings. *SIAM Journal of Applied Mathematics*, 35:68–82, 1978.

[Rab89] L. R. Rabiner. A tutorial on hidden Markov models and selected applications in speech recognition. *Proc IEEE*, 77(2):257–286, 1989.

[Sae84] W. Saenger. *Principles of nucleic acid structure*. Springer Advanced Texts in Chemistry. Springer-Verlag, New York, 1984.

[Sak92] Y. Sakakibara. Efficient learning of context-free grammars from positive structural examples. *Information and Computation*, 97:23–60, 1992.

[San85] D. Sankoff. Simultaneous solution of the RNA folding, alignment and protosequence problems. *SIAM J. Appl. Math.*, 45:810–825, 1985.

[SBH+93] Y. Sakakibara, M. Brown, R. Hughey, I. S. Mian, K. Sjölander, R. Underwood, and D. Haussler. The application of stochastic context-free grammars to folding, aligning and modeling homologous RNA sequences. Submitted for publication, 1993.

[SBM+94] Y. Sakakibara, M. Brown, I. S. Mian, R. Underwood, and D. Haussler. Stochastic context-free grammars for modeling RNA. In *Proceedings of the Hawaii International Conference on System Sciences*, Los Alamitos, CA, 1994. IEEE Computer Society Press.

[SBU+93] Y. Sakakibara, M. Brown, R. Underwood, I. S. Mian, and D. Haussler. Stochastic context-free grammars for modeling RNA. Technical Report UCSC-CRL-93-16, UC Santa Cruz, Computer and Information Sciences Dept., Santa Cruz, CA 95064, 1993.

[SD89] T. J. Santner and D. E. Duffy. *The Statistical Analysis of Discrete Data*. Springer Verlag, New York, 1989.

[SD93] D. B. Searls and S. Dong. A syntactic pattern recognition system for DNA sequences. In *Proc. 2nd Int. Conf. on Bioinformatics, Supercomputing and complex genome analysis*. World Scientific, 1993. In press.

[Sea92] David B. Searls. The linguistics of DNA. *American Scientist*, 80:579–591, November–December 1992.

[Sea93a] D. B. Searls. The computational linguistics of biological sequences. In *Artificial Intelligence and Molecular Biology*, chapter 2, pages 47–120. AAAI Press, 1993.

[Sea93b] D. B. Searls. String variable grammar: a logic grammar formalism for DNA sequences, 1993. Unpublished.

[SZ90] B. A. Shapiro and K. Zhang. Comparing multiple RNA secondary structures using tree comparisons. *CABIOS*, 6(4):309–318, 1990.

[TE93]   A. J. Tranguch and D. R. Engelke. Comparative structural analysis of nuclear RNase P RNAs from yeast. *Journal of Biological Chemistry*, 268:14045–1455, 1993.

[TSF88]   D. H. Turner, N. Sugimoto, and S. M. Freier. RNA structure prediction. *Annual Review of Biophysics and Biophysical Chemistry*, 17:167–192, 1988.

[TUL71]   I. Tinoco Jr., O. C. Uhlenbeck, and M. D. Levine. Estimation of secondary structure in ribonucleic acids. *Nature*, 230:363–367, 1971.

[TW68]   J. W. Thatcher and J. B. Wright. Generalized finite automata theory with an application to a decision problem of second-order logic. *Mathematical Systems Theory*, 2:57–81, 1968.

[Wat88]   M. S. Waterman. Computer analysis of nucleic acid sequences. *Methods in Enzymology*, 164:765–792, 1988.

[Wat89]   M. S. Waterman. Consensus methods for folding single-stranded nucleic acids. In M. S. Waterman, editor, *Mathematical Methods for DNA Sequences*, chapter 8. CRC Press, 1989.

[WGGN83]   C. R. Woese, R. R. Gutell, R. Gupta, and H. F. Noller. Detailed analysis of the higher-order structure of 16S-like ribosomal ribonucleic acids. *Microbiology Reviews*, 47(4):621–669, 1983.

[WOW⁺90]   S. Winker, R. Overbeek, C.R. Woese, G.J. Olsen, and N. Pfluger. Structure detection through automated covariance search. *Computer Applications in the Biosciences*, 6:365–371, 1990.

[WPT89]   J. R. Wyatt, J. D. Puglisi, and I. Tinoco Jr. RNA folding: pseudoknots, loops and bulges. *BioEssays*, 11(4):100–106, 1989.

[Zuk89]   M. Zuker. On finding all suboptimal foldings of an RNA molecule. *Science*, 244:48–52, 1989.

[Zwi89]   C. Zwieb. Structure and function of signal recognition particle RNA. *Progress in Nucleic Acid Research and Molecular Biology*, 37:207–234, 1989.

Efficient bounds for oriented chromosome inversion distance

John Kececioglu*   David Sankoff[†]

Abstract We study the problem of comparing two circular chromosomes that have evolved by chromosome inversion, assuming that the order of corresponding genes is known, as well as their orientation. Determining the minimum number of inversions is equivalent to finding the minimum of reversals to sort a signed circular permutation, where a reversal takes an arbitrary substring of elements and reverses their order, as well as flipping their sign. We show that tight bounds on the minimum number of reversals can be found by simple and efficient algorithms.

Keywords Genome rearrangements, chromosome inversion, reversal distance, sorting by signed reversals

1 Introduction

With the advent of genome sequencing in molecular biology, there is an increasing interest in the development of algorithms for comparing chromosomes in terms of high-level mutational events. In this paper we consider the comparison of two circular chromosomes from two related organisms on the basis of the order of their common genes, and in terms of chromosome inversions. An *inversion* replaces an arbitrary region of the chromosome with the reverse complement sequence. This has the effect of reversing the order of the genes within the region, as well as complementing the sequence for each gene.

We model this comparison problem as that of determining the minimum number of reversals to transform one signed circular permutation into another. The permutations represent the order of corresponding genes, and the sign of an element represents whether or not the sequence for the gene in one chromosome is reverse complemented in the other chromosome.

---

*Department of Computer Science, University of California, Davis, CA 95616, USA. Electronic mail: kece@cs.ucdavis.edu. Research supported by a U.S. Department of Energy Human Genome Distinguished Postdoctoral Fellowship.

†Centre de recherches mathématiques, Université de Montréal, C.P. 6128, succ. A, Montréal, Québec H3C 3J7, Canada. Electronic mail: sankoff@ere.umontreal.ca. Research supported by grants from the Natural Sciences and Engineering Research Council of Canada, and the Fonds pour la formation de chercheurs et l'aide à la recherche (Québec). David Sankoff is a fellow of the Canadian Institute for Advanced Research.

We write a permutation π as a list $(\pi_1\ \pi_2\ \cdots\ \pi_n)$ of elements, where $\pi_i = \pi(i)$. A *signed permutation* is just an ordinary permutation, except that elements may be positively or negatively signed. In a *circular permutation*, any cyclic shift of a permutation is considered to be equivalent. We adopt the convention that for a signed circular permutation, element 1 is always in the first position, and is positive. A *reversal* ρ of interval $[i, j]$ is the permutation

$$\rho = (1\ 2\ \cdots\ i{-}1\ -(j)\ -(j{-}1)\ \cdots\ -(i)\ j{+}1\ \cdots\ n).$$

Applying ρ to a permutation π by the composition $\pi \circ \rho$ has the effect of reversing the order of the elements in π in interval $[i, j]$, and flipping their sign.

The *reversal distance* between two n element permutations σ and τ, denoted by $d(\sigma, \tau)$, is the length ℓ of a shortest series of reversals $\rho_1, \rho_2, \ldots, \rho_\ell$ such that

$$\sigma \circ \rho_1 \circ \rho_2 \circ \cdots \circ \rho_\ell = \tau.$$

Since we can always take one permutation to be the *identity permutation* $\imath = (1\ 2\ \cdots\ n)$, the problem is equivalent to finding the minimum number of reversals to transform a permutation π into $\imath$. This formulation is called *sorting by reversals*. We denote the minimum number of reversals to sort π by $d(\pi)$.

Kececioglu and Sankoff [7, 8] studied the problem of sorting an unsigned linear permutation by reversals, and developed the first approximation algorithm for the problem. Their algorithm runs in $O(n^2)$ time and is guaranteed to use no more than twice the minimum number of reversals. They also developed a family of lower bounds on $d(\pi)$ in terms of the cycles of a graph whose evaluation required linear programming. Using these bounds in the context of a branch-and-bound procedure, they were able to solve to optimality permutations of up to 30 elements, and bound the exact value to within 2 reversals for permutations of up to 50 elements.

Bafna and Pevzer [1] subsequently found an improved approximation algorithm with a performance guarantee of $\frac{7}{4}$, and presented an approximation algorithm for signed reversals with a guarantee of $\frac{3}{2}$. They also resolved a conjecture due to Holger Gollan [7] that for every n there is an n-element permutation requiring $n - 1$ reversals.

This paper applies the techniques developed in [8] to the case of signed circular permutations. This is important for the study of organisms with circular chromosomes when the orientation of genes, as well as their order, can be determined. In this context a tremendous simplification occurs. The lower bound can be evaluated in linear time, without the need for linear programming, and it can be given a simple combinatorial proof. The greedy approximation algorithm is also shown to perform remarkably well in this context. We first review the greedy algorithm in Section 2, and then develop a simplified lower bound in Section 3. Section 4 develops a more parsimonious branch-and-bound algorithm. We also prove in this section one of the first theorems concerning an optimal solution, namely that there is always an optimal series in which a quantity called the number of breakpoints is non-increasing. In Section 5 we show that for every n there is a signed permutation requiring at least $n - 1$ reversals, which gives a tight bound on the diameter for the signed problem and proves that the greedy algorithm is essentially worst-case

309

optimal. Section 6 studies the empirical performance of these algorithms on random permutations, and permutations generated by random reversals. Suprisingly, the observed average difference between the greedy approximation and the lower bound in these experiments remained less than 1 reversal for signed permutations with up to 10,000 elements, and the maximum observed difference was less than 4 reversals. The tightness of these bounds allowed us to solve most problems on permutations of up to 250 elements to optimality in less than an hour of computation, and failing that, to determine the minimum to within 1 reversal.

Throughout the paper we use n to denote the number of elements in a permutation. Unless otherwise stated, the term "permutation" will refer to signed circular permutations.

2   Upper bounding the distance

In this section we review for completeness the greedy algorithm of Kececioglu and Sankoff [8] for sorting by reversals, and adapt it to the context of signed circular permutations.

The key idea is that of a *breakpoint*. A breakpoint of a permutation π is a pair $(i, i \oplus 1)$ of consecutive positions in π such that $\pi_{i\oplus 1} \neq \pi_i \oplus 1$. Here $\oplus$ is the usual operation of addition, except that $n \oplus 1 = 1$, and $(-1) \oplus 1 = (-n)$. Similarly we use $\ominus$ for ordinary subtraction, except that $1 \ominus 1 = n$, and $(-n) \ominus 1 = (-1)$. A *strip* is a maximal run of elements between breakpoints.

We use $\Phi(\pi)$ to denote the *number of breakpoints* of π. A reversal can change $\Phi(\pi)$ by removing 2, 1, or 0 breakpoints, or by creating 1 or 2 breakpoints. We call a reversal that removes i breakpoints an *i-reversal*, where $i \in \{2, 1, 0, -1, -2\}$.

To sort a permutation π, we must reduce the number of breakpoints from $\Phi(\pi)$ to $\Phi(\iota) = 0$. This suggests the following *greedy heuristic*: at any step, choose a reversal that removes the most breakpoints. The algorithm of Figure 1 uses this rule, with the twist that it favors reversals that leave negative elements.

```
procedure Greedy(π) begin
    i := 0
    while π contains a breakpoint do begin
        i := i + 1
        Let ρi be a reversal that removes the most breakpoints of π,
            resolving ties in favor of reversals that leave negative elements.
        π := π ∘ ρi
    end
    return i, ρ1ρ2 ··· ρi
end
```

Figure 1   The greedy algorithm.

The following results were proved in [8] for the greedy algorithm in the context of unsigned permutations. They carry over to signed permutations, and we state them without proof.

Lemma 1 *Every permutation with a negative element has a 1-reversal or a 2-reversal. Consequently the greedy algorithm performs a 0-reversal only when the permutation has no negative elements. Furthermore, if every reversal that removes a breakpoint of π leaves a permutation with no negative elements, then π has a 2-reversal.*

Theorem 1 *The greedy algorithm sorts any permutation π with a negative element in at most $\Phi(\pi) - 1$ reversals, and any permutation without a negative element in at most $\Phi(\pi)$ reversals. Consequently*

$$d(\pi) \leq \Phi(\pi).$$

A consequence of Theorem 1 [7] is that the greedy algorithm is an approximation algorithm for sorting by reversals with a worst-case peformance ratio of 2. The algorithm can be implemented to run in time $O(n + \Phi^2(\pi)) = O(n^2)$, using $O(n)$ space [8]. Bafna and Pevzner [1] subsequently found a more elaborate algorithm that uses the greedy algorithm as a subroutine to guarantee a performance ratio of $\frac{3}{2}$.

In Section 6 we present extensive experiments indicating that, in contrast to what was observed for unsigned permutations [8], for *signed* permutations the simple greedy algorithm may yield satisfactory bounds, as it appears to have good performance in terms of its *difference* from the optimum, even for very large problems.

3 Lower bounding the distance

In this section we derive a lower bound on $d(\pi)$ using the notion of a cycle graph introduced in [7]. We show that, in contrast to unsigned permutations, the bound has a remarkably simple proof, and can be evaluated in $O(n)$ time.

With each permutation π we associate an undirected graph $G(\pi)$. Its vertices are as follows. For every pair $(i, i \oplus 1)$ of consecutive positions in π, there is a vertex v_i in G. In effect, vertices lie between positions of π. We call π_i the left value for v_i, and $\pi_{i\oplus 1}$ the right value for v_i. Similarly we call v_i the right vertex for value π_i, and v_i the left vertex for value $\pi_{i\oplus 1}$.

Figure 2 shows the construction of edges for G. Let x and y be the left and right values of vertex v_i. There is an edge from v_i to a vertex associated with $x \oplus 1$ and to a vertex associated with $y \ominus 1$. Consider the vertex associated with $x \oplus 1$. Either value $x \oplus 1$ or $-(x \oplus 1)$ appears in π. If value $x \oplus 1$ appears, v_i is joined by an edge to the vertex on the left of $x \oplus 1$. If value $(x \oplus 1)$ appears, v_i is joined to the vertex on the right of $-(x \oplus 1)$. In a similar manner v_i is joined to a vertex associated with $y \ominus 1$, as shown in the figure. We note that a pair of vertices may

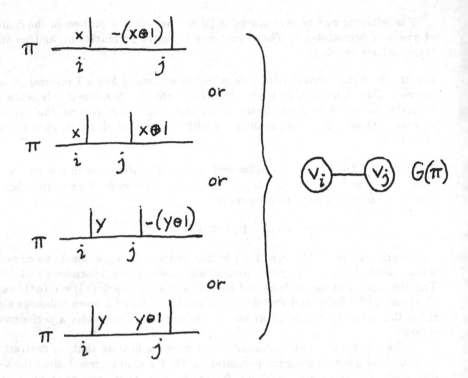

Figure 2 Construction of graph $G(\pi)$.

have two parallel edges between them, which we treat as a cycle of length two, and that a vertex may have a self-loop.

A significant difference between this construction and the one given in [7] for unsigned permutations is that every vertex has degree two. This implies G has a particularly simple structure: it consists of vertex-disjoint cycles. This radically simplifies the form of our lower bound, as well as its proof.

In the following we denote the *number of cycles* of $G(\pi)$ by $\Psi(\pi)$.

Theorem 2 *For every signed circular permutation π on n elements,*

$$d(\pi) \geq n - \Psi(\pi).$$

Proof Consider the effect on $G(\pi)$ of an arbitrary reversal $[i+1, j]$. This changes values only at vertices $v_i, v_{i+1}, \ldots, v_j$.

At an interior vertex, a vertex other than v_i and v_j, the reversal exchanges the left value with the right value and negates their signs. In our construction, this does not affect the cycle passing through the vertex.

At the end vertices v_i and v_j, the reversal exchanges the right value of v_i with the left value of v_j, and negates their sign. This has three possible effects, as shown

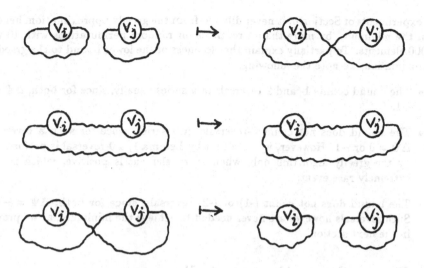

Figure 3 The effect of a reversal $[i+1, j]$ on $G(\pi)$.

in Figure 3. In each case the number of cycles in G either decreases by one, remains the same, or increases by one.

A series of reversals that sorts π must change the number of cycles from $\Psi(\pi)$ to $\Psi(\iota)$. A reversal can change $\Psi(\pi)$ by at most one. Hence any series that sorts π requires at least $|\Psi(\pi) - \Psi(\iota)|$ reversals, which is $n - \Psi(\pi)$. □

Bafna and Pevzner [1] have also observed that their lower bound, which is in terms of edge-disjoint Eulerian cycles that alternate in color in an edge-colored graph, simplifies as well in the context of signed permutations [1].

Since $G(\pi)$ consists of vertex-disjoint cycles, $\Psi(\pi)$ is simply the number of connected components, which yields the algorithm of Figure 4.

procedure LowerBound(π) **begin**
    Let n be the number of elements of π.
    Construct $G(\pi)$ and count the number k of connected components.
    return $n - k$
end

Figure 4 The lower bound algorithm.

This trivial $O(n)$ time algorithm gives an astonishingly tight lower bound. In our experiments of Section 6, it never differed from the greedy approximation, hence from the minimum, by more than 4 reversals on random permutations with 10 to 10,000 elements. To partially explain the closeness of the lower bound to the greedy approximation, we note the following:

- The bound counts 1- and 2-reversals in a series equally, since for both, $\Delta \Psi = +1$.

- The bound does not count 0-reversals in a series, since for such a reversal $\Delta \Psi = 0$ or -1. However, we note that by Lemma 1, a 0-reversal is performed by the greedy algorithm only when every element is positive, which is an extremely rare event.

- The bound does not count (-1) or (-2)-reversals, since for both, $\Delta \Psi = -1$. Such reversals however are never needed to achieve the minimum, as we prove in the next section.

4 Determining the exact distance

We can use these bounds to determine the minimum number of reversals by the branch-and-bound algorithm of Figure 5. The algorithm explores a tree of reversals depth-first, using the lower bound of Section 3 to prune subtrees, and can be implemented to run in $O(mn)$ time for a tree of m nodes, using $O(\Phi^2(\pi)) = O(n^2)$ space. It differs from the branch-and-bound algorithm for unsigned permutations [7] in that the tree does not contain reversals that cut strips. This reduces the branching factor significantly, from $\binom{n}{2}$ to $\binom{\Phi(\pi)}{2}$.

We now prove that the algorithm is correct, and note that the following is one of the first theorems regarding the structure of an optimal solution.

Theorem 3 *Every permutation has an optimal solution that does not cut strips.*

Proof We show that any series of reversals that cuts strips while sorting a permutation π, can be transformed into an equivalent series that also sorts π, but does not cut strips, without increasing the length of the series. Since we can apply this to a shortest series, there is an optimal solution that does not cut strips. The argument uses induction on both the length of the series and the number of reversals that cut strips. A series of one reversal that sorts a permutation cannot cut any strips, so the basis holds.

In general, consider a series that sorts π and cuts strips. Let ρ be the last reversal that cuts a strip and σ be the permutation to which ρ applies. To simplify matters, we assume that only the left end of ρ cuts a strip. (If ρ cuts a strip on the right, we can apply the argument to the right end as well.) Reversal ρ then has the

```
global d*, r*[1..n], r[1..n]

procedure BranchAndBound(π) begin
    d*, r* := UpperBound(π)
    Search(π, 0)
    return d*, r*
end

procedure Search(π, d) begin
    if π is the identity permutation then
        d*, r* := d, r
    else
        for every reversal ρ that does not cut a strip,
            considering reversals in order of decreasing ΔΦ, and
            resolving ties in favor of reversals that leave negative strips,
        do
            if d + 1 + LowerBound(π ∘ ρ) < d* then begin
                r[d + 1] := ρ
                Search(π ∘ ρ, d + 1)
            end
    end
```

Figure 5 The branch-and-bound algorithm.

following form.

$$\sigma \quad \cdots \boxed{A \mid B \mid W} \cdots \underbrace{}_{\rho} \quad \vdash \quad \sigma \circ \rho \quad \cdots \boxed{A \mid W^R \mid B^R} \cdots$$

Here A, B, and W represent substrings of σ, and W^R represents the string formed by reversing W and negating its elements. Strings A and B form a strip, while W is an arbitrary substring of strips.

Instead of cutting strip AB with ρ, we perform the following reversal $\widetilde{\rho}$.

$$\sigma \quad \cdots \boxed{AB \mid W} \cdots \underbrace{}_{\widetilde{\rho}} \quad \vdash \quad \sigma \circ \widetilde{\rho} \quad \cdots \boxed{\overset{\widetilde{A}}{\overset{\shortparallel}{AB}} \mid W^R} \cdots$$

All reversals after ρ are then simulated by the following scheme. No reversal after ρ cuts a strip, so B^R remains intact in all subsequent permutations. Delete B^R from these permutations, and give strip AB in $\sigma \circ \tilde{\rho}$ the new name $\tilde{A}$. If after deleting B^R we identify $\tilde{A}$ in $\sigma \circ \tilde{\rho}$ with strip A in $\sigma \circ \rho$, permutations $\sigma \circ \tilde{\rho}$ and $\sigma \circ \rho$ are identical. Thus, after performing this deletion and renaming, the endpoints of the reversal following ρ can be mapped onto $\sigma \circ \tilde{\rho}$. Continuing in this manner, we can map all subsequent reversals onto the transformed permutations.

In the identity permutation, either configuration AB or $B^R A^R$ appears. In the first case, A must be reversed an even number of times by reversals following ρ, and in the second case, A must be reversed an odd number of times. Thus, when we finish simulating the remaining reversals on $\sigma \circ \tilde{\rho}$, the strip $\tilde{A}$, which will be reversed the same number of times as A, will be oriented at the conclusion as it should be in the identity permutation. In short, the transformed series sorts π.

It may happen in the course of the simulation that a reversal which formerly did not cut a strip is transformed into a reversal that now cuts a strip. The length of the suffix of the series that contains such a reversal has decreased by at least one, however, since $\tilde{\rho}$ does not cut a strip. By induction, this suffix can be transformed into a series that does not cut strips. At this point the total number of reversals that cut strips in the series has decreased by one. By induction on the number of such reversals, the entire series can be transformed into a series free of reversals that cut strips. □

To compute the initial upper bound, we improve the greedy approximation using look-ahead. Instead of constructing a series by choosing a reversal that immediately removes the most breakpoints, we construct a series by looking ahead k reversals, finding a series of length k that removes the maximum number of breakpoints, and performing these k reversals. The best series of length k can be found by branch-and-bound just as in Figure 5. The only difference is that the search is stopped at a depth of k, and we are maximizing the number of breakpoints eliminated, rather than minimizing the length of a solution.

To carry this out, we need an upper bound on the total number of breakpoints that can be eliminated in a given number of reversals, and a way of obtaining a good initial series of k reversals. We find a good initial series by calling the algorithm recursively: to find a series of length k that eliminates a lot of breakpoints, we find a best series of length $\lceil \frac{k}{2} \rceil$ and follow it by a best series of length $\lfloor \frac{k}{2} \rfloor$. At the bottom of the recursion, we use the simple greedy algorithm. We call this approach logarithmic *bootstrapping*, as it bootstraps itself to a good initial solution in a logarithmic number of levels.

An upper bound on the number of breakpoints that can be eliminated in k reversals is obtained using the idea of a *cycle packing*, introduced in [8] in the context of unsigned permutations. A cycle packing for a permutation π is a set of cycles of $G(\pi)$ that are vertex-disjoint. We say the size of a cycle is its number of vertices minus one, and the size of a packing is the sum of the sizes of its cycles. A *k-packing* is a packing of size at most k.

Theorem 4 *Let c be the size of a maximum k-packing for π. Then the number* of breakpoints of π that can be eliminated in k reversals is at most

$$\min\{k + c, \Phi(\pi)\}.$$

Since $G(\pi)$ consists of vertex-disjoint cycles, a maximum k-packing for a signed permutation can be found in $O(n)$ time with a simple greedy procedure. Again this in contrast to the situation for unsigned permutations, where bounding the size of a maximum packing required linear programming [7].

5 Bounding the diameter

Much of the theoretical work related to sorting by reversals has been concerned with bounds on the diameter [5, 3, 6]. The *diameter* $D(n)$ of the set S_n of n-element permutations, with respect to reversal distance, is the maximum number of reversals required to sort an n-element permutation:

$$D(n) = \max_{\pi \in S_n} d(\pi).$$

We now show that the lower bound of Section 3, together with the greedy algorithm, give a tight bound on the diameter.

Theorem 5 *For signed circular permutations, and all n,*

$$n-1 \leq D(n) \leq n.$$

Proof By Theorems 1 and 2 we know that for any permutation π on n elements,

$$n - \Psi(\pi) \leq d(\pi) \leq \Phi(\pi).$$

For circular permutations, $\Phi(\pi)$ attains a maximum value of n, which gives the upper bound on $D(n)$. We prove the lower bound on $D(n)$ by demonstrating a permutation π for every n for which $\Psi(\pi)$ is 1. The form of the permutation depends on whether n is odd or even.

For odd n, consider

$$\omega_n = (n \ n-1 \ \cdots \ 2 \ 1).$$

Recall that $G(\omega_n)$ has a vertex between every consecutive pair of positions in ω_n. Number the vertices so that vertex v_i contains values $(i \oplus 1, i)$. In $G(\omega_n)$, v_i is adjacent to two vertices: those containing values $(\cdot, i \oplus 2)$ and $(i \ominus 1, \cdot)$, which are $v_{i \oplus 2}$ and $v_{i \ominus 2}$. Consider following edges from v_2. We will visit

$$v_2, v_4, v_6, \ldots, v_{n-1}, v_1, v_3, v_5, \ldots, v_n, v_2.$$

This is every vertex of $G(\omega_n)$. Hence $G(\omega_n)$ consists of a single cycle when n is odd.

For even n, consider

$$\zeta_n = \left(-(\tfrac{n}{2}+1) \ \tfrac{n}{2}+2 \ -(\tfrac{n}{2}) \ \tfrac{n}{2}+3 \ \cdots \ n-2 \ -4 \ n-1 \ -3 \ n \ -2 \ 1\right).$$

We group the vertices of $G(\zeta_n)$ into two sets and name them $\{v_2, v_3, \ldots, v_{\frac{n}{2}+1}\}$ and $\{w_2, w_3, \ldots, w_{\frac{n}{2}+1}\}$. In general vertex v_i contains values $(\cdot, -i)$, and vertex w_i contains values $(-i, \cdot)$. We note the following three properties of $G(\zeta_n)$:

(i) w_i is joined by an edge to v_i for $2 \leq i \leq \frac{n}{2}+1$,
(ii) v_i is joined by an edge to w_{i+1} for $2 \leq i \leq \frac{n}{2}$, and
(iii) $v_{\frac{n}{2}+1}$ is joined by an edge to w_2.

Together these imply $G(\zeta_n)$ is a single cycle. $\qquad\square$

Theorem 5 implies that for signed linear permutations, $n - 2 \leq D(n) \leq n - 1$. Moreover these bounds are tight, as shown in the next section.

6 Computational results

We studied the behavior of the bounds and $d(\pi)$ in experiments on all permutations of a fixed size, random permutations, and permutations generated by a fixed number of random reversals. We now summarize the results.

6.1 Exact distances for small permutations

Table 1 gives the distribution of $d(\pi)$ for small n, obtained by running our exact algorithm on all n-element permutations. The distribution differs from that of unsigned permutations [8] in two respects.

Table 1 The number of permutations on n elements at distance d from the identity.

d	n=2	3	4	5	6	7	8	9
0	1	1	1	1	1	1	1	1
1	1	3	6	10	15	21	28	36
2		3	16	50	120	245	448	756
3		1	25	170	700	2,170	5,586	12,600
4				145	1,554	8,820	35,612	115,254
5				8	1,447	19,495	138,229	684,525
6					3	15,148	262,688	2,295,786
7						180	202,464	4,198,049
8							64	3,006,846
9								8,067

First, we note that the extremal permutations are numerous (those π for which $d(\pi) = D(n)$) in contrast to the unsigned case. For unsigned permutations, Holger Gollan conjectured in a talk in 1992 that the extremal permutation is unique up to taking its inverse, and took a large step towards a proof by positing its general form (see [7] for the statement of the conjecture and the form of Gollan's permutation). Bafna and Pevzner [1] established the conjecture by applying their lower bound to this permutation.

Second, the diameter does not grow uniformly, again in contrast to the unsigned case [1]. The table shows that the bounds of Section 5 on the diameter are tight. We suspect, however, that the diameter does not hit the lower bound infinitely often, and conjecture that for all sufficiently large n, $D(n) = n$.

To aid characterization of an extremal permutation, Table 2 lists the extremal permutations consisting only of positive elements for $n \leq 6$. Since any permutation containing a negative element can be sorted in $n - 1$ steps by Theorem 1, only positive permutations are candidates for showing $D(n) = n$.

Table 2   Extremal positive permutations on n elements.

		n		
3	4	5	6
1 3 2	1 2 4 3	1 5 2 4 3	1 6 2 4 3 5
	1 4 2 3	1 4 2 5 3	1 3 5 4 6 2
	1 4 3 2	1 5 4 3 2	1 3 2 4 6 5
	1 3 4 2	1 4 3 5 2	
	1 3 2 4	1 3 5 4 2	
		1 5 3 2 4	
		1 3 5 2 4	
		1 3 2 5 4	

6.2 Bounds for random permutations

To study the quality of the greedy approximation we compared it to the lower bound on random permutations. The results, shown in Table 3, are striking.

Unexpectedly, the *difference* remained bounded for n ranging from 10 to 10,000. This is quite different from the behavior observed for unsigned permutations. For random unsigned permutations the average *ratio* A/L varied between 1.2 and 1.3 for n from 10 to 100 [8].

Table 4 records the variance in L for the same range of n. The variance is small, and slow growing, but it is not sufficiently small to completely account for the tightness of the bounds. In the next experiments on randomly reversed permutations, for example, the difference is small even when the standard deviation in the lower bound exceeds 8 reversals.

Moreover, the concentration of distance about the mean for random permutations may be useful for detecting when the gene order between two organisms is sufficiently scrambled to suggest that they are unrelated. For example, Table 4 indicates that a measured distance of 45 inversions for 50 genes is around what

Table 3 Difference between the approximation A and the lower bound L for random permutations. For each n the sample size is 100.

n	$A - L$ mean	$A - L$ max
10	0.05	1
25	0.12	2
50	0.14	2
100	0.06	1
250	0.10	2
500	0.11	2
1,000	0.08	1
2,500	0.15	1
5,000	0.11	2
10,000	0.11	1

Table 4 Growth of the lower bound L for random permutations. For each n the sample size is 100.

n	L mean	L min	L max	L dev	$n - L$ mean
10	7.82	5	9	0.91	2.18
25	22.38	17	24	1.32	2.62
50	47.02	43	49	1.36	2.98
100	96.64	91	99	1.65	3.36
250	246.25	241	249	1.64	3.75
500	496.12	492	499	1.73	3.88
1,000	995.56	990	999	1.92	4.44
2,500	2,494.94	2,488	2,498	2.14	5.06
5,000	4,994.53	4,988	4,999	2.13	5.47
10,000	9,994.23	9,988	9,998	2.15	5.77

one would expect for completely unrelated organisms. We note that the expected difference between n and $d(\pi)$ appears to be proportional to roughly $\log n$.

6.3 Bounds for permutations generated by random reversals

An input more typical than a random permutation would be one generated from the identity by a fixed number k of random reversals. Table 5 shows the observed difference between the approximation and the lower bound for a range of k and n. As can be seen, the difference remains quite small.

It is interesting that the difference observed is generally greatest when the permutation is large but the number of reversals is small. At this extreme, the ends of the reversals are likely to cut at disjoint locations, which causes a preponderance of 2-reversals in a solution. As the greedy algorithm does not distinguish between the 2-reversals available, in such a situation it is more likely to choose a "bad" 2-reversal—one that prevents other 2-reversals, or fails to set them up, by reversing one of their endpoints. Bafna and Pevzner [1] improved the performance ratio of the greedy algorithm by refining its choice in this situation, and it would be interesting to see if their improvement smooths out the behavior of the algorithm for the full range of k.

Comparison of L and k in Table 5 reveals that as k gets large relative to n, reversal distance underestimates the true number of reversals. What percentage of reversals can we recover by measuring reversal distance?

Table 6 indicates that for roughly $k < .5n$ reversal distance is a good measure of the true number of reversals. We have observed the same transition point for $n \leq 1,000$, and for unsigned permutations as well [7]. This suggests the following rule of thumb: a measurement of inversion distance d between two organisms should be based on a sample of more than $2d$ approximately equally-spaced markers.

6.4 Exact distances for large permutations

Our last experiments studied the behavior of the exact algorithm on large permutations. Table 7 summarizes the results for both permutations generated by k random reversals, and completely random permutations ($k = \infty$).

Except for one problem on 100 elements, all problems of up to 250 elements were solved to optimality. As can be seen from the table, whenever optimal solution was possible, the initial lower bound was tight.

Generally when there was a gap between the lower bound and the upper bound obtained with look-ahead, the branch-and-bound algorithm was unable to close the gap within the search limit. The one exception is a random problem on 500 elements where the exact algorithm closed the gap of 1 reversal in a search of around 29,000 nodes.

For all problems the series obtained with look-ahead was known at termination to be within 1 reversal of the optimum. The largest improvement obtained by look-ahead was a savings of 4 reversals on a series of length 104. Running times on a standard workstation varied from less than 1 minute for 50 element problems, to around 45 minutes for 500 element problems.

Table 5 Difference between the approximation A and the lower bound L for permutations generated by k random reversals. For each n and k the sample size is 100.

n	k	L		$A-L$	
		mean	dev	mean	max
25	5	4.95	0.30	0.32	2
	10	9.68	0.75	0.16	1
	25	18.66	1.74	0.11	2
50	5	5.00	0.00	0.19	2
	10	9.88	0.46	0.17	2
	25	24.18	1.06	0.25	2
	50	39.83	2.21	0.10	2
100	5	5.00	0.00	0.17	2
	10	10.00	0.00	0.44	4
	25	24.87	0.56	0.24	2
	50	48.90	1.24	0.25	2
	100	81.16	3.03	0.07	1
250	10	10.00	0.00	0.37	2
	25	24.98	0.20	0.28	2
	50	49.92	0.37	0.48	3
	100	99.49	0.93	0.27	3
	250	207.22	4.07	0.11	2
500	10	10.00	0.00	0.23	2
	25	25.00	0.00	0.43	4
	50	49.98	0.20	0.51	3
	100	99.93	0.33	0.54	3
	250	248.48	1.64	0.40	3
	500	415.71	6.72	0.16	2
1000	10	10.00	0.00	0.56	4
	25	25.00	0.00	0.54	4
	50	50.00	0.00	0.49	2
	100	100.00	0.00	0.50	3
	250	249.78	0.60	0.60	3
	500	498.12	1.67	0.44	3
	1000	835.58	8.66	0.09	1

Table 6   Difference between the true number of reversals and the lower bound L for permutations generated by k random reversals. The sample for each k is 100 permutations of 1,000 elements.

	L		$k - L$
k	mean	dev	mean
50	50.00	0.00	0.00
100	100.00	0.00	0.00
150	149.96	0.28	0.04
200	199.85	0.61	0.15
250	249.78	0.60	0.22
300	299.75	0.66	0.25
350	349.51	0.89	0.49
400	399.49	0.89	0.51
450	449.11	1.19	0.89
500	498.12	1.67	1.88
550	546.85	2.56	3.15
600	593.30	3.88	6.70
650	636.81	5.60	13.19
700	674.42	6.06	25.58
750	709.36	6.38	40.64
800	741.44	8.44	58.56
850	768.54	8.58	81.46
900	793.02	8.63	106.98
950	815.46	8.59	134.54
1000	835.58	8.66	164.42

450	449.11	1.19	0.89
460	459.08	1.33	0.92
470	468.60	1.55	1.40
480	478.78	1.29	1.22
490	488.33	1.69	1.67
500	498.12	1.67	1.88

Table 7 Behavior of the exact algorithm on permutations of n elements with k random reversals. Parameters are the exact algorithm value E, upper bound U, approximation A, lower bound L, and tree sizes T_E and T_U for the exact and upper bound algorithms. The sample size for each n and k is 10, with look-ahead 5 and maximum tree sizes of 10,000 and 100,000 for the upper bound and exact algorithms, respectively.

n	k	$E - L$ max	$U - E$ max	$A - U$ max	T_U mean	T_E max
25	5	0	0	1	20	0
	10	0	0	1	322	0
	25	0	0	0	1,243	0
	∞	0	0	2	1,232	0
50	10	0	0	0	264	0
	25	0	0	1	7,526	0
	50	0	0	0	2,095	0
	∞	0	0	2	6,534	0
100	25	1	0	2	6,020	100,000
	50	0	0	1	9,593	0
	100	0	0	0	7,207	0
	∞	0	0	1	3,020	0
250	25	0	0	2	2,148	0
	50	0	0	2	9,510	0
	100	0	0	2	10,000	0
	∞	0	0	0	9,483	0
500	25	1	1	1	105	100,000
	50	0	0	3	3,604	0
	100	1	0	4	10,000	100,000
	∞	0	1	1	9,000	28,931

7 Conclusions

While computing the reversal distance between signed permutations appears to be hard, good bounds can be obtained quite efficiently. The greedy algorithm yields an upper bound in $O(n^2)$ time, and the lower bound can be evaluated in $O(n)$ time.

As well as being simple and easy to implement, these methods yield bounds that are extremely tight. For random and randomly reversed permutations we have not observed them to differ by more than 4 reversals in extensive trials involving permutations of up to 10,000 elements. Coupled with a branch-and-bound algorithm using look-ahead, we could solve most problems of up to 250 elements to optimality in less than an hour, and failing this, determine the reversal distance to within 1 reversal.

This success is due by and large to the tightness of the lower bound, and suggests that if one is simply interested in the reversal distance between two organisms and not in an actual series of reversals, for instance when constructing an evolutionary tree, one might simply evaluate the lower bound, in linear time. This phenomena has been observed for other optimization problems, such as in the traveling salesman problem, where the *value* of an optimal solution can be determined quite closely and easily by a suitable relaxation, such as to a matching problem, yet finding a *solution* near that value requires a tremendous amount of effort. Indeed, it would be interesting to examine how well a shortest series of reversals recovers the actual reversals in a random series.

Given that the lower bound relaxes all information concerning how reversals in a series overlap, and that the greedy algorithm forms a series based on extremely local and naive decisions, it is a mystery why the bounds they yield are so tight, and in particular why their *difference* is so small. Can one prove that the expected difference between the greedy approximation and the lower bound is a slowly growing function? Our experimental results suggest that this difference grows more slowly than $\log n$, and may in fact be bounded by a constant. Fortunately the lower bound and the greedy algorithm have a simple form, which provides some hope for a satisfying analysis.

References

[1] Bafna, Vineet and Pavel A. Pevzner. Genome rearrangements and sorting by reversals. In Proceedings of the 34th Annual IEEE *Symposium on Foundations of Computer Science*, 148–157, November 1993.

[2] Chrobak, M., T. Szymacha, and A. Krawczyk. A data structure useful for finding Hamiltonian cycles. *Theoretical Computer Science* 71, 419–424, 1990.

[3] Cohen, David S. and Manuel Blum. On the problem of sorting burnt pancakes. Manuscript, Computer Science Division, University of California at Berkeley, 1993.

[4] Fredman, M.L., D.S. Johnson, L.A. McGeoch, and G. Ostheimer. Data structures for traveling salesmen. In Proceedings of the 4th Annual ACM-SIAM *Symposium on Discrete Algorithms*, 145–154, 1993.

[5] Gates, William H. and Christos H. Papadimitriou. Bounds for sorting by prefix reversals. *Discrete Mathematics* 27, 47–57, 1979.

[6] Heydari, Mohammad H. *The Pancake Problem*. PhD dissertation, Department of Computer Science, University of Texas at Dallas, 1993.

[7] Kececioglu, John and David Sankoff. Exact and approximation algorithms for the inversion distance between two chromosomes. In Proceedings of the 4th Annual *Symposium on Combinatorial Pattern Matching*, Lecture Notes in Computer Science 684, Springer-Verlag, 87–105, June 1993. (An earlier version appeared as "Exact and approximation algorithms for the reversal distance between two permutations," Technical Report 1824, Centre de recherches mathématiques, Université de Montréal, Montréal, Canada, July 1992.)

[8] Kececioglu, John and David Sankoff. Exact and approximation algorithms for sorting by reversals, with application to genome rearrangement. To appear in *Algorithmica*, 1993.

[9] Sankoff, David, Guillame Leduc, Natalie Antoine, Bruno Paquin, B. Franz Lang, and Robert Cedergren.   Gene order comparisons for phylogenetic inference: evolution of the mitochondrial genome.   *Proceedings of the National Academy of Science USA* 89, 6575–6579, 1992.

[10] Schöniger, Michael and Michael S. Waterman.   A local algorithm for DNA sequence alignment with inversions.   *Bulletin of Mathematical Biology* 54, 521–536, 1992.

[11] Watterson, G.A., W.J. Ewens, T.E. Hall, and A. Morgan.   The chromosome inversion problem.   *Journal of Theoretical Biology* 99, 1–7, 1982.

Index of Authors

Lecture Notes in Computer Science

For information about Vols. 1–729
please contact your bookseller or Springer-Verlag

Lecture Notes in Computer Science

This series reports new developments in computer science research and teaching, quickly, informally, and at a high level. The timeliness of a manuscript is more important than its form, which may be unfinished or tentative. The type of material considered for publication includes

– drafts of original papers or monographs,

– technical reports of high quality and broad interest,

– advanced-level lectures,

– reports of meetings, provided they are of exceptional interest and focused on a single topic.

Publication of Lecture Notes is intended as a service to the computer science community in that the publisher Springer-Verlag offers global distribution of documents which would otherwise have a restricted readership. Once published and copyrighted they can be cited in the scientific literature.

Manuscripts

Lecture Notes are printed by photo-offset from the master copy delivered in camera-ready form. Manuscripts should be no less than 100 and preferably no more than 500 pages of text. Authors of monographs and editors of proceedings volumes receive 50 free copies of their book. Manuscripts should be printed with a laser or other high-resolution printer onto white paper of reasonable quality. To ensure that the final photo-reduced pages are easily readable, please use one of the following formats:

Font size	Printing area		Final size
(points)	(cm)	(inches)	(%)
10	12.2 x 19.3	4.8 x 7.6	100
12	15.3 x 24.2	6.0 x 9.5	80

On request the publisher will supply a leaflet with more detailed technical instructions or a T$_E$X macro package for the preparation of manuscripts.

Manuscripts should be sent to one of the series editors or directly to:

Springer-Verlag, Computer Science Editorial I, Tiergartenstr. 17, D-69121 Heidelberg, Germany

ISBN 3-540-58094-8
ISBN 0-387-58094-8